Stochastic Network Optimization with Application to Communication and Queueing Systems

Stochastic Network Optimization with Application to Communication and Queueing Systems
Michael J. Neely

ISBN: 978-3-031-79994-5 paperback
ISBN: 978-3-031-79995-2 ebook

DOI 10.1007/978-3-031-79995-2

A Publication in the Springer series
SYNTHESIS LECTURES ON COMMUNICATION NETWORKS

Lecture #7
Series Editor: Jean Walrand, *University of California, Berkeley*
Series ISSN
Synthesis Lectures on Communication Networks
Print 1935-4185 Electronic 1935-4193

This material is supported in part by one or more of the following: the DARPA IT-MANET program grant W911NF-07-0028, the NSF Career grant CCF-0747525, and continuing through participation in the Network Science Collaborative Technology Alliance sponsored by the U.S. Army Research Laboratory.

Synthesis Lectures on Communication Networks

Editor
Jean Walrand, ***University of California, Berkeley***

Synthesis Lectures on Communication Networks is an ongoing series of 50- to 100-page publications on topics on the design, implementation, and management of communication networks. Each lecture is a self-contained presentation of one topic by a leading expert. The topics range from algorithms to hardware implementations and cover a broad spectrum of issues from security to multiple-access protocols. The series addresses technologies from sensor networks to reconfigurable optical networks. The series is designed to:

- Provide the best available presentations of important aspects of communication networks.
- Help engineers and advanced students keep up with recent developments in a rapidly evolving technology.
- Facilitate the development of courses in this field.

Stochastic Network Optimization with Application to Communication and Queueing Systems
Michael J. Neely
2010

Scheduling and Congestion Control for Wireless and Processing Networks
Libin Jiang and Jean Walrand
2010

Performance Modeling of Communication Networks with Markov Chains
Jeonghoon Mo
2010

Communication Networks: A Concise Introduction
Jean Walrand and Shyam Parekh
2010

Path Problems in Networks
John S. Baras and George Theodorakopoulos
2010

Stochastic Network Optimization with Application to Communication and Queueing Systems

Michael J. Neely
University of Southern California

SYNTHESIS LECTURES ON COMMUNICATION NETWORKS #7

ABSTRACT

This text presents a modern theory of analysis, control, and optimization for dynamic networks. Mathematical techniques of Lyapunov drift and Lyapunov optimization are developed and shown to enable constrained optimization of time averages in general stochastic systems. The focus is on communication and queueing systems, including wireless networks with time-varying channels, mobility, and randomly arriving traffic. A simple drift-plus-penalty framework is used to optimize time averages such as throughput, throughput-utility, power, and distortion. Explicit performance-delay tradeoffs are provided to illustrate the cost of approaching optimality. This theory is also applicable to problems in operations research and economics, where energy-efficient and profit-maximizing decisions must be made without knowing the future.

Topics in the text include the following:

- Queue stability theory
- Backpressure, max-weight, and virtual queue methods
- Primal-dual methods for non-convex stochastic utility maximization
- Universal scheduling theory for arbitrary sample paths
- Approximate and randomized scheduling theory
- Optimization of renewal systems and Markov decision systems

Detailed examples and numerous problem set questions are provided to reinforce the main concepts.

KEYWORDS

dynamic scheduling, decision theory, wireless networks, Lyapunov optimization, congestion control, fairness, network utility maximization, multi-hop, mobile networks, routing, backpressure, max-weight, virtual queues

Contents

Preface

This text is written to teach the theory of Lyapunov drift and Lyapunov optimization for stochastic network optimization. It assumes only that the reader is familiar with basic probability concepts (such as expectations and the law of large numbers). Familiarity with Markov chains and with standard (non-stochastic) optimization is useful but not required. A variety of examples and simulation results are given to illustrate the main concepts. Diverse problem set questions (several with example solutions) are also given. These questions and examples were developed over several years for use in the stochastic network optimization course taught by the author. They include topics of wireless opportunistic scheduling, multi-hop routing, network coding for maximum throughput, distortion-aware data compression, energy-constrained and delay-constrained queueing, dynamic decision making for maximum profit, and more.

The Lyapunov theory for optimizing network time averages was described collectively in our previous text (22). The current text is significantly different from (22). It has been reorganized with many more examples to help the reader. This is done while still keeping all of the details for a complete and self-contained exposition of the material. This text also provides many recent topics not covered in (22), including:

- A more detailed development of queue stability theory (Chapter 2).
- Variable-V algorithms that provide exact optimality of time averages subject to a weaker form of stability called "mean rate stability" (Section 4.7).
- Place-holder bits for delay improvement (Sections 3.2.4 and 4.8).
- Universal scheduling for non-ergodic sample paths (Section 4.9).
- Worst case delay bounds (Sections 5.6 and 7.6.1).
- Non-convex stochastic optimization (Section 5.5).
- Approximate scheduling and full throughput scheduling in interference networks via the Jiang-Walrand theorem (Chapter 6).
- Optimization of renewal systems and Markov decision examples (Chapter 7).
- Treatment of problems with equality constraints and abstract set constraints (Section 5.4).

Finally, this text emphasizes the simplicity of the Lyapunov method, showing how all of the results follow directly from four simple concepts: (i) telescoping sums, (ii) iterated expectations, (iii) opportunistically minimizing an expectation, and (iv) Jensen's inequality.

Michael J. Neely
September 2010

CHAPTER 1
Introduction

This text considers the analysis and control of stochastic networks, that is, networks with random events, time variation, and uncertainty. Our focus is on communication and queueing systems. Example applications include wireless mesh networks with opportunistic scheduling, cognitive radio networks, ad-hoc mobile networks, internets with peer-to-peer communication, and sensor networks with joint compression and transmission. The techniques are also applicable to stochastic systems that arise in operations research, economics, transportation, and smart-grid energy distribution. These problems can be formulated as problems that optimize the time averages of certain quantities subject to time average constraints on other quantities, and they can be solved with a common mathematical framework that is intimately connected to queueing theory.

1.1 EXAMPLE OPPORTUNISTIC SCHEDULING PROBLEM

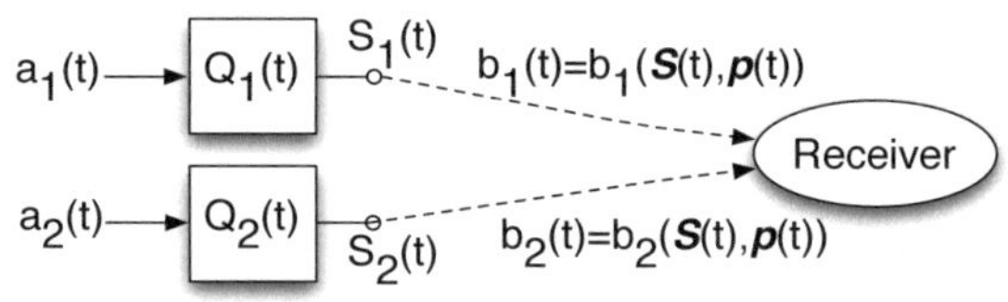

Figure 1.1: The 2-user wireless system for the example of Section 1.1.

Here we provide a simple wireless example to illustrate how the theory for optimizing time averages can be used. Consider a 2-user wireless uplink that operates in slotted time $t \in \{0, 1, 2, \dots\}$. Every slot new data randomly arrives to each user for transmission to a common receiver. Let $(a_1(t), a_2(t))$ be the vector of new arrivals on slot t, in units of bits. The data is stored in queues $Q_1(t)$ and $Q_2(t)$ to await transmission (see Fig. 1.1). We assume the receiver coordinates network decisions every slot.

Channel conditions are assumed to be constant for the duration of a slot, but they can change from slot to slot. Let $\boldsymbol{S}(t) = (S_1(t), S_2(t))$ denote the channel conditions between users and the receiver on slot t. The channel conditions represent any information that affects the channel on slot t, such as fading coefficients and/or noise ratios. We assume the network controller can observe $\boldsymbol{S}(t)$ at the beginning of each slot t before making a transmission decision. This channel-aware scheduling is called *opportunistic scheduling*. Every slot t, the network controller observes the current $\boldsymbol{S}(t)$

and chooses a *power allocation vector* $\boldsymbol{p}(t) = (p_1(t), p_2(t))$ within some set $\mathcal{P}$ of possible power allocations. This decision, together with the current $\boldsymbol{S}(t)$, determines the *transmission rate vector* $(b_1(t), b_2(t))$ for slot t, where $b_k(t)$ represents the transmission rate (in bits/slot) from user $k \in \{1, 2\}$ to the receiver on slot t. Specifically, we have general transmission rate functions $\hat{b}_k(\boldsymbol{p}(t), \boldsymbol{S}(t))$:

$$b_1(t) = \hat{b}_1(\boldsymbol{p}(t), \boldsymbol{S}(t)) \ , \ b_2(t) = \hat{b}_2(\boldsymbol{p}(t), \boldsymbol{S}(t))$$

The precise form of these functions depends on the modulation and coding strategies used for transmission. The queueing dynamics are then:

$$Q_k(t+1) = \max[Q_k(t) - \hat{b}_k(\boldsymbol{p}(t), \boldsymbol{S}(t)), 0] + a_k(t) \quad \forall k \in \{1, 2\}, \forall t \in \{0, 1, 2, \ldots\}$$

Several types of optimization problems can be considered for this simple system.

1.1.1 EXAMPLE PROBLEM 1: MINIMIZING TIME AVERAGE POWER SUBJECT TO STABILITY

Let $\overline{p}_k$ be the time average power expenditure of user k under a particular power allocation algorithm (for $k \in \{1, 2\}$):

$$\overline{p}_k \triangleq \lim_{t\rightarrow\infty} \frac{1}{t} \sum_{\tau=0}^{t-1} p_k(\tau)$$

The problem of designing an algorithm to minimize time average power expenditure subject to queue stability can be written mathematically as:

$$\begin{array}{lll} \text{Minimize:} & & \overline{p}_1 + \overline{p}_2 \\ \text{Subject to:} & 1) & \text{Queues } Q_k(t) \text{ are stable } \forall k \in \{1, 2\} \\ & 2) & \boldsymbol{p}(t) \in \mathcal{P} \ \forall t \in \{0, 1, 2, \ldots\} \end{array}$$

where queue stability is defined in the next chapter. It is shown in the next chapter that queue stability ensures the time average output rate of the queue is equal to the time average input rate. Our theory will allow the design of a simple algorithm that makes decisions $\boldsymbol{p}(t) \in \mathcal{P}$ every slot t, without requiring a-priori knowledge of the probabilities associated with the arrival and channel processes $\boldsymbol{a}(t)$ and $\boldsymbol{S}(t)$. The algorithm meets all desired constraints in the above problem whenever it is possible to do so. Further, the algorithm is parameterized by a constant $V \geq 0$ that can be chosen as desired to yield time average power within $O(1/V)$ from the minimum possible time average power required for queue stability. Choosing a large value of V can thus push average power arbitrarily close to optimal. However, this comes with a tradeoff in average queue backlog and delay that is $O(V)$.

1.1.2 EXAMPLE PROBLEM 2: MAXIMIZING THROUGHPUT SUBJECT TO TIME AVERAGE POWER CONSTRAINTS

Consider the same system, but now assume the arrival process $\boldsymbol{a}(t) = (a_1(t), a_2(t))$ can be *controlled* by a flow control mechanism. We thus have two decision vectors: $\boldsymbol{p}(t)$ (the power allocation vector) and $\boldsymbol{a}(t)$ (the data admission vector). The admission vector $\boldsymbol{a}(t)$ is chosen within some set $\mathcal{A}$ every slot t. Let $\overline{a}_k$ be the time average admission rate (in bits/slot) for user k, which is the same as the time average throughput of user k if its queue is stable (as shown in the next chapter). We have the following problem of maximizing a weighted sum of throughput subject to average power constraints:

$$
\begin{array}{lll}
\text{Maximize:} & & w_1\overline{a}_1 + w_2\overline{a}_2 \\
\text{Subject to:} & 1) & \overline{p}_k \leq p_{k,av} \ \forall k \in \{1, 2\} \\
& 2) & \text{Queues } Q_k(t) \text{ are stable } \forall k \in \{1, 2\} \\
& 3) & \boldsymbol{p}(t) \in \mathcal{P} \ \forall t \in \{0, 1, 2, \ldots\} \\
& 4) & \boldsymbol{a}(t) \in \mathcal{A} \ \forall t \in \{0, 1, 2, \ldots\}
\end{array}
$$

where w_1, w_2 are given positive weights that define the relative importance of user 1 traffic and user 2 traffic, and $p_{1,av}$, $p_{2,av}$ are given constants that represent desired average power constraints for each user. Again, our theory leads to an algorithm that meets all desired constraints and comes within $O(1/V)$ of the maximum throughput possible under these constraints, with an $O(V)$ tradeoff in average backlog and delay.

1.1.3 EXAMPLE PROBLEM 3: MAXIMIZING THROUGHPUT-UTILITY SUBJECT TO TIME AVERAGE POWER CONSTRAINTS

Consider the same system as Example Problem 2, but now assume the objective is to maximize a concave function of throughput, rather than a linear function of throughput (the definition of "concave" is given in footnote 1 in the next subsection). Specifically, let $g_1(a)$ and $g_2(a)$ be continuous, concave, and non-decreasing functions of a over the range $a \geq 0$. Such functions are called *utility functions*. The value $g_1(\overline{a}_1)$ represents the *utility* (or *satisfaction*) that user 1 gets by achieving a throughput of $\overline{a}_1$. Maximizing $g_1(\overline{a}_1) + g_2(\overline{a}_2)$ can provide a more "fair" throughput vector $(\overline{a}_1, \overline{a}_2)$. Indeed, maximizing a linear function often yields a vector with one component that is very high and the other component very low (possibly 0). We then have the problem:

$$
\begin{array}{lll}
\text{Maximize:} & & g_1(\overline{a}_1) + g_2(\overline{a}_2) \\
\text{Subject to:} & 1) & \overline{p}_k \leq p_{k,av} \ \forall k \in \{1, 2\} \\
& 2) & \text{Queues } Q_k(t) \text{ are stable } \forall k \in \{1, 2\} \\
& 3) & \boldsymbol{p}(t) \in \mathcal{P} \ \forall t \in \{0, 1, 2, \ldots\} \\
& 4) & \boldsymbol{a}(t) \in \mathcal{A} \ \forall t \in \{0, 1, 2, \ldots\}
\end{array}
$$

Typical utility functions are $g_1(a) = g_2(a) = \log(a)$, or $g_1(a) = g_2(a) = \log(1 + a)$. These functions are non-decreasing and strictly concave, so that $g_1(\overline{a}_1)$ has a *diminishing returns* property with each incremental increase in throughput $\overline{a}_1$. This means that if $\overline{a}_1 < \overline{a}_2$, the sum utility $g_1(\overline{a}_1) + g_2(\overline{a}_2)$ would be improved more by increasing $\overline{a}_1$ than by increasing $\overline{a}_2$. This creates a more evenly distributed throughput vector. The $\log(a)$ utility functions provide a type of fairness called *proportional fairness* (see (1)(2)). Fairness properties of different types of utility functions are considered in (3)(4)(5)(6).

For any given continuous and concave utility functions, our theory enables the design of an algorithm that meets all desired constraints and provides throughput-utility within $O(1/V)$ of optimality, with a tradeoff in average backlog and delay that is $O(V)$.

We emphasize that these three problems are just examples. The general theory can treat many more types of networks. Indeed, the examples and problem set questions provided in this text include networks with probabilistic channel errors, network coding, data compression, multi-hop communication, and mobility. The theory is also useful for problems within operations research and economics.

1.2 GENERAL STOCHASTIC OPTIMIZATION PROBLEMS

The three example problems considered in the previous section all involved optimizing a time average (or a function of time averages) subject to time average constraints. Here we state the general problems of this type. Consider a stochastic network that operates in discrete time with unit time slots $t \in \{0, 1, 2, \ldots\}$. The network is described by a collection of *queue backlogs*, written in vector form $\boldsymbol{Q}(t) = (Q_1(t), \ldots, Q_K(t))$, where K is a non-negative integer. The case $K = 0$ corresponds to a system without queues. Every slot t, a control action is taken, and this action affects arrivals and departures of the queues and also creates a collection of real valued *attribute vectors* $\boldsymbol{x}(t)$, $\boldsymbol{y}(t)$, $\boldsymbol{e}(t)$:

$$\begin{aligned} \boldsymbol{x}(t) &= (x_1(t), \ldots, x_M(t)) \\ \boldsymbol{y}(t) &= (y_0(t), y_1(t), \ldots, y_L(t)) \\ \boldsymbol{e}(t) &= (e_1(t), \ldots, e_J(t)) \end{aligned}$$

for some non-negative integers M, L, J (used to distinguish between equality constraints and two types of inequality constraints). The attributes can be positive or negative, and they represent penalties or rewards associated with the network on slot t, such as power expenditures, distortions, or packet drops/admissions. These attributes are given by general functions:

$$\begin{aligned} x_m(t) &= \hat{x}_m(\alpha(t), \omega(t)) \ \forall m \in \{1, \ldots, M\} \\ y_l(t) &= \hat{y}_l(\alpha(t), \omega(t)) \ \forall l \in \{0, 1, \ldots, L\} \\ e_j(t) &= \hat{e}_j(\alpha(t), \omega(t)) \ \forall j \in \{1, \ldots, J\} \end{aligned}$$

where $\omega(t)$ is a random event observed on slot t (such as new packet arrivals or channel conditions) and $\alpha(t)$ is the control action taken on slot t (such as packet admissions or transmissions). The action $\alpha(t)$ is chosen within an abstract set $\mathcal{A}_{\omega(t)}$ that possibly depends on $\omega(t)$. Let $\overline{x}_m, \overline{y}_l, \overline{e}_j$ represent the *time average* of $x_m(t), y_l(t), e_j(t)$ under a particular control algorithm. Our first objective is to design an algorithm that solves the following problem:

$$
\begin{array}{lll}
\text{Minimize:} & & \overline{y}_0 \quad (1.1)\\
\text{Subject to:} & 1) & \overline{y}_l \leq 0 \text{ for all } l \in \{1, \ldots, L\} \quad (1.2)\\
& 2) & \overline{e}_j = 0 \text{ for all } j \in \{1, \ldots, J\} \quad (1.3)\\
& 3) & \alpha(t) \in \mathcal{A}_{\omega(t)} \ \forall t \quad (1.4)\\
& 4) & \text{Stability of all Network Queues} \quad (1.5)
\end{array}
$$

Our second objective, more general than the first, is to optimize *convex functions* of time averages.[1] Specifically, let $f(\boldsymbol{x}), g_1(\boldsymbol{x}), \ldots, g_L(\boldsymbol{x})$ be convex functions from $\mathbb{R}^M$ to $\mathbb{R}$, and let $\mathcal{X}$ be a closed and convex subset of $\mathbb{R}^M$. Let $\overline{\boldsymbol{x}} = (\overline{x}_1, \ldots, \overline{x}_M)$ be the vector of time averages of the $x_m(t)$ attributes under a given control algorithm. We desire a solution to the following problem:

$$
\begin{array}{lll}
\text{Minimize:} & & \overline{y}_0 + f(\overline{\boldsymbol{x}}) \quad (1.6)\\
\text{Subject to:} & 1) & \overline{y}_l + g_l(\overline{\boldsymbol{x}}) \leq 0 \text{ for all } l \in \{1, \ldots, L\} \quad (1.7)\\
& 2) & \overline{e}_j = 0 \text{ for all } j \in \{1, \ldots, J\} \quad (1.8)\\
& 3) & \overline{\boldsymbol{x}} \in \mathcal{X} \quad (1.9)\\
& 4) & \alpha(t) \in \mathcal{A}_{\omega(t)} \ \forall t \quad (1.10)\\
& 5) & \text{Stability of all Network Queues} \quad (1.11)
\end{array}
$$

These problems (1.1)-(1.5) and (1.6)-(1.11) can be viewed as *stochastic programs*, and are analogues of the classic linear programs and convex programs of static optimization theory. A *solution* is an algorithm for choosing control actions over time in reaction to the existing network state, such that all of the constraints are satisfied and the quantity to be minimized is as small as possible. These problems have wide applications, and they are of interest even when there is no underlying queueing network to be stabilized (so that the "Stability" constraints in (1.5) and (1.11) are removed). However, it turns out that queueing theory plays a central role in this type of stochastic optimization. Indeed, even if there are no underlying queues in the original problem, we can introduce *virtual queues* as a strong method for ensuring that the required time average constraints are satisfied. Inefficient control actions incur larger backlog in certain queues. These backlogs act as "sufficient statistics" on which to base the next control decision. This enables algorithms that do not require knowledge of the probabilities associated with the random network events $\omega(t)$.

1.3 LYAPUNOV DRIFT AND LYAPUNOV OPTIMIZATION

We solve the problems described above with a simple and elegant theory of *Lyapunov drift* and *Lyapunov optimization*. While this theory is presented in detail in future chapters, we briefly describe it here. The first step is to look at the constraints of the problem to be solved. For example, for the

[1] A set $\mathcal{X} \subseteq \mathbb{R}^M$ is *convex* if the line segment formed by any two points in $\mathcal{X}$ is also in $\mathcal{X}$. A function $f(\boldsymbol{x})$ defined over a convex set $\mathcal{X}$ is a *convex function* if for any two points $\boldsymbol{x}_1, \boldsymbol{x}_2 \in \mathcal{X}$ and any two probabilities $p_1, p_2 \geq 0$ such that $p_1 + p_2 = 1$, we have $f(p_1\boldsymbol{x}_1 + p_2\boldsymbol{x}_2) \leq p_1 f(\boldsymbol{x}_1) + p_2 f(\boldsymbol{x}_2)$. A function $f(\boldsymbol{x})$ is *concave* if $-f(\boldsymbol{x})$ is convex. A function $f(\boldsymbol{x})$ is *affine* if it is linear plus a constant, having the form: $f(\boldsymbol{x}) = c_0 + \sum_{m=1}^{M} c_m x_m$.

problem (1.1)-(1.5), the constraints are (1.2)-(1.5). Then construct *virtual queues* (in a way to be specified) that help to meet the desired constraints. Next, define a function $L(t)$ as the sum of squares of backlog in all virtual and actual queues on slot t. This is called a *Lyapunov function*, and it is a scalar measure of network congestion. Intuitively, if $L(t)$ is "small," then all queues are small, and if $L(t)$ is "large," then at least one queue is large. Define $\Delta(t) = L(t+1) - L(t)$, being the difference in the Lyapunov function from one slot to the next.[2] If control decisions are made every slot t to greedily minimize $\Delta(t)$, then backlogs are consistently pushed towards a lower congestion state, which intuitively maintains network stability (where "stability" is precisely defined in the next chapter).

Minimizing $\Delta(t)$ every slot is called *minimizing the Lyapunov drift*. Chapter 3 shows this method provides queue stability for a particular example network, and Chapter 4 shows it also stabilizes general networks. However, at this point, the problem is only half solved: The virtual queues and Lyapunov drift help only to ensure the desired time average constraints are met. The objective function to be minimized has not yet been incorporated. For example, $y_0(t)$ is the objective function for the problem (1.1)-(1.5). The objective function is mapped to an appropriate function $penalty(t)$. Instead of taking actions to greedily minimize $\Delta(t)$, actions are taken every slot t to greedily minimize the following *drift-plus-penalty* expression:

$$\Delta(t) + V \times penalty(t)$$

where V is a non-negative control parameter that is chosen as desired. Choosing $V = 0$ corresponds to the original algorithm of minimizing the drift alone. Choosing $V > 0$ includes the weighted penalty term in the control decision and allows a smooth tradeoff between backlog reduction and penalty minimization. We show that the time average objective function deviates by at most $O(1/V)$ from optimality, with a time average queue backlog bound of $O(V)$.

While Lyapunov techniques have a long history in the field of control theory, this form of Lyapunov drift was perhaps first used to construct stable routing and scheduling policies for queueing networks in the pioneering works (7)(8) by Tassiulas and Ephremides. These works used the technique of minimizing $\Delta(t)$ every slot, resulting in *backpressure routing* and *max-weight scheduling* algorithms that stabilize the network whenever possible. The algorithms are particularly interesting because they only require knowledge of the current network state, and they do not require knowledge of the probabilities associated with future random events. Minimizing $\Delta(t)$ has had wide success for stabilizing many other types of networks, including packet switch networks (9)(10)(11), wireless systems (7)(8)(12)(13)(14), and ad-hoc mobile networks (15). A related technique was used for computing multi-commodity network flows in (16).

We introduced the $V \times penalty(t)$ term to the drift minimization in (17)(18)(19) to solve problems of joint network stability and stochastic utility maximization, and we introduced the virtual queue technique in (20)(21) to solve problems of maximizing throughput in a wireless network

[2]The notation used in later chapters is slightly different. Simplified notation is used here to give the main ideas.

subject to individual average power constraints at each node. Our previous text (22) unified these ideas for application to general problems of the type described in Section 1.2.

1.4 DIFFERENCES FROM OUR EARLIER TEXT

The theory of Lyapunov drift and Lyapunov optimization is described collectively in our previous text (22). The current text is different from (22) in that we emphasize the general optimization problems first, showing how the problem (1.6)-(1.11) can be solved directly by using the solution to the simpler problem (1.1)-(1.5). We also provide a variety of examples and problem set questions to help the reader. These have been developed over several years for use in the stochastic network optimization course taught by the author. This text also provides many new topics not covered in (22), including:

- A more detailed development of queue stability theory (Chapter 2).
- Variable-V algorithms that provide exact optimality of time averages subject to a weaker form of stability called "mean rate stability" (Section 4.7).
- Place-holder bits for delay improvement (Sections 3.2.4 and 4.8).
- Universal scheduling for non-ergodic sample paths (Section 4.9).
- Worst case delay bounds (Sections 5.6 and 7.6.1).
- Non-convex stochastic optimization (Section 5.5).
- Approximate scheduling and full throughput scheduling in interference networks via the Jiang-Walrand theorem (Chapter 6).
- Optimization of renewal systems and Markov decision examples (Chapter 7).
- Treatment of problems with equality constraints (1.3) and abstract set constraints (1.9) (Section 5.4).

1.5 ALTERNATIVE APPROACHES

The relationship between network utility maximization, Lagrange multipliers, convex programming, and duality theory is developed for static wireline networks in (2)(23)(24) and for wireless networks in (25)(26)(27)(28)(29) where the goal is to converge to a static flow allocation and/or resource allocation over the network. Scheduling in wireless networks with static channels is considered from a duality perspective in (30)(31). Primal-dual techniques for maximizing utility in a stochastic wireless downlink are developed in (32)(33) for systems without queues. The primal-dual technique is extended in (34)(35) to treat networks with queues and to solve problems similar to (1.6)-(1.11) in a fluid limit sense. Specifically, the work (34) shows the primal-dual technique leads to a fluid limit with an optimal utility, and it conjectures that the utility of the actual network is close to this fluid limit when an exponential averaging parameter is scaled. It makes a statement concerning weak limits of scaled systems. A related primal-dual algorithm is used in (36) and shown to converge to utility-optimality as a parameter is scaled.

Our drift-plus-penalty approach can be viewed as a dual-based approach to the stochastic problem (rather than a primal-dual approach), and it reduces to the well known dual subgradient algorithm for linear and convex programs when applied to non-stochastic problems (see (37)(22)(17) for discussions on this). One advantage of the drift-plus-penalty approach is the explicit convergence analysis and performance bounds, resulting in the $[O(1/V), O(V)]$ performance-delay tradeoff. This tradeoff is not shown in the alternative approaches described above. The dual approach is also robust to non-ergodic variations and has "universal scheduling" properties, i.e., properties that hold for systems with arbitrary sample paths, as shown in Section 4.9 (see also (38)(39)(40)(41)(42)). However, one advantage of the primal-dual approach is that it provides *local optimum* guarantees for problems of minimizing $f(\overline{\boldsymbol{x}})$ for *non-convex functions* $f(\cdot)$ (see Section 5.5 and (43)). Related dual-based approaches are used for "infinitely backlogged" systems in (31)(44)(45)(46) using static optimization, fluid limits, and stochastic gradients, respectively. Related algorithms for channel-aware scheduling in wireless downlinks with different analytical techniques are developed in (47)(48)(49).

We note that the $[O(1/V), O(V)]$ performance-delay tradeoff achieved by the drift-plus-penalty algorithm on general systems is not necessarily the optimal tradeoff for particular networks. An optimal $[O(1/V), O(\sqrt{V})]$ energy-delay tradeoff is shown by Berry and Gallager in (50) for a single link with known channel statistics, and optimal performance-delay tradeoffs for multi-queue systems are developed in (51)(52)(53) and shown to be achievable even when channel statistics are unknown. This latter work builds on the Lyapunov optimization method, but it uses a more aggressive drift steering technique. A *place-holder* technique for achieving near-optimal delay tradeoffs is developed in (37) and related implementations are in (54)(55).

1.6 ON GENERAL MARKOV DECISION PROBLEMS

The penalties $\hat{x}_m(\alpha(t), \omega(t))$, described in Section 1.2, depend only on the network control action $\alpha(t)$ and the random event $\omega(t)$ (where $\omega(t)$ is generated by "nature" and is not influenced by past control actions). In particular, the queue backlogs $\boldsymbol{Q}(t)$ are not included in the penalties. A more advanced penalty structure would be $\hat{x}_m(\alpha(t), \omega(t), z(t))$, where $z(t)$ is a controlled Markov chain (possibly related to the queue backlog) with transition probabilities that depend on control actions. Extensions of Lyapunov optimization for this case are developed in Chapter 7 using a drift-plus-penalty metric defined over renewal frames (56)(57)(58).

A related 2-timescale approach to learning optimal decisions in Markov decision problems is developed in (59), and learning approaches to power-aware scheduling in single queues are developed in (60)(61)(62)(63). Background on dynamic programming and Markov decision problems can be found in (64)(65)(66), and approximate dynamic programming, neuro-dynamic programming, and Q-learning theory can be found in (67)(68)(69). All of these approaches may suffer from large convergence times, high complexity, or inaccurate approximation when applied to large networks. This is due to the *curse of dimensionality* for Markov decision problems. This problem does not arise when using the Lyapunov optimization technique and when penalties have the structure given in Section 1.2.

1.7 ON NETWORK DELAY

This text develops general $[O(1/V), O(V)]$ tradeoffs, giving explicit bounds on average queue backlog and delay that grow linearly with V. We also provide examples of exact delay analysis for randomized algorithms (Exercises 2.6-2.10), delay-limited transmission (Exercises 5.13-5.14 and Section 7.6.1), worst case delay (Section 5.6), and average delay constraints (Section 7.6.2). Further work on delay-limited transmission is found in (70)(71), and Lyapunov drift algorithms that use delays as weights, rather than queue backlogs, are considered in (72)(73)(74)(75)(76). There are many additional interesting topics on network delay that we do not cover in this text. We briefly discuss some of those topics in the following sub-sections, with references given for further reading.

1.7.1 DELAY AND DYNAMIC PROGRAMMING

Dynamic programming and Markov decision frameworks are considered for one-queue energy and delay optimality problems in (77)(78)(79)(80)(81). One-queue problems with strict deadlines and a-priori knowledge of future events are treated in (82)(83)(84)(85)(86), and filter theory is used to establish delay bounds in (87). Control rules for two interacting service stations are given in (88). Optimal scheduling in a finite buffer 2×2 packet switch is treated in (89).

Minimum energy problems with delay deadlines are considered for multi-queue wireless systems in (90). In the case when channels are static, the work (90) maps the problem to a shortest path problem. In the case when channels are varying but rate-power functions are linear, (90) shows the optimal multi-dimensional dynamic program has a very simple threshold structure. Heuristic approximations are given for more general rate-power curves. Related work in (91) considers delay optimal scheduling in multi-queue systems and derives structural results of the dynamic programs, resulting in efficient approximation algorithms. These approximations are shown to have optimal decay exponents for sum queue backlog in (92), which relies on techniques developed in (93) for optimal max-queue exponents. A mixed Lyapunov optimization and dynamic programming approach is given in (56) for networks with a small number of delay-constrained queues and an arbitrarily large number of other queues that only require stability. Approximate dynamic programs and q-learning type algorithms, which attempt to learn optimal decision strategies, are considered in (61)(60)(56)(57)(62)(63).

1.7.2 OPTIMAL $O(\sqrt{V})$ AND $O(\log(V))$ DELAY TRADEOFFS

The $[O(1/V), O(V)]$ performance-delay tradeoffs we derive for general networks in this text are not necessarily the optimal tradeoffs for particular networks. The work (50) considers the optimal energy-delay tradeoff for a one-queue wireless system with a fading channel. It shows that no algorithm can do better than an $[O(1/V), O(\sqrt{V})]$ tradeoff, and it proposes a buffer-partitioning algorithm that can be shown to come within a logarithmic factor of this tradeoff. This optimal $[O(1/V), O(\sqrt{V})]$ tradeoff is extended to multi-queue systems in (51), and an algorithm with an exponential Lyapunov function and aggressive drift steering is shown to meet this tradeoff to within a logarithmic factor. The work (51) also shows an improved $[O(1/V), O(\log(V))]$ tradeoff is achievable in certain exceptional cases with piecewise linear structure.

Optimal $[O(1/V), O(\log(V))]$ energy-delay tradeoffs are shown in (53) in cases when packet dropping is allowed, and optimal $[O(1/V), O(\log(V))]$ utility-delay tradeoffs are shown for flow control problems in (52). Near-optimal $[O(1/V), O(\log^2(V))]$ tradeoffs are shown for the basic quadratic Lyapunov drift-plus-penalty method in (37)(55) using *place-holders* and *Last-In-First-Out (LIFO) scheduling*, described in more detail in Section 4.8, and related implementations are in (54).

1.7.3 DELAY-OPTIMAL ALGORITHMS FOR SYMMETRIC NETWORKS

The works (8)(94)(95)(96)(97) treat multi-queue wireless systems with "symmetry," where arrival rates and channel probabilities are the same for all queues. They use stochastic coupling theory to prove delay optimality for particular algorithms. The work (8) proves delay optimality of the *longest connected queue first* algorithm for ON/OFF channels with a single server, the work (94)(97) considers multi-server systems, and the work (95)(96) considers wireless problems under the information theoretic multi-access capacity region. Related work in (98) proves delay optimality of the *join the shortest queue* strategy for routing packets to two queues with identical exponential service.

1.7.4 ORDER-OPTIMAL DELAY SCHEDULING AND QUEUE GROUPING

The work (99) shows that delay is at least linear in N for $N \times N$ packet switches that use queue-unaware scheduling, and it develops a simple queue-aware scheduling algorithm that gives $O(\log(N))$ delay whenever rates are within the capacity region. Related work in (100) considers scheduling in N-user wireless systems with ON/OFF channels and shows that delay is at least linear in N if queue-unaware algorithms are used, but it can be made $O(1)$ with a simple queue-aware *queue grouping* algorithm. This $O(1)$ delay, independent of the number of users, is called *order optimal* because it differs from optimal only in a constant coefficient that does not depend on N. Order optimality of the simple *longest connected queue first* rule (simpler than the algorithm of (100)) is proven in (101) via a queue grouping analysis.

Order-optimal delay for 1-hop switch scheduling under *maximal scheduling* (which provides stability only when rates are within a constant factor of the capacity boundary) are developed in (102)(103), again using queue grouping theory. In particular, it is shown that $N \times N$ packet switches can provide $O(1)$ delay (order-optimal) if they are at most half-loaded. The best known delay bound beyond the half-loaded region is the $O(\log(N))$ delay result of (99), and it is not known if it is possible to achieve $O(1)$ delay in this region. Time-correlated "bursty" traffic is considered in (103). The queue grouping results in (101)(103) are inspired by queue-grouped Lyapunov functions developed in (104)(105) for stability analysis.

1.7.5 HEAVY TRAFFIC AND DECAY EXPONENTS

A line of work addresses asymptotic delay optimality in a "heavy traffic" regime where input rates are pushed very close to the capacity region boundary. Delay is often easier to understand in this heavy traffic regime due to a phenomenon of *state space collapse* (106). Of course, delay grows to infinity if input rates are pushed toward the capacity boundary, but the goal is to design an algorithm that minimizes an asymptotic growth coefficient. Heavy traffic analysis is considered in (107) for wireless scheduling and (108)(109) for packet switches.

The work (108)(109) suggests that delay in packet switches can be improved by changing the well-known max-weight rule, which seeks to maximize a weighted sum of queue backlog and service rates every slot t ($\sum_i Q_i(t)\mu_i(t)$), to an α-max weight rule that seeks to maximize $\sum_i Q_i(t)^\alpha \mu_i(t)$, where $0 < \alpha \leq 1$. Simulations on $N \times N$ packet switches in (110) show that delay is improved when α is positive but small. A discussion of this in the context of heavy traffic theory is given in (111), along with some counterexamples. It is interesting to note that α-max weight policies with small but positive α make matching decisions that are similar to the max-size matches used in the frame-based algorithm of (99), which achieves $O(\log(N))$ delay. This may be a reason why the delay of α-max weight policies is also small. Large deviation theory is often used to analyze queue backlog and delay, and this is considered for α-max weight policies in (112), for delay-based scheduling in (73), and for processor sharing queues in (113)(114). Algorithms that optimize the exponent of queue backlog are considered in (93) for optimizing the max-queue exponent and in (92) for the sum-queue exponent. These consider analysis of queue backlog when the queue is very large. An analysis of backlog distributions that are valid also in the small buffer regime is given in (115) for the case when the number of network channels is scaled to infinity.

1.7.6 CAPACITY AND DELAY TRADEOFFS FOR MOBILE NETWORKS

Work by Gupta and Kumar in (116) shows that per-node capacity of ad-hoc wireless networks with N nodes and with random source-destination pairings is roughly $\Theta(1/\sqrt{N})$ (neglecting logarithmic factors in N for simplicity). Grossglauser and Tse show in (117) that mobility increases per-node capacity to $\Theta(1)$, which does not vanish with N. However, the algorithm in (117) uses a 2-hop relay algorithm that creates a large delay. The exact capacity and average end-to-end delay are computed in (118)(17) for a cell-partitioned network with a simplified i.i.d. mobility model. The work (118)(17) also shows for this simple model that the average delay $\overline{W}$ of *any* scheduling and routing protocol, possibly one that uses redundant packet transfers, must satisfy:

$$\frac{\overline{W}}{\lambda} \geq \frac{N-d}{4d}(1-\log(2))$$

where λ is the per-user throughput, C is the number of cells, $d = N/C$ is the node/cell density, and $\log(\cdot)$ denotes the natural logarithm. Thus, if the node/cell density $d = \Theta(1)$, then $\overline{W}/\lambda \geq \Omega(N)$. The 2-hop relay algorithm meets this bound with $\lambda = \Theta(1)$ and $\overline{W} = \Theta(N)$, and a relay algorithm that redundantly transmits packets over multiple paths meets this bound with $\lambda = \Theta(1/\sqrt{N})$ and $\overline{W} = \Theta(\sqrt{N})$. Similar i.i.d. mobility models are considered in (119)(120)(121). The work (119) shows that improved tradeoffs are possible if the transmission radius of each node can be scaled to include a large amount of users in each transmission (so that the $d = \Theta(1)$ assumption is relaxed). The work (120)(121) quantifies the optimal tradeoff achievable under this type of radius scaling, and it also shows improved tradeoffs are possible if the model is changed to allow time slot scaling and network bit-pipelining. Related delay tradeoffs via transmission radius scaling for non-mobile networks are in (122). Analysis of non-i.i.d. mobility models is more complex and considered in (123)(124)(122)(125). Recent network coding approaches are in (126)(127)(128).

1.8 PRELIMINARIES

We assume the reader is comfortable with basic concepts of probability and random processes (such as expectations, the law of large numbers, etc.) and with basic mathematical analysis. Familiarity with queueing theory, Markov chains, and convex functions is useful but not required as we present or derive results in these areas as needed in the text. For additional references on queueing theory and Markov chains, including discussions of Little's Theorem and the renewal-reward theorem, see (129)(66)(130)(131)(132). For additional references on convex analysis, including discussions of convex hulls, Caratheodory's theorem, and Jensen's inequality, see (133)(134)(135).

All of the major results of this text are derived directly from one or more of the following four key concepts:

- *Law of Telescoping Sums:* For any function $f(t)$ defined over integer times $t \in \{0, 1, 2, \ldots\}$, we have for any integer time $t > 0$:

$$\sum_{\tau=0}^{t-1}[f(\tau+1) - f(\tau)] = f(t) - f(0)$$

  The proof follows by a simple cancellation of terms. This is the main idea behind Lyapunov drift arguments: Controlling the change in a function at every step allows one to control the ending value of the function.

- *Law of Iterated Expectations:* For any random variables X and Y, we have:[3]

$$\mathbb{E}\{X\} = \mathbb{E}\{\mathbb{E}\{X|Y\}\}$$

[3] Strictly speaking, the law of iterated expectations holds whenever the result of *Fubini's Theorem* holds (which allows one to switch the integration order of a double integral). This holds whenever any one of the following hold: (i) $\mathbb{E}\{|X|\} < \infty$, (ii) $\mathbb{E}\{\max[X, 0]\} < \infty$, (iii) $\mathbb{E}\{\min[X, 0]\} > -\infty$.

where the outer expectation is with respect to the distribution of Y, and the inner expectation is with respect to the conditional distribution of X given Y.

- *Opportunistically Minimizing an Expectation:* Consider a game we play against nature, where nature generates a random variable ω with some (possibly unknown) probability distribution. We look at nature's choice of ω and then choose a control action α within some action set $\mathcal{A}_\omega$ that possibly depends on ω. Let $c(\alpha, \omega)$ represent a general *cost function*. Our goal is to design a (possibly randomized) policy for choosing $\alpha \in \mathcal{A}_\omega$ to minimize the expectation $\mathbb{E}\{c(\alpha, \omega)\}$, where the expectation is taken with respect to the distribution of ω and the distribution of our action α that possibly depends on ω. Assume for simplicity that for any given outcome ω, there is at least one action α_ω^{min} that minimizes the function $c(\alpha, \omega)$ over all $\alpha \in \mathcal{A}_\omega$. Then, not surprisingly, the policy that minimizes $\mathbb{E}\{c(\alpha, \omega)\}$ is the one that observes ω and selects a minimizing action α_ω^{min}.

  This is easy to prove: If α_ω^* represents any random control action chosen in the set $\mathcal{A}_\omega$ in response to the observed ω, we have: $c(\alpha_\omega^{min}, \omega) \leq c(\alpha_\omega^*, \omega)$. This is an inequality relationship concerning the random variables ω, α_ω^{min}, α_ω^*. Taking expectations yields $\mathbb{E}\left\{c(\alpha_\omega^{min}, \omega)\right\} \leq \mathbb{E}\left\{c(\alpha_\omega^*, \omega)\right\}$, showing that the expectation under the policy α_ω^{min} is less than or equal to the expectation under any other policy. This is useful for designing *drift minimizing algorithms*.

- *Jensen's Inequality (not needed until Chapter 5):* Let $\mathcal{X}$ be a convex subset of $\mathbb{R}^M$ (possibly being the full space $\mathbb{R}^M$ itself), and let $f(\boldsymbol{x})$ be a convex function over $\mathcal{X}$. Let $\boldsymbol{X}$ be any random vector that takes values in $\mathcal{X}$, and assume that $\mathbb{E}\{\boldsymbol{X}\}$ is well defined and finite (where the expectation is taken entrywise). Then:

$$\mathbb{E}\{\boldsymbol{X}\} \in \mathcal{X} \quad \text{and} \quad f(\mathbb{E}\{\boldsymbol{X}\}) \leq \mathbb{E}\{f(\boldsymbol{X})\} \tag{1.12}$$

This text also uses, in addition to regular limits of functions, the lim sup and lim inf. Using (or not using) these limits does not impact any of the main ideas in this text, and readers who are not familiar with these limits can replace all instances of "lim sup" and "lim inf" with regular limits "lim," without loss of rigor, under the additional assumption that the regular limit exists. For readers interested in more details on this, note that a function $f(t)$ may or may not have a well defined limit as $t \to \infty$ (consider, for example, a cosine function). We define $\limsup_{t\to\infty} f(t)$ as the largest possible limiting value of $f(t)$ over any subsequence of times t_k that increase to infinity, and for which the limit of $f(t_k)$ exists. Likewise, $\liminf_{t\to\infty} f(t)$ is the smallest possible limiting value. It can be shown that these limits always exist (possibly being ∞ or $-\infty$). For example, the lim sup and lim inf of the cosine function are 1 and -1, respectively. The main properties of lim sup and lim inf that we use in this text are:

- If $f(t)$, $g(t)$ are functions that satisfy $f(t) \leq g(t)$ for all t, then $\limsup_{t\to\infty} f(t) \leq \limsup_{t\to\infty} g(t)$. Likewise, $\liminf_{t\to\infty} f(t) \leq \liminf_{t\to\infty} g(t)$.

- For any function $f(t)$, we have $\liminf_{t\to\infty} f(t) \leq \limsup_{t\to\infty} f(t)$, with equality if and only if the regular limit exists. Further, whenever the regular limit exists, we have $\liminf_{t\to\infty} f(t) = \limsup_{t\to\infty} f(t) = \lim_{t\to\infty} f(t)$.
- For any function $f(t)$, we have $\limsup_{t\to\infty} f(t) = -\liminf_{t\to\infty}[-f(t)]$ and $\liminf_{t\to\infty} f(t) = -\limsup[-f(t)]$.
- If $f(t)$ and $g(t)$ are functions such that $\lim_{t\to\infty} g(t) = g^*$, where g^* is a finite constant, then $\limsup_{t\to\infty}[g(t) + f(t)] = g^* + \limsup_{t\to\infty} f(t)$.

CHAPTER 2

Introduction to Queues

Let $Q(t)$ represent the contents of a single-server discrete time queueing system defined over integer time slots $t \in \{0, 1, 2, \ldots\}$. Specifically, the initial state $Q(0)$ is assumed to be a non-negative real valued random variable. Future states are driven by stochastic arrival and server processes $a(t)$ and $b(t)$ according to the following dynamic equation:

$$Q(t+1) = \max[Q(t) - b(t), 0] + a(t) \quad \text{for } t \in \{0, 1, 2, \ldots\} \tag{2.1}$$

We call $Q(t)$ the *backlog* on slot t, as it can represent an amount of work that needs to be done. The stochastic processes $\{a(t)\}_{t=0}^{\infty}$ and $\{b(t)\}_{t=0}^{\infty}$ are sequences of real valued random variables defined over slots $t \in \{0, 1, 2, \ldots\}$.

The value of $a(t)$ represents the amount of new work that arrives on slot t, and it is assumed to be non-negative. The value of $b(t)$ represents the amount of work the server of the queue can process on slot t. For most physical queueing systems, $b(t)$ is assumed to be non-negative, although it is sometimes convenient to allow $b(t)$ to take negative values. This is useful for the *virtual queues* defined in future sections where $b(t)$ can be interpreted as a (possibly negative) attribute.[1] Because we assume $Q(0) \geq 0$ and $a(t) \geq 0$ for all slots t, it is clear from (2.1) that $Q(t) \geq 0$ for all slots t.

The units of $Q(t)$, $a(t)$, and $b(t)$ depend on the context of the system. For example, in a communication system with fixed size data units, these quantities might be integers with units of *packets*. Alternatively, they might be real numbers with units of *bits*, *kilobits*, or some other unit of unfinished work relevant to the system.

We can equivalently re-write the dynamics (2.1) without the non-linear $\max[\cdot, 0]$ operator as follows:

$$Q(t+1) = Q(t) - \tilde{b}(t) + a(t) \quad \text{for } t \in \{0, 1, 2, \ldots\} \tag{2.2}$$

where $\tilde{b}(t)$ is the actual work processed on slot t (which may be less than the offered amount $b(t)$ if there is little or no backlog in the system on slot t). Specifically, $\tilde{b}(t)$ is mathematically defined:

$$\tilde{b}(t) \triangleq \min[b(t), Q(t)]$$

[1] Assuming that the $b(t)$ value in (2.1) is possibly negative also allows treatment of modified queueing models that place new arrivals inside the $\max[\cdot, 0]$ operator. For example, a queue with dynamics $\hat{Q}(t+1) = \max[\hat{Q}(t) - \beta(t) + \alpha(t), 0]$ is the same as (2.1) with $a(t) = 0$ and $b(t) = \beta(t) - \alpha(t)$ for all t. Leaving $a(t)$ outside the $\max[\cdot, 0]$ is crucial for treatment of multi-hop networks, where $a(t)$ can be a sum of exogenous and endogenous arrivals.

Note by definition that $\tilde{b}(t) \leq b(t)$ for all t. The dynamic equation (2.2) yields a simple but important property for all sample paths, described in the following lemma.

Lemma 2.1 *(Sample Path Property) For any discrete time queueing system described by (2.1), and for any two slots t_1 and t_2 such that $0 \leq t_1 < t_2$, we have:*

$$Q(t_2) - Q(t_1) = \sum_{\tau=t_1}^{t_2-1} a(\tau) - \sum_{\tau=t_1}^{t_2-1} \tilde{b}(\tau) \tag{2.3}$$

Therefore, for any $t > 0$, we have:

$$\frac{Q(t)}{t} - \frac{Q(0)}{t} = \frac{1}{t}\sum_{\tau=0}^{t-1} a(\tau) - \frac{1}{t}\sum_{\tau=0}^{t-1} \tilde{b}(\tau) \tag{2.4}$$

$$\frac{Q(t)}{t} - \frac{Q(0)}{t} \geq \frac{1}{t}\sum_{\tau=0}^{t-1} a(\tau) - \frac{1}{t}\sum_{\tau=0}^{t-1} b(\tau) \tag{2.5}$$

Proof. By (2.2), we have for any slot $\tau \geq 0$:

$$Q(\tau + 1) - Q(\tau) = a(\tau) - \tilde{b}(\tau)$$

Summing the above over $\tau \in \{t_1, \ldots, t_2 - 1\}$ and using the law of telescoping sums yields:

$$Q(t_2) - Q(t_1) = \sum_{\tau=t_1}^{t_2-1} a(\tau) - \sum_{\tau=t_1}^{t_2-1} \tilde{b}(\tau)$$

This proves (2.3). Inequality (2.4) follows by substituting $t_1 = 0, t_2 = t$, and dividing by t. Inequality (2.5) follows because $\tilde{b}(\tau) \leq b(\tau)$ for all τ. □

An important application of Lemma 2.1 to power-aware systems is treated in Exercise 2.11. The equality (2.4) is illuminating. It shows that $\lim_{t\to\infty} Q(t)/t = 0$ if and only if the time average of the process $a(t) - \tilde{b}(t)$ is zero (where the time average of $a(t) - \tilde{b}(t)$ is the limit of the right-hand-side of (2.4)). This happens when the time average rate of arrivals $a(t)$ is equal to the time average rate of actual departures $\tilde{b}(t)$. This motivates the definitions of *rate stability* and *mean rate stability*, defined in the next section.

2.1 RATE STABILITY

Let $Q(t)$ be a real valued stochastic process that evolves in discrete time over slots $t \in \{0, 1, 2, \ldots\}$ according to some probability law.

Definition 2.2 A discrete time process $Q(t)$ is *rate stable* if:

$$\lim_{t\to\infty} \frac{Q(t)}{t} = 0 \text{ with probability 1}$$

Definition 2.3 A discrete time process $Q(t)$ is *mean rate stable* if:

$$\lim_{t\to\infty} \frac{\mathbb{E}\{|Q(t)|\}}{t} = 0$$

We use an absolute value of $Q(t)$ in the mean rate stability definition, even though our queue in (2.1) is non-negative, because later it will be useful to define mean rate stability for virtual queues that can be possibly negative.

Theorem 2.4 *(Rate Stability Theorem) Suppose $Q(t)$ evolves according to (2.1), with $a(t) \geq 0$ for all t, and with $b(t)$ real valued (and possibly negative) for all t. Suppose that the time averages of the processes $a(t)$ and $b(t)$ converge with probability 1 to finite constants a^{av} and b^{av}, so that:*

$$\lim_{t\to\infty} \frac{1}{t}\sum_{\tau=0}^{t-1} a(\tau) = a^{av} \quad \textit{with probability } 1 \tag{2.6}$$

$$\lim_{t\to\infty} \frac{1}{t}\sum_{\tau=0}^{t-1} b(\tau) = b^{av} \quad \textit{with probability } 1 \tag{2.7}$$

Then:

(a) $Q(t)$ is rate stable if and only if $a^{av} \leq b^{av}$.
(b) If $a^{av} > b^{av}$, then:

$$\lim_{t\to\infty} \frac{Q(t)}{t} = a^{av} - b^{av} \quad \textit{with probability 1}$$

(c) Suppose there are finite constants $\epsilon > 0$ and $C > 0$ such that $\mathbb{E}\left\{[a(t) + b^-(t)]^{1+\epsilon}\right\} \leq C$ for all t, where $b^-(t) \triangleq -\min[b(t), 0]$. Then $Q(t)$ is mean rate stable if and only if $a^{av} \leq b^{av}$.

Proof. Here we prove only the necessary condition of part (a). Suppose that $Q(t)$ is rate stable, so that $Q(t)/t \rightarrow 0$ with probability 1. Because (2.5) holds for all slots $t > 0$, we can take limits in (2.5) as $t \rightarrow \infty$ and use (2.6)-(2.7) to conclude that $0 \geq a^{av} - b^{av}$. Thus, $a^{av} \leq b^{av}$ is *necessary* for rate stability. The proof for sufficiency in part (a) and the proof of part (b) are developed in Exercises 2.3 and 2.4. The proof of part (c) is more complex and is omitted (see (136)). □

The following theorem presents a more general necessary condition for rate stability that does not require the arrival and server processes to have well defined limits.

Theorem 2.5 *(Necessary Condition for Rate Stability) Suppose $Q(t)$ evolves according to (2.1), with any general processes $a(t)$ and $b(t)$ such that $a(t) \geq 0$ for all t. Then:*

(a) If $Q(t)$ is rate stable, then:

$$\limsup_{t\rightarrow\infty} \frac{1}{t}\sum_{\tau=0}^{t-1}[a(\tau) - b(\tau)] \leq 0 \quad \textit{with probability 1} \tag{2.8}$$

(b) If $Q(t)$ is mean rate stable and if $\mathbb{E}\{Q(0)\} < \infty$, then:

$$\limsup_{t\rightarrow\infty} \frac{1}{t}\sum_{\tau=0}^{t-1}\mathbb{E}\{a(\tau) - b(\tau)\} \leq 0 \tag{2.9}$$

Proof. The proof of (a) follows immediately by taking a lim sup of both sides of (2.5) and noting that $Q(t)/t \rightarrow 0$ because $Q(t)$ is rate stable. The proof of (b) follows by first taking an expectation of (2.5) and then taking limits. □

2.2 STRONGER FORMS OF STABILITY

Rate stability and mean rate stability only describe the long term average rate of arrivals and departures from the queue, and do not say anything about the fraction of time the queue backlog exceeds a certain value, or about the time average expected backlog. The stronger stability definitions given below are thus useful.

Definition 2.6 A discrete time process $Q(t)$ is *steady state stable* if:

$$\lim_{M\rightarrow\infty} g(M) = 0$$

where for each $M \geq 0$, $g(M)$ is defined:

$$g(M) \triangleq \limsup_{t\rightarrow\infty} \frac{1}{t}\sum_{\tau=0}^{t-1} Pr[|Q(\tau)| > M] \tag{2.10}$$

Definition 2.7 A discrete time process $Q(t)$ is *strongly stable* if:

$$\limsup_{t\to\infty} \frac{1}{t}\sum_{\tau=0}^{t-1} \mathbb{E}\{|Q(\tau)|\} < \infty \tag{2.11}$$

Under mild boundedness assumptions, strong stability implies all of the other forms of stability, as specified in Theorem 2.8 below.

Theorem 2.8 *(Strong Stability Theorem) Suppose $Q(t)$ evolves according to (2.1) for some general stochastic processes $\{a(t)\}_{t=0}^{\infty}$ and $\{b(t)\}_{t=0}^{\infty}$, where $a(t) \geq 0$ for all t, and $b(t)$ is real valued for all t. Suppose $Q(t)$ is strongly stable. Then:*

(a) $Q(t)$ is steady state stable.

(b) If there is a finite constant C such that either $a(t) + b^-(t) \leq C$ with probability 1 for all t (where $b^-(t) \triangleq -\min[b(t), 0]$), or $b(t) - a(t) \leq C$ with probability 1 for all t, then $Q(t)$ is rate stable, so that $Q(t)/t \to 0$ with probability 1.

(c) If there is a finite constant C such that either $\mathbb{E}\left\{a(t) + b^-(t)\right\} \leq C$ for all t, or $\mathbb{E}\{b(t) - a(t)\} \leq C$ for all t, then $Q(t)$ is mean rate stable.

Proof. Part (a) is given in Exercise 2.5. Parts (b) and (c) are omitted (see (136)). □

Readers familiar with discrete time Markov chains (DTMCs) may be interested in the following connection: For processes $Q(t)$ defined over an ergodic DTMC with a finite or countably infinite state space and with the property that, for each real value M, the event $\{|Q(t)| \leq M\}$ corresponds to only a finite number of states, steady state stability implies the existence of a steady state distribution, and strong stability implies finite average backlog and (by Little's theorem (129)) finite average delay.

2.3 RANDOMIZED SCHEDULING FOR RATE STABILITY

The Rate Stability Theorem (Theorem 2.4) suggests the following simple method for stabilizing a multi-queue network: Make scheduling decisions so that the time average service and arrival rates are well defined and satisfy $a_i^{av} \leq b_i^{av}$ for each queue i. This method typically requires perfect knowledge of the arrival and channel probabilities so that the desired time averages can be achieved. Some representative examples are provided below. A better method that does not require a-priori statistical knowledge is developed in Chapters 3 and 4.

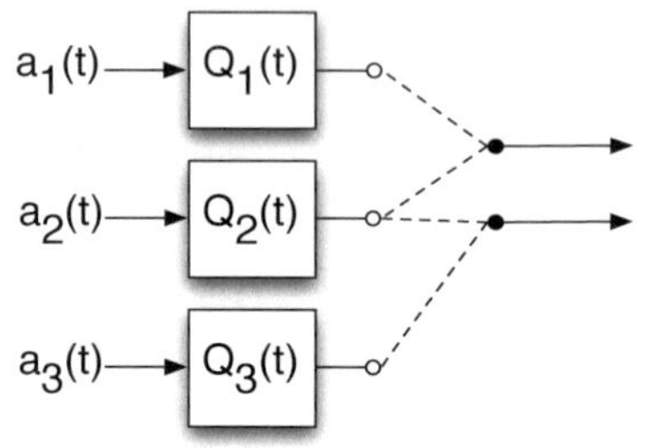

Figure 2.1: A 3-queue, 2-server system. Every slot the network controller decides which 2 queues receive servers. A single queue cannot receive 2 servers on the same slot.

2.3.1 A 3-QUEUE, 2-SERVER EXAMPLE

Example Problem: Consider the 3-queue, 2-server system of Fig. 2.1. All packets have fixed length, and a queue that is allocated a server on a given slot can serve exactly one packet on that slot. Every slot we choose which 2 queues to serve. The service is given for $i \in \{1, 2, 3\}$ by:

$$b_i(t) = \begin{cases} 1 & \text{if a server is connected to queue } i \text{ on slot } t \\ 0 & \text{otherwise} \end{cases}$$

Assume the arrival processes have well defined time average rates $(a_1^{av}, a_2^{av}, a_3^{av})$, in units of packets/slot. Design a server allocation algorithm to make all queues rate stable when arrival rates are given as follows:

a) $(a_1^{av}, a_2^{av}, a_3^{av}) = (0.5, 0.5, 0.9)$
b) $(a_1^{av}, a_2^{av}, a_3^{av}) = (2/3, 2/3, 2/3)$
c) $(a_1^{av}, a_2^{av}, a_3^{av}) = (0.7, 0.9, 0.4)$
d) $(a_1^{av}, a_2^{av}, a_3^{av}) = (0.65, 0.5, 0.75)$
e) Use (2.5) to prove that the constraints $0 \leq a_i^{av} \leq 1$ for all $i \in \{1, 2, 3\}$, and $a_1^{av} + a_2^{av} + a_3^{av} \leq 2$, are necessary for the existence of a rate stabilizing algorithm.

Solution:

a) Choose the service vector $(b_1(t), b_2(t), b_3(t))$ to be independent and identically distributed (i.i.d.) every slot, choosing $(0, 1, 1)$ with probability $1/2$ and $(1, 0, 1)$ with probability $1/2$. Then $\{b_1(t)\}_{t=0}^{\infty}$ is i.i.d. over slots with $b_1^{av} = 0.5$ by the law of large numbers. Likewise, $b_2^{av} = 0.5$ and $b_3^{av} = 1$. Then clearly $a_i^{av} \leq b_i^{av}$ for all $i \in \{1, 2, 3\}$, and so the Rate Stability Theorem ensures all queues are rate stable. While this is a *randomized* scheduling algorithm, one could also design a deterministic algorithm, such as one that alternates between $(0, 1, 1)$ (on odd slots) and $(1, 0, 1)$ (on even slots).

b) Choose $(b_1(t), b_2(t), b_3(t))$ i.i.d. over slots, equally likely over the three options $(1, 1, 0)$, $(1, 0, 1)$, and $(0, 1, 1)$. Then $b_i^{av} = 2/3 = a_i^{av}$ for all $i \in \{1, 2, 3\}$, and so by the Rate Stability Theorem all queues are rate stable.

c) Every slot, independently choose the service vector $(0, 1, 1)$ with probability p_1, $(1, 0, 1)$ with probability p_2, and $(1, 1, 0)$ with probability p_3, so that p_1, p_2, p_3 satisfy:

$$p_1(0,1,1) + p_2(1,0,1) + p_3(1,1,0) \geq (0.7, 0.9, 0.4) \tag{2.12}$$
$$p_1 + p_2 + p_3 = 1 \tag{2.13}$$
$$p_i \geq 0 \ \forall i \in \{1, 2, 3\} \tag{2.14}$$

where the inequality (2.12) is taken entrywise. This is an example of a *linear program*. Linear programs are typically difficult to solve by hand, but this one can be solved easily by guessing that the constraint in (2.12) can be solved with equality. One can verify the following (unique) solution: $p_1 = 0.3$, $p_2 = 0.1$, $p_3 = 0.6$. Thus, $b_1^{av} = p_2 + p_3 = 0.7$, $b_2^{av} = p_1 + p_3 = 0.9$, $b_3^{av} = p_1 + p_2 = 0.4$, and so all queues are rate stable by the Rate Stability Theorem. It is an interesting exercise to design an alternative *deterministic* algorithm that uses a periodic schedule to produce the same time averages.

d) Use the same linear program (2.12)-(2.14), but replace the constraint (2.12) with the following:

$$p_1(0,1,1) + p_2(1,0,1) + p_3(1,1,0) \geq (0.65, 0.5, 0.75)$$

This can be solved by hand by trial-and-error. One simplifying trick is to replace the above inequality constraint with the following equality constraint:

$$p_1(0,1,1) + p_2(1,0,1) + p_3(1,1,0) = (0.7, 0.5, 0.8)$$

Then we can use $p_1 = 0.3$, $p_2 = 0.5$, $p_3 = 0.2$.

e) Consider any algorithm that makes all queues rate stable, and let $b_i(t)$ be the queue-i decision made by the algorithm on slot t. For each queue i, we have for all $t > 0$:

$$\begin{aligned}\frac{Q_i(t)}{t} - \frac{Q_i(0)}{t} &\geq \frac{1}{t}\sum_{\tau=0}^{t-1} a_i(\tau) - \frac{1}{t}\sum_{\tau=0}^{t-1} b_i(\tau)\\ &\geq \frac{1}{t}\sum_{\tau=0}^{t-1} a_i(\tau) - 1\end{aligned}$$

where the first inequality follows by (2.5) and the final inequality holds because $b_i(\tau) \leq 1$ for all τ. The above holds for all $t > 0$. Taking a limit as $t \to \infty$ and using the fact that queue i is rate stable yields, with probability 1:

$$0 \geq a_i^{av} - 1$$

and so we find that, for each $i \in \{1, 2, 3\}$, the condition $a_i^{av} \leq 1$ is *necessary* for the existence of an algorithm that makes all queues rate stable. Similarly, we have:

$$\begin{aligned}
\frac{Q_1(t) + Q_2(t) + Q_3(t)}{t} - \frac{Q_1(0) + Q_2(0) + Q_3(0)}{t} \\
&\geq \frac{1}{t}\sum_{\tau=0}^{t-1}[a_1(\tau) + a_2(\tau) + a_3(\tau)] - \frac{1}{t}\sum_{\tau=0}^{t-1}[b_1(\tau) + b_2(\tau) + b_3(\tau)] \\
&\geq \frac{1}{t}\sum_{\tau=0}^{t-1}[a_1(\tau) + a_2(\tau) + a_3(\tau)] - 2
\end{aligned}$$

where the final inequality holds because $b_1(\tau) + b_2(\tau) + b_3(\tau) \leq 2$ for all τ. Taking limits shows that $0 \geq a_1^{av} + a_2^{av} + a_3^{av} - 2$ is also a necessary condition.

Discussion: Define Λ as the set of all rate vectors $(a_1^{av}, a_2^{av}, a_3^{av})$ that satisfy the constraints in part (e) of the above example problem. We know from part (e) that $(a_1^{av}, a_2^{av}, a_3^{av}) \in \Lambda$ is a *necessary* condition for existence of an algorithm that makes all queues rate stable. Further, it can be shown that for any vector $(a_1^{av}, a_2^{av}, a_3^{av}) \in \Lambda$, there exist probabilities p_1, p_2, p_3 that solve the following linear program:

$$\begin{aligned}
p_1(0, 1, 1) + p_2(1, 0, 1) + p_3(1, 1, 0) &\geq (a_1^{av}, a_2^{av}, a_3^{av}) \\
p_1 + p_2 + p_3 &= 1 \\
p_i &\geq 0 \;\; \forall i \in \{1, 2, 3\}
\end{aligned}$$

Showing this is not trivial and is left as an advanced exercise. However, this fact, together with the Rate Stability Theorem, shows that it is possible to design an algorithm to make all queues rate stable whenever $(a_1^{av}, a_2^{av}, a_3^{av}) \in \Lambda$. That is, $(a_1^{av}, a_2^{av}, a_3^{av}) \in \Lambda$ is *necessary and sufficient* for the existence of an algorithm that makes all queues rate stable. The set Λ is called the *capacity region* for the network. Exercises 2.7 and 2.8 provide additional practice questions about scheduling and delay in this system.

2.3.2 A 2-QUEUE OPPORTUNISTIC SCHEDULING EXAMPLE

Example Problem: Consider a 2-queue wireless downlink that operates in discrete time (Fig. 2.2a). All data consists of fixed length packets. The arrival process $(a_1(t), a_2(t))$ represents the (integer) number of packets that arrive to each queue on slot t. There are two wireless channels, and packets in queue i must be transmitted over channel i, for $i \in \{1, 2\}$. At the beginning ofeach slot, the network controller observes the *channel state vector* $\boldsymbol{S}(t) = (S_1(t), S_2(t))$, where $S_i(t) \in \{ON, OFF\}$, so that there are four possible channel state vectors. The controller can transmit at most one packet per slot, and it can only transmit a packet over a channel that is ON. Thus, for each channel $i \in \{1, 2\}$, we have:

$$b_i(t) = \begin{cases} 1 & \text{if } S_i(t) = ON \text{ and channel } i \text{ is chosen for transmission on slot } t \\ 0 & \text{otherwise} \end{cases}$$

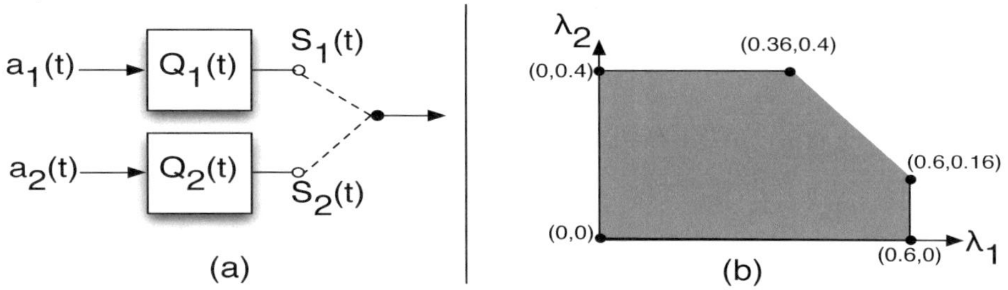

Figure 2.2: (a) The 2-queue, 1-server opportunistic scheduling system with ON/OFF channels. (b) The capacity region for the specific channel probabilities given below.

If $\boldsymbol{S}(t) = (OFF, OFF)$, then $b_1(t) = b_2(t) = 0$. If exactly one channel is ON, then clearly the controller should choose to transmit over that channel. The only decision is which channel to use when $\boldsymbol{S}(t) = (ON, ON)$. Suppose that $(a_1(t), a_2(t))$ is i.i.d. over slots with $\mathbb{E}\{a_1(t)\} = \lambda_1$ and $\mathbb{E}\{a_2(t)\} = \lambda_2$. Suppose that $\boldsymbol{S}(t)$ is i.i.d. over slots with $Pr[(OFF, OFF)] \triangleq p_{00}$, $Pr[(OFF, ON)] = p_{01}$, $Pr[(ON, OFF)] = p_{10}$, $Pr[ON, ON] = p_{11}$.

a) Define Λ as the set of all vectors (λ_1, λ_2) that satisfy the constraints $0 \leq \lambda_1 \leq p_{10} + p_{11}, 0 \leq \lambda_2 \leq p_{01} + p_{11}, \lambda_1 + \lambda_2 \leq p_{01} + p_{10} + p_{11}$. Show that $(\lambda_1, \lambda_2) \in \Lambda$ is necessary for the existence of a rate stabilizing algorithm.

b) Plot the 2-dimensional region Λ for the special case when $p_{00} = 0.24$, $p_{10} = 0.36$, $p_{01} = 0.16$, $p_{11} = 0.24$.

c) For the system of part (b): Use a randomized algorithm that independently transmits over channel 1 with probability β whenever $\boldsymbol{S}(t) = (ON, ON)$. Choose β to make both queues rate stable when $(\lambda_1, \lambda_2) = (0.6, 0.16)$.

d) For the system of part (b): Choose β to make both queues rate stable when $(\lambda_1, \lambda_2) = (0.5, 0.26)$.

Solution:

a) Let $b_1(t), b_2(t)$ be the decisions made by a particular algorithm that makes both queues rate stable. From (2.5), we have for queue 1 and for all slots $t > 0$:

$$\frac{Q_1(t)}{t} - \frac{Q_1(0)}{t} \geq \frac{1}{t}\sum_{\tau=0}^{t-1} a_1(\tau) - \frac{1}{t}\sum_{\tau=0}^{t-1} b_1(\tau)$$

Because $b_1(\tau) \leq 1_{\{S_1(\tau)=ON\}}$, where the latter is an indicator function that is 1 if $S_1(\tau) = ON$, and 0 else, we have:

$$\frac{Q_1(t)}{t} - \frac{Q_1(0)}{t} \geq \frac{1}{t}\sum_{\tau=0}^{t-1} a_1(\tau) - \frac{1}{t}\sum_{\tau=0}^{t-1} 1_{\{S_1(\tau)=ON\}} \tag{2.15}$$

However, we know that $Q_1(t)/t \to 0$ with probability 1. Further, by the law of large numbers, we have (with probability 1):

$$\lim_{t\to\infty} \frac{1}{t}\sum_{\tau=0}^{t-1} a_1(\tau) = \lambda_1 \ , \ \lim_{t\to\infty} \frac{1}{t}\sum_{\tau=0}^{t-1} 1_{\{S_1(\tau)=ON\}} = p_{10} + p_{11}$$

Thus, taking a limit as $t \to \infty$ in (2.15) yields:

$$0 \geq \lambda_1 - (p_{10} + p_{11})$$

and hence $\lambda_1 \leq p_{10} + p_{11}$ is a necessary condition for any rate stabilizing algorithm. A similar argument shows that $\lambda_2 \leq p_{01} + p_{11}$ is a necessary condition. Finally, note that for all $t > 0$:

$$\frac{Q_1(t) + Q_2(t)}{t} - \frac{Q_1(0) + Q_2(0)}{t} \geq \frac{1}{t}\sum_{\tau=0}^{t-1}[a_1(\tau) + a_2(\tau)] - \frac{1}{t}\sum_{\tau=0}^{t-1} 1_{\{\{S_1(\tau)=ON\}\cup\{S_2(\tau)=ON\}\}}$$

Taking a limit of the above proves that $\lambda_1 + \lambda_2 \leq p_{01} + p_{10} + p_{11}$ is necessary.

b) See Fig. 2.2b.

c) If $\boldsymbol{S}(t) = (OFF, OFF)$ then don't transmit. If $\boldsymbol{S}(t) = (ON, OFF)$ or (ON, ON) then transmit over channel 1. If $\boldsymbol{S}(t) = (OFF, ON)$, then transmit over channel 2. Then by the law of large numbers, we have $b_1^{av} = p_{10} + p_{11} = 0.6$, $b_2^{av} = p_{01} = 0.16$, and so both queues are rate stable (by the Rate Stability Theorem).

d) Choose $\beta = 0.14/0.24$. Then $b_1^{av} = 0.36 + 0.24\beta = 0.5$, and $b_2^{av} = 0.16 + 0.24(1 - \beta) = 0.26$.

Discussion: Exercise 2.9 treats scheduling and delay issues in this system. It can be shown that the set Λ given in part (a) above is the *capacity region*, so that $(\lambda_1, \lambda_2) \in \Lambda$ is *necessary and sufficient* for the existence of a rate stabilizing policy. See (8) for the derivation of the capacity region for ON/OFF opportunistic scheduling systems with K queues (with $K \geq 2$). See also (8) for optimal delay scheduling in symmetric systems of this type (where all arrival rates are the same, as are all ON/OFF probabilities), and (101)(100) for "order-optimal" delay in general (possibly asymmetric) situations.

It is possible to support any point in Λ using a *stationary randomized policy* that makes a scheduling decision as a random function of the observed channel state $\boldsymbol{S}(t)$. Such policies are called $\boldsymbol{S}$*-only policies*. The solutions given in parts (c) and (d) above use $\boldsymbol{S}$-only policies. Further, the randomized server allocation policies considered in the 3-queue, 2-server example of Section 2.3.1 can be viewed as "degenerate" $\boldsymbol{S}$-only policies, because, in that case, there is only one "channel state" (i.e., (ON, ON, ON)). It is known that the capacity region of general single-hop and multi-hop networks with time varying channels $\boldsymbol{S}(t)$ can be described in terms of $\boldsymbol{S}$-only policies (15)(22) (see also Theorem 4.5 of Chapter 4 for a related result for more general systems).

Note that $\boldsymbol{S}$-only policies do not consider queue backlog information, and thus they may serve a queue that is empty, which is clearly inefficient. Thus, one might wonder how $\boldsymbol{S}$-only policies can stabilize queueing networks whenever traffic rates are inside the capacity region. Intuitively, the reason is that inefficiency only arises when a queue becomes empty, a rare event when traffic rates are near the boundary of the capacity region.[2] Thus, using queue backlog information cannot "enlarge" the region of supportable rates. However, Chapter 3 shows that queue backlogs are extremely useful for designing dynamic algorithms that do not require a-priori knowledge of channel statistics or a-priori computation of a randomized policy with specific time averages.

2.4 EXERCISES

Exercise 2.1. (Queue Sample Path) Fill in the missing entries of the table in Fig. 2.3 for a queue $Q(t)$ that satisfies (2.1).

	t	0	1	2	3	4	5	6	7	8	9	10
Arrivals	$a(t)$	3	3	0	2	1	0	0	2		0	0
Current Rate	$b(t)$	4	2	1	3	3	2	2	4	0	2	1
Backlog	$Q(t)$	0	3	4	3						2	
Transmitted	$\tilde{b}(t)$	0	2	1		2	1					

Figure 2.3: An example sample path for the queueing system of Exercise 2.1.

Exercise 2.2. (Inequality comparison) Let $Q(t)$ satisfy (2.1) with server process $b(t)$ and arrival process $a(t)$. Let $\tilde{Q}(t)$ be another queueing system with the same server process $b(t)$ but with an arrival process $\tilde{a}(t) = a(t) + z(t)$, where $z(t) \geq 0$ for all $t \in \{0, 1, 2, \ldots\}$. Assuming that $Q(0) = \tilde{Q}(0)$, prove that $Q(t) \leq \tilde{Q}(t)$ for all $t \in \{0, 1, 2, \ldots\}$.

Exercise 2.3. (Proving sufficiency for Theorem 2.4a) Let $Q(t)$ satisfy (2.1) with arrival and server processes with well defined time averages a^{av} and b^{av}. Suppose that $a^{av} \leq b^{av}$. Fix $\epsilon > 0$, and define $Q_\epsilon(t)$ as a queue with $Q_\epsilon(0) = Q(0)$, and with the same server process $b(t)$ but with an arrival process $\tilde{a}(t) = a(t) + (b^{av} - a^{av}) + \epsilon$ for all t.

a) Compute the time average of $\tilde{a}(t)$.

b) Assuming the result of Theorem 2.4b, compute $\lim_{t\to\infty} Q_\epsilon(t)/t$.

c) Use the result of part (b) and Exercise 2.2 to prove that $Q(t)$ is rate stable. Hint: I am thinking of a non-negative number x. My number has the property that $x \leq \epsilon$ for all $\epsilon > 0$. What is my number?

[2] For example, in the GI/B/1 queue of Exercise 2.6, it can be shown by Little's Theorem (129) that the fraction of time the queue is empty is $1 - \lambda/\mu$ (assuming $\lambda \leq \mu$), which goes to zero when $\lambda \to \mu$.

Exercise 2.4. (Proof of Theorem 2.4b) Let $Q(t)$ be a queue that satisfies (2.1). Assume time averages of $a(t)$ and $b(t)$ are given by finite constants a^{av} and b^{av}, respectively.

a) Use the following equation to prove that $\lim_{t\to\infty} a(t)/t = 0$ with probability 1:

$$\frac{1}{t+1}\sum_{\tau=0}^{t} a(\tau) = \left(\frac{t}{t+1}\right)\frac{1}{t}\sum_{\tau=0}^{t-1} a(\tau) + \left(\frac{t}{t+1}\right)\frac{a(t)}{t}$$

b) Suppose that $\tilde{b}(t_i) < b(t_i)$ for some slot t_i (where we recall that $\tilde{b}(t_i) \triangleq \min[b(t_i), Q(t_i)]$). Use (2.1) to compute $Q(t_i + 1)$.

c) Use part (b) and (2.5) to show that if $\tilde{b}(t_i) < b(t_i)$, then:

$$a(t_i) \geq Q(0) + \sum_{\tau=0}^{t_i} [a(\tau) - b(\tau)]$$

Conclude that if $\tilde{b}(t_i) < b(t_i)$ for an infinite number of slots t_i, then $a^{av} \leq b^{av}$.

d) Use part (c) to conclude that if $a^{av} > b^{av}$, there is some slot $t^* \geq 0$ such that for all $t > t^*$, we have:

$$Q(t) = Q(t^*) + \sum_{\tau=t^*}^{t-1} [a(\tau) - b(\tau)]$$

Use this to prove the result of Theorem 2.4b.

Exercise 2.5. (Strong stability implies steady state stability) Prove that strong stability implies steady state stability using the fact that $\mathbb{E}\{|Q(\tau)|\} \geq M Pr[|Q(\tau)| > M]$.

Exercise 2.6. (Discrete time GI/B/1 queue) Consider a queue $Q(t)$ with dynamics (2.1). Assume that $a(t)$ is i.i.d. over slots with non-negative integer values, with $\mathbb{E}\{a(t)\} = \lambda$ and $\mathbb{E}\left\{a(t)^2\right\} = \mathbb{E}\left\{a^2\right\}$. Assume that $b(t)$ is independent of the arrivals and is i.i.d. over slots with $Pr[b(t) = 1] = \mu$, $Pr[b(t) = 0] = 1 - \mu$. Thus, $Q(t)$ is always integer valued. Suppose that $\lambda < \mu$, and that there are finite values $\mathbb{E}\{Q\}$, $\overline{Q}$, Q^{av}, $\mathbb{E}\left\{Q^2\right\}$ such that:

$$\lim_{t\to\infty} \frac{1}{t}\sum_{\tau=0}^{t-1} \mathbb{E}\{Q(\tau)\} = \overline{Q} \ , \ \lim_{t\to\infty} \frac{1}{t}\sum_{\tau=0}^{t-1} Q(\tau) = Q^{av} \text{ with prob. 1}$$

$$\lim_{t\to\infty} \mathbb{E}\{Q(t)\} = \mathbb{E}\{Q\} \ , \ \lim_{t\to\infty} \mathbb{E}\left\{Q(t)^2\right\} = \mathbb{E}\left\{Q^2\right\}$$

Using ergodic Markov chain theory, it can be shown that $\overline{Q} = Q^{av} = \mathbb{E}\{Q\}$ (see also Exercise 7.9). Here we want to compute $\mathbb{E}\{Q\}$, using the magic of a quadratic.

a) Take expectations of equation (2.2) to find $\lim_{t\to\infty} \mathbb{E}\left\{\tilde{b}(t)\right\}$.

b) Explain why $\tilde{b}(t)^2 = \tilde{b}(t)$ and $Q(t)\tilde{b}(t) = Q(t)b(t)$.
c) Square equation (2.2) and use part (b) to prove:

$$Q(t+1)^2 = Q(t)^2 + \tilde{b}(t) + a(t)^2 - 2Q(t)(b(t) - a(t)) - 2\tilde{b}(t)a(t)$$

d) Take expectations in (c) and let $t \to \infty$ to conclude that:

$$\mathbb{E}\{Q\} = \frac{\mathbb{E}\{a^2\} + \lambda - 2\lambda^2}{2(\mu - \lambda)}$$

We have used the fact that $Q(t)$ is independent of $b(t)$, even though it is *not* independent of $\tilde{b}(t)$. This establishes the average backlog for an integer-based GI/B/1 queue (where "GI" means the arrivals are general and i.i.d. over slots, "B" means the service is i.i.d. Bernoulli, and "1" means there is a single server). By Little's Theorem (129), it follows that average delay (in units of slots) is $\overline{W} = \overline{Q}/\lambda$. When the arrival process is Bernoulli, these formulas simplify to $\overline{Q} = \lambda(1-\lambda)/(\mu - \lambda)$ and $\overline{W} = (1-\lambda)/(\mu-\lambda)$. Using reversible Markov chain theory (130)(66)(131), it can be shown that the steady state output process of a B/B/1 queue is also i.i.d. Bernoulli with rate λ (regardless of μ, provided that $\lambda < \mu$), which makes analysis of tandems of B/B/1 queues very easy.

Exercise 2.7. (Server Scheduling) Consider the 3-queue, 2-server system example of Section 2.3.1 (Fig. 2.1). Assume the arrival vector $(a_1(t), a_2(t), a_3(t))$ is i.i.d. over slots with $\mathbb{E}\{a_i(t)\} = \lambda_i$ for $i \in \{1, 2, 3\}$. Design a randomized server allocation algorithm to make all queues rate stable when:
a) $(\lambda_1, \lambda_2, \lambda_3) = (0.2, 0.9, 0.6)$
b) $(\lambda_1, \lambda_2, \lambda_3) = (3/4, 3/4, 1/2)$
c) $(\lambda_1, \lambda_2, \lambda_3) = (0.6, 0.5, 0.9)$
d) $(\lambda_1, \lambda_2, \lambda_3) = (0.7, 0.6, 0.5)$
e) Give a deterministic algorithm that uses a periodic schedule to support the rates in part (b).
f) Give a deterministic algorithm that uses a periodic schedule to support the rates in part (c).

Exercise 2.8. (Delay for Server Scheduling) Consider the 3-queue, 2-server system of Fig. 2.1 that operates according to the randomized schedule of the solution given in part (d) of Section 2.3.1, so that $p_1 = 0.3$, $p_2 = 0.5$, $p_3 = 0.2$. Suppose $a_1(t)$ is i.i.d. over slots and Bernoulli, with $Pr[a_1(t) = 0] = 0.35$, $Pr[a_1(t) = 1] = 0.65$. Use the formula of Exercise 2.6 to compute the average backlog $\overline{Q}_1$ and average delay $\overline{W}_1$ in queue 1. (First, you must convince yourself that queue 1 is indeed a discrete time GI/B/1 queue).

Exercise 2.9. (Delay for Opportunistic Scheduling) Consider the 2-queue wireless downlink with ON/OFF channels as described in the example of Section 2.3.2 (Fig. 2.2). The channel probabilities are given as in that example: $p_{00} = 0.24$, $p_{10} = 0.36$, $p_{01} = 0.16$, $p_{11} = 0.24$. Suppose the arrival process $a_1(t)$ is i.i.d. Bernoulli with rate $\lambda_1 = 0.4$, so that $Pr[a_1(t) = 1] = 0.4$, $Pr[a_1(t) = 0] = 0.6$. Suppose $a_2(t)$ is i.i.d. Bernoulli with rate $\lambda_2 = 0.3$. Design a randomized algorithm, using parameter β as the probability that we transmit over channel 1 when $\boldsymbol{S}(t) = (ON, ON)$, that ensures the average delay satisfies $\overline{W}_1 \leq 25$ slots and $\overline{W}_2 \leq 25$ slots. You should use the delay formula in Exercise 2.6 (first convincing yourself that each queue is indeed a GI/B/1 queue) along with an educated guess for β and/or trial and error for β.

Exercise 2.10. (Simulation of a B/B/1 queue) Write a computer program to simulate a Bernoulli/Bernoulli/1 (B/B/1) queue. Specifically, we have $Q(0) = 0$, $\{a(t)\}_{t=0}^{\infty}$ is i.i.d over slots with $Pr[a(t) = 1] = \lambda$, $Pr[a(t) = 0] = 1 - \lambda$, and $\{b(t)\}_{t=0}^{\infty}$ is independent of the arrival process and is i.i.d. over slots with $Pr[b(t) = 1] = \mu$, $Pr[b(t) = 0] = 1 - \mu$. Assume that $\mu = 0.7$, run the experiment over 10^6 slots, and give the empirical time average Q^{av} and the value of $Q(t)/t$ for $t = 10^6$, for λ values of $0.4, 0.5, 0.6, 0.7, 0.8$. Compare these to the exact value (given in Exercise 2.6) for $t \to \infty$.

Exercise 2.11. (Virtual Queues) Suppose we have a system that operates in discrete time with slots $t \in \{0, 1, 2, \ldots\}$. A controller makes decisions every slot t about how to operate the system, and these decisions incur power $p(t)$. The controller wants to ensure the time average power expenditure is no more than 12.3 power units per slot. Define a *virtual queue* $Z(t)$ with $Z(0) = 0$, and with update equation:

$$Z(t+1) = \max[Z(t) - 12.3, 0] + p(t) \tag{2.16}$$

The controller keeps the value of $Z(t)$ as a state variable, and updates $Z(t)$ at the end of each slot via (2.16) using the power $p(t)$ that was spent on that slot.

a) Use Lemma 2.1 to prove that if $Z(t)$ is rate stable, then:[3]

$$\lim_{t\to\infty} \frac{1}{t} \sum_{\tau=0}^{t-1} p(\tau) \leq 12.3 \text{ with probability 1}$$

b) Suppose there is a positive constant Z_{max} such that $Z(t) \leq Z_{max}$ for all $t \in \{0, 1, 2, \ldots\}$. Use (2.3) to show that for any integer $T > 0$ and any interval of T slots, defined by $\{t_1, \ldots, t_1 + T - 1\}$ (where $t_1 \geq 0$), we have:

$$\sum_{\tau=t_1}^{t_1+T-1} p(\tau) \leq 12.3T + Z_{max}$$

This idea is used in (21) to ensure the total power used in a communication system over any interval is less than or equal to the desired per-slot average power constraint multiplied by the interval size, plus a constant allowable "power burst" Z_{max}. A variation of this technique is used in (137) to bound the worst-case number of collisions with a primary user in a cognitive radio network.

[3] For simplicity, we have implicitly assumed the limit $\lim_{t\to\infty} \frac{1}{t} \sum_{\tau=0}^{t-1} p(\tau)$ in Exercise 2.11(a) exists. More generally, the result holds when "lim" is replaced with "lim sup."

CHAPTER 3

Dynamic Scheduling Example

The dynamic scheduling algorithms developed in this text use powerful techniques of *Lyapunov drift* and *Lyapunov optimization.* To build intuition, this chapter introduces the main concepts for a simple 2-user wireless downlink example, similar to the example given in Section 2.3.2 of the previous chapter. First, the problem is formulated in terms of known arrival rates and channel state probabilities. However, rather than using a randomized scheduling algorithm that bases decisions only on the current channel states (as considered in the previous chapter), we use an alternative approach based on minimizing the drift of a *Lyapunov function.* The advantage is that the drift-minimizing approach uses both current channel states and current queue backlogs to stabilize the system, and it does not require a-priori knowledge of traffic rates or channel probabilities. This Lyapunov drift technique is extended at the end of the chapter to allow for joint stability and average power minimization.

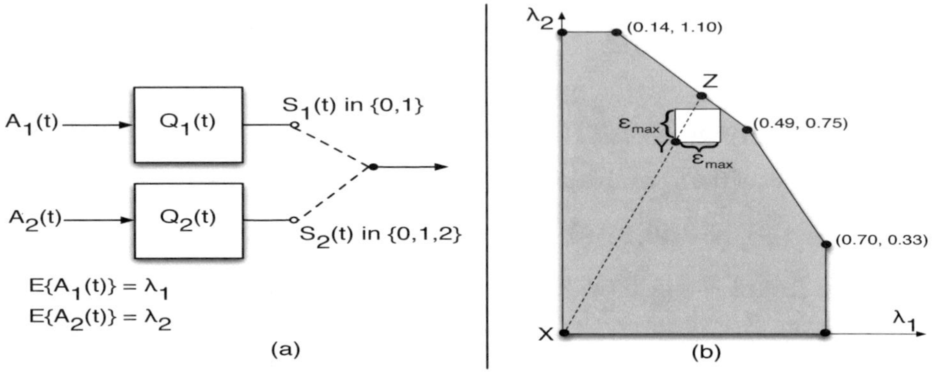

Figure 3.1: (a) The 2-queue wireless downlink example with time-varying channels. (b) The capacity region Λ. For $\boldsymbol{\lambda} = (0.3, 0.7)$ (i.e., point Y illustrated), we have $\epsilon_{max}(\boldsymbol{\lambda}) = 0.12$.

3.1 SCHEDULING FOR STABILITY

Consider a slotted system with two queues, as shown in Fig. 3.1(a). The arrival vector $(A_1(t), A_2(t))$ is i.i.d. over slots, where $A_1(t)$ and $A_2(t)$ take integer units of packets. The arrival rates are given by $\lambda_1 \triangleq \mathbb{E}\{A_1(t)\}$ and $\lambda_2 \triangleq \mathbb{E}\{A_2(t)\}$. The second moments $\mathbb{E}\{A_1^2\} \triangleq \mathbb{E}\{A_1(t)^2\}$ and $\mathbb{E}\{A_2^2\} \triangleq \mathbb{E}\{A_2(t)^2\}$ are assumed to be finite. The wireless channels are time varying, and every slot t we have a channel vector $\boldsymbol{S}(t) = (S_1(t), S_2(t))$, where $S_i(t)$ is a non-negative integer that represents the number of packets that can be transmitted over channel i on slot t (for $i \in \{1, 2\}$), provided that the scheduler decides to transmit over that channel. The channel state processes $S_1(t)$ and $S_2(t)$ are independent of each other and are i.i.d. over slots, with:

- $Pr[S_1(t) = 0] = 0.3$, $Pr[S_1(t) = 1] = 0.7$
- $Pr[S_2(t) = 0] = 0.2$, $Pr[S_2(t) = 1] = 0.5$, $Pr[S_2(t) = 2] = 0.3$

Every slot t the network controller observes the current channel state vector $\boldsymbol{S}(t)$ and chooses a single channel over which to transmit. Let $\alpha(t)$ be the *transmission decision* on slot t, taking three possible values:

$$\alpha(t) \in \{\text{"Transmit over channel 1"}, \text{"Transmit over channel 2"}, \text{"Idle"}\}$$

where $\alpha(t) =$ "Idle" means that no transmission takes place on slot t. The queueing dynamics are given by:

$$Q_i(t+1) = \max[Q_i(t) - b_i(t), 0] + A_i(t) \ \ \forall i \in \{1, 2\}, \forall t \in \{0, 1, 2, \ldots\} \tag{3.1}$$

where $b_i(t)$ represents the amount of service offered to channel i on slot t (for $i \in \{1, 2\}$), defined by a function $\hat{b}_i(\alpha(t), \boldsymbol{S}(t))$:

$$b_i(t) = \hat{b}_i(\alpha(t), \boldsymbol{S}(t)) \triangleq \begin{cases} S_i(t) & \text{if } \alpha(t) = \text{"Transmit over channel } i\text{"} \\ 0 & \text{otherwise} \end{cases} \tag{3.2}$$

3.1.1 THE S-ONLY ALGORITHM AND ϵ_{max}

Let $\mathcal{S}$ represent the set of the 6 possible outcomes for channel state vector $\boldsymbol{S}(t)$ in the above system:

$$\mathcal{S} \triangleq \{(0, 0), (0, 1), (0, 2), (1, 0), (1, 1), (1, 2)\}$$

Consider first the class of $\boldsymbol{S}$*-only scheduling algorithms* that make independent, stationary, and randomized transmission decisions every slot t based only on the observed $\boldsymbol{S}(t)$ (and hence independent of queue backlog). A particular $\boldsymbol{S}$-only algorithm for this system is characterized by probabilities $q_1(S_1, S_2)$ and $q_2(S_1, S_2)$ for all $(S_1, S_2) \in \mathcal{S}$, where $q_i(S_1, S_2)$ is the probability of transmitting over channel i if $\boldsymbol{S}(t) = (S_1, S_2)$. These probabilities must satisfy $q_1(S_1, S_2) + q_2(S_1, S_2) \leq 1$ for all $(S_1, S_2) \in \mathcal{S}$, where we use inequality to allow the possibility of transmitting over *neither* channel (useful for the power minimization problem considered later). Let $\alpha^*(t)$ represent the transmission decisions under a particular $\boldsymbol{S}$-only policy, and define $b_1^*(t) \triangleq \hat{b}_1(\alpha^*(t), \boldsymbol{S}(t)), b_2^*(t) \triangleq \hat{b}_2(\alpha^*(t), \boldsymbol{S}(t))$ as the resulting transmission rates offered by this policy on slot t. We thus have for every slot t:

$$\begin{aligned} \mathbb{E}\left\{b_1^*(t)\right\} &= \sum_{(S_1, S_2) \in \mathcal{S}} Pr[S_1, S_2] S_1 q_1(S_1, S_2) \\ \mathbb{E}\left\{b_2^*(t)\right\} &= \sum_{(S_1, S_2) \in \mathcal{S}} Pr[S_1, S_2] S_2 q_2(S_1, S_2) \end{aligned}$$

where we have used $Pr[S_1, S_2]$ as short-hand notation for $Pr[(S_1(t), S_2(t)) = (S_1, S_2)]$.

Note that the above expectations are over the random channel state vector $\boldsymbol{S}(t)$ and the random transmission decision in reaction to this vector. Under this $\boldsymbol{S}$-only algorithm, $b_1^*(t)$ is i.i.d. over slots with mean $\mathbb{E}\left\{b_1^*(t)\right\}$, and thus the time average of $b_1^*(t)$ is equal to $\mathbb{E}\left\{b_1^*(t)\right\}$ with probability 1 (by the law of large numbers). It follows by the Rate Stability Theorem (Theorem 2.4) that queue 1 is rate stable if and only if $\lambda_1 \leq \mathbb{E}\left\{b_1^*(t)\right\}$. Likewise, queue 2 is rate stable if and only if $\lambda_2 \leq \mathbb{E}\left\{b_2^*(t)\right\}$. However, for finite delay, it is useful to design the transmission rates to be strictly larger than the arrival rates (see Exercises 2.6, 2.8, 2.9, 2.10). The following linear program seeks to design an $\boldsymbol{S}$-only policy that maximizes the value of ϵ for which $\lambda_1 + \epsilon \leq \mathbb{E}\left\{b_1^*(t)\right\}$ and $\lambda_2 + \epsilon \leq \mathbb{E}\left\{b_2^*(t)\right\}$:

$$\begin{aligned} &\text{Maximize:} && \epsilon && (3.3)\\ &\text{Subject to:} && \lambda_1 + \epsilon \leq \textstyle\sum_{(S_1,S_2)\in\mathcal{S}} Pr[S_1, S_2] S_1 q_1(S_1, S_2) && (3.4)\\ &&& \lambda_2 + \epsilon \leq \textstyle\sum_{(S_1,S_2)\in\mathcal{S}} Pr[S_1, S_2] S_2 q_2(S_1, S_2) && (3.5)\\ &&& q_1(S_1, S_2) + q_2(S_1, S_2) \leq 1 \;\; \forall (S_1, S_2) \in \mathcal{S} && (3.6)\\ &&& q_1(S_1, S_2) \geq 0, q_2(S_1, S_2) \geq 0 \;\; \forall (S_1, S_2) \in \mathcal{S} && (3.7) \end{aligned}$$

There are 8 known parameters that appear as constants in the above linear program:

$$\lambda_1, \lambda_2, Pr[S_1, S_2] \;\; \forall (S_1, S_2) \in \mathcal{S} \tag{3.8}$$

There are 13 unknowns that act as variables to be optimized in the above linear program:

$$\epsilon, q_1(S_1, S_2), q_2(S_1, S_2) \;\; \forall (S_1, S_2) \in \mathcal{S} \tag{3.9}$$

Define $\boldsymbol{\lambda} \triangleq (\lambda_1, \lambda_2)$, and define $\epsilon_{max}(\boldsymbol{\lambda})$ as the maximum value of ϵ in the above problem. It can be shown that the *network capacity region* is the set Λ of all non-negative rate vectors $\boldsymbol{\lambda}$ for which $\epsilon_{max}(\boldsymbol{\lambda}) \geq 0$. The value of ϵ_{max} represents a measure of the distance between the rate vector $\boldsymbol{\lambda}$ and the capacity region boundary. If the rate vector $\boldsymbol{\lambda}$ is *interior* to the capacity region Λ, then $\epsilon_{max}(\boldsymbol{\lambda}) > 0$. In this simple example, it is possible to compute the capacity region explicitly, and that is shown in Fig. 3.1(b). The figure also illustrates an example arrival rate vector $(\lambda_1, \lambda_2) = (0.3, 0.7)$ (shown as point Y in the figure), for which we have $\epsilon_{max}(0.3, 0.7) = 0.12$.

It follows that for any rate vector $\boldsymbol{\lambda} = (\lambda_1, \lambda_2)$ that is interior to the capacity region Λ, we have $\epsilon_{max}(\boldsymbol{\lambda}) > 0$, and there exists an $\boldsymbol{S}$-only algorithm that yields transmission variables $(b_1^*(t), b_2^*(t))$ that satisfy:

$$\mathbb{E}\left\{b_1^*(t)\right\} \geq \lambda_1 + \epsilon_{max}(\boldsymbol{\lambda}) \;\;,\;\; \mathbb{E}\left\{b_2^*(t)\right\} \geq \lambda_2 + \epsilon_{max}(\boldsymbol{\lambda}) \tag{3.10}$$

3.1.2 LYAPUNOV DRIFT FOR STABLE SCHEDULING

Rather than trying to solve the linear program of the preceding sub-section (which would require a-priori knowledge of the arrival rates and channel probabilities specified in (3.8)), here we pursue queue stability via an algorithm that makes decisions based on both the current channel states and the current queue backlogs. Thus, the algorithm we present is *not* an $\boldsymbol{S}$-only algorithm. Remarkably, the proof that it provides strong stability whenever the arrival rate vector is interior to the capacity region will use the *existence* of the $\boldsymbol{S}$-only algorithm that satisfies (3.10), without ever needing to solve for the 13 variables in (3.9) that define this $\boldsymbol{S}$-only algorithm.

Let $\boldsymbol{Q}(t) = (Q_1(t), Q_2(t))$ be the vector of current queue backlogs, and define a *Lyapunov function* $L(\boldsymbol{Q}(t))$ as follows:

$$L(\boldsymbol{Q}(t)) \triangleq \frac{1}{2}[Q_1(t)^2 + Q_2(t)^2] \tag{3.11}$$

This represents a scalar measure of queue congestion in the network, and has the following properties:

- $L(\boldsymbol{Q}(t)) \geq 0$ for all backlog vectors $\boldsymbol{Q}(t) = (Q_1(t), Q_2(t))$, with equality if and only if the network is empty on slot t.
- $L(\boldsymbol{Q}(t))$ being "small" implies that both queue backlogs are "small."
- $L(\boldsymbol{Q}(t))$ being "large" implies that at least one queue backlog is "large."

For example, if $L(\boldsymbol{Q}(t)) \leq 32$, then $Q_1(t)^2 + Q_2(t)^2 \leq 64$, and thus we know that both $Q_1(t) \leq 8$ and $Q_2(t) \leq 8$.

If there is a finite constant M such that $L(\boldsymbol{Q}(t)) \leq M$ for all t, then clearly all queue backlogs are always bounded by $\sqrt{2M}$, and so all queues are trivially strongly stable. While we usually cannot guarantee that the Lyapunov function is deterministically bounded, it is intuitively clear that designing an algorithm to consistently push the queue backlog towards a region such that $L(\boldsymbol{Q}(t)) \leq M$ (for some finite constant M) will help to control congestion and stabilize the queues.

One may wonder why we use a *quadratic* Lyapunov function, when another function, such as a linear function, would satisfy properties similar to those stated above. When computing the change in the Lyapunov function from one slot to the next, we will find that the quadratic has important *dominant cross terms* that include an inner product of queue backlogs and transmission rates. This is important for the same reason that it was important to use a quadratic function in the delay computation of Exercise 2.6, and readers seeking more intuition on the "magic" of the quadratic function are encouraged to review that exercise.

To understand how we can consistently push the Lyapunov function towards a low congestion region, we first use (3.1) to compute a bound on the change in the Lyapunov function from one slot to the next:

$$\begin{aligned} L(\boldsymbol{Q}(t+1)) - L(\boldsymbol{Q}(t)) &= \frac{1}{2}\sum_{i=1}^{2}[Q_i(t+1)^2 - Q_i(t)^2] \\ &= \frac{1}{2}\sum_{i=1}^{2}\left[(\max[Q_i(t) - b_i(t), 0] + A_i(t))^2 - Q_i(t)^2\right] \\ &\leq \sum_{i=1}^{2}\frac{[A_i(t)^2 + b_i(t)^2]}{2} + \sum_{i=1}^{2} Q_i(t)[A_i(t) - b_i(t)] \end{aligned} \tag{3.12}$$

where in the final inequality we have used the fact that for any $Q \geq 0, b \geq 0, A \geq 0$, we have:

$$(\max[Q - b, 0] + A)^2 \leq Q^2 + A^2 + b^2 + 2Q(A - b)$$

Now define $\Delta(\boldsymbol{Q}(t))$ as the *conditional Lyapunov drift* for slot t:

$$\Delta(\boldsymbol{Q}(t)) \triangleq \mathbb{E}\left\{L(\boldsymbol{Q}(t+1) - L(\boldsymbol{Q}(t))|\boldsymbol{Q}(t)\right\} \tag{3.13}$$

where the expectation depends on the control policy, and is with respect to the random channel states and the (possibly random) control actions made in reaction to these channel states. From (3.12), we have that $\Delta(\boldsymbol{Q}(t))$ for a general control policy satisfies:

$$\Delta(\boldsymbol{Q}(t)) \leq \mathbb{E}\left\{\sum_{i=1}^{2} \frac{A_i(t)^2 + b_i(t)^2}{2} \mid \boldsymbol{Q}(t)\right\} + \sum_{i=1}^{2} Q_i(t)\lambda_i - \mathbb{E}\left\{\sum_{i=1}^{2} Q_i(t)b_i(t)|\boldsymbol{Q}(t)\right\} \tag{3.14}$$

where we have used the fact that arrivals are i.i.d. over slots and hence independent of current queue backlogs, so that $\mathbb{E}\{A_i(t)|\boldsymbol{Q}(t)\} = \mathbb{E}\{A_i(t)\} = \lambda_i$. Now define B as a finite constant that bounds the first term on the right-hand-side of the above drift inequality, so that for all t, all possible $\boldsymbol{Q}(t)$, and all possible control actions that can be taken, we have:

$$\mathbb{E}\left\{\sum_{i=1}^{2} \frac{A_i(t)^2 + b_i(t)^2}{2} \mid \boldsymbol{Q}(t)\right\} \leq B$$

For our system, we have that at most one $b_i(t)$ value can be non-zero on a given slot t. The probability that the non-zero $b_i(t)$ (if any) is equal to 2 is at most 0.3 (because $Pr[S_2(t) = 2] = 0.3$), and if it is not equal to 2, then it is at most 1. Hence:

$$\frac{1}{2}\mathbb{E}\left\{\sum_{i=1}^{2} b_i(t)^2|\boldsymbol{Q}(t)\right\} \leq \frac{2^2(0.3) + 1^2(0.7)}{2} = 0.95$$

and thus we can define B as:

$$B \triangleq 0.95 + \frac{1}{2}\sum_{i=1}^{2} \mathbb{E}\left\{A_i^2\right\} \tag{3.15}$$

Using this in (3.14) yields:

$$\Delta(\boldsymbol{Q}(t)) \leq B + \sum_{i=1}^{2} Q_i(t)\lambda_i - \mathbb{E}\left\{\sum_{i=1}^{2} Q_i(t)b_i(t)|\boldsymbol{Q}(t)\right\}$$

To emphasize how the right-hand-side of the above inequality depends on the transmission decision $\alpha(t)$, we use the identity $b_i(t) = \hat{b}_i(\alpha(t), \boldsymbol{S}(t))$ to yield:

$$\Delta(\boldsymbol{Q}(t)) \leq B + \sum_{i=1}^{2} Q_i(t)\lambda_i - \mathbb{E}\left\{\sum_{i=1}^{2} Q_i(t)\hat{b}_i(\alpha(t), \boldsymbol{S}(t))|\boldsymbol{Q}(t)\right\} \tag{3.16}$$

3.1.3 THE "MIN-DRIFT" OR "MAX-WEIGHT" ALGORITHM

Our dynamic algorithm is designed to observe the current queue backlogs $(Q_1(t), Q_2(t))$ and the current channel states $(S_1(t), S_2(t))$ and to make a transmission decision $\alpha(t)$ to minimize the right-hand-side of the drift bound (3.16). Note that the transmission decision on slot t only affects the final term on the right-hand-side. Thus, we seek to design an algorithm that maximizes the following expression:

$$\mathbb{E}\left\{\sum_{i=1}^{2} Q_i(t)\hat{b}_i(\alpha(t), \boldsymbol{S}(t))|\boldsymbol{Q}(t)\right\}$$

The above conditional expectation is with respect to the randomly observed channel states $\boldsymbol{S}(t) = (S_1(t), S_2(t))$ and the (possibly random) control decision $\alpha(t)$. We now use the concept of *opportunistically maximizing an expectation*: The above expression is maximized by the algorithm that observes the current queues $(Q_1(t), Q_2(t))$ and channel states $(S_1(t), S_2(t))$ and chooses $\alpha(t)$ to maximize:

$$\sum_{i=1}^{2} Q_i(t)\hat{b}_i(\alpha(t), \boldsymbol{S}(t)) \tag{3.17}$$

This is often called the "max-weight" algorithm, as it seeks to maximize a weighted sum of the transmission rates, where the weights are queue backlogs. As there are only three decisions (transmit over channel 1, transmit over channel 2, or don't transmit), it is easy to evaluate the weighted sum (3.17) for each option:

- $\sum_{i=1}^{2} Q_i(t)\hat{b}_i(\alpha(t), \boldsymbol{S}(t)) = Q_1(t)S_1(t)$ if we choose to transmit over channel 1.
- $\sum_{i=1}^{2} Q_i(t)\hat{b}_i(\alpha(t), \boldsymbol{S}(t)) = Q_2(t)S_2(t)$ if we choose to transmit over channel 2.
- $\sum_{i=1}^{2} Q_i(t)\hat{b}_i(\alpha(t), \boldsymbol{S}(t)) = 0$ if we choose to remain idle.

It follows that the max-weight algorithm chooses to transmit over the channel i with the largest (positive) value of $Q_i(t)S_i(t)$, and remains idle if this value is 0 for both channels. This simple algorithm just makes decisions based on the current queue states and channel states, and it does not need knowledge of the arrival rates or channel probabilities.

Because this algorithm maximizes the weighted sum (3.17) over all alternative decisions, we have:

$$\sum_{i=1}^{2} Q_i(t)\hat{b}_i(\alpha(t), \boldsymbol{S}(t)) \geq \sum_{i=1}^{2} Q_i(t)\hat{b}_i(\alpha^*(t), \boldsymbol{S}(t))$$

where $\alpha^*(t)$ represents *any alternative (possibly randomized) transmission decision that can be made on slot t*. This includes the case when $\alpha^*(t)$ is an $\boldsymbol{S}$-only decision that randomly chooses one of the three transmit options (transmit 1, transmit 2, or idle) with a distribution that depends on the observed $\boldsymbol{S}(t)$. Fixing a particular alternative (possibly randomized) decision $\alpha^*(t)$ for comparison and taking a conditional expectation of the above inequality (given $\boldsymbol{Q}(t)$) yields:

$$\mathbb{E}\left\{\sum_{i=1}^{2} Q_i(t)\hat{b}_i(\alpha(t), \boldsymbol{S}(t))|\boldsymbol{Q}(t)\right\} \geq \mathbb{E}\left\{\sum_{i=1}^{2} Q_i(t)\hat{b}_i(\alpha^*(t), \boldsymbol{S}(t))|\boldsymbol{Q}(t)\right\}$$

where the decision $\alpha(t)$ on the left-hand-side of the above inequality represents the max-weight decision made on slot t, and the decision $\alpha^*(t)$ represents any other particular decision that could have been made. Plugging the above directly into (3.16) yields:

$$\Delta(\boldsymbol{Q}(t)) \leq B + \sum_{i=1}^{2} Q_i(t)\lambda_i - \mathbb{E}\left\{\sum_{i=1}^{2} Q_i(t)\hat{b}_i(\alpha^*(t), \boldsymbol{S}(t))|\boldsymbol{Q}(t)\right\} \tag{3.18}$$

where the left-hand-side represents the drift under the max-weight decision $\alpha(t)$, and the final term on the right-hand-side involves any other decision $\alpha^*(t)$. It is remarkable that the inequality (3.18) holds true for all of the (infinite) number of possible randomized alternative decisions that can be plugged into the final term on the right-hand-side. However, this should not be too surprising, as we designed the max-weight policy to have exactly this property! Rearranging the terms in (3.18) yields:

$$\Delta(\boldsymbol{Q}(t)) \leq B - \sum_{i=1}^{2} Q_i(t)[\mathbb{E}\left\{b_i^*(t)|\boldsymbol{Q}(t)\right\} - \lambda_i] \tag{3.19}$$

where we have used the identity $b_i^*(t) \triangleq \hat{b}_i(\alpha^*(t), \boldsymbol{S}(t))$ to represent the transmission rate that would be offered over channel i if decision $\alpha^*(t)$ were made.

Now suppose the arrival rates (λ_1, λ_2) are interior to the capacity region Λ, and consider the particular $\boldsymbol{S}$-only decision $\alpha^*(t)$ that chooses a transmit option independent of queue backlog to yield (3.10). Because channel states are i.i.d. over slots, the resulting rates $(b_1^*(t), b_2^*(t))$ are independent of current queue backlog, and so by (3.10), we have for $i \in \{1, 2\}$:

$$\mathbb{E}\left\{b_i^*(t)|\boldsymbol{Q}(t)\right\} = \mathbb{E}\left\{b_i^*(t)\right\} \geq \lambda_i + \epsilon_{max}(\boldsymbol{\lambda})$$

Plugging this directly into (3.19) yields:

$$\Delta(\boldsymbol{Q}(t)) \leq B - \sum_{i=1}^{2} Q_i(t)\epsilon_{max}(\boldsymbol{\lambda}) \tag{3.20}$$

where we recall that $\epsilon_{max}(\boldsymbol{\lambda}) > 0$. The above is a drift inequality concerning the max-weight algorithm on slot t, and it is now in terms of a value $\epsilon_{max}(\boldsymbol{\lambda})$ associated with the linear program (3.3)-(3.7). However, we did not need to solve the linear program to obtain this inequality or to implement the algorithm! It was enough to know that the solution to the linear program *exists*!

3.1.4 ITERATED EXPECTATIONS AND TELESCOPING SUMS

Taking an expectation of (3.20) over the randomness of the $Q_1(t)$ and $Q_2(t)$ values yields:

$$\mathbb{E}\{\Delta(\boldsymbol{Q}(t))\} \leq B - \epsilon_{max}(\boldsymbol{\lambda}) \sum_{i=1}^{2} \mathbb{E}\{Q_i(t)\} \tag{3.21}$$

Using the definition of $\Delta(\boldsymbol{Q}(t))$ in (3.13) with the law of iterated expectations yields:

$$\mathbb{E}\{\Delta(\boldsymbol{Q}(t))\} = \mathbb{E}\{\mathbb{E}\{L(\boldsymbol{Q}(t+1)) - L(\boldsymbol{Q}(t))|\boldsymbol{Q}(t)\}\} = \mathbb{E}\{L(\boldsymbol{Q}(t+1))\} - \mathbb{E}\{L(\boldsymbol{Q}(t))\}$$

Substituting this identity into (3.21) yields:

$$\mathbb{E}\{L(\boldsymbol{Q}(t+1))\} - \mathbb{E}\{L(\boldsymbol{Q}(t))\} \leq B - \epsilon_{max}(\boldsymbol{\lambda}) \sum_{i=1}^{2} \mathbb{E}\{Q_i(t)\}$$

The above holds for all $t \in \{0, 1, 2, \ldots\}$. Summing over $t \in \{0, 1, \ldots, T-1\}$ for some integer $T > 0$ yields (by telescoping sums):

$$\mathbb{E}\{L(\boldsymbol{Q}(T))\} - \mathbb{E}\{L(\boldsymbol{Q}(0))\} \leq BT - \epsilon_{max}(\boldsymbol{\lambda}) \sum_{t=0}^{T-1} \sum_{i=1}^{2} \mathbb{E}\{Q_i(t)\}$$

Rearranging terms, dividing by $\epsilon_{max}(\boldsymbol{\lambda})T$, and using the fact that $L(\boldsymbol{Q}(T)) \geq 0$ yields:

$$\frac{1}{T} \sum_{t=0}^{T-1} \sum_{i=1}^{2} \mathbb{E}\{Q_i(t)\} \leq \frac{B}{\epsilon_{max}(\boldsymbol{\lambda})} + \frac{\mathbb{E}\{L(\boldsymbol{Q}(0))\}}{\epsilon_{max}(\boldsymbol{\lambda})T}$$

Assuming that $\mathbb{E}\{L(\boldsymbol{Q}(0))\} < \infty$ and taking a lim sup yields:

$$\limsup_{T\to\infty} \frac{1}{T} \sum_{t=0}^{T-1} \sum_{i=1}^{2} \mathbb{E}\{Q_i(t)\} \leq \frac{B}{\epsilon_{max}(\boldsymbol{\lambda})}$$

Thus, all queues are strongly stable, and the total average backlog (summed over both queues) is less than or equal to $B/\epsilon_{max}(\boldsymbol{\lambda})$. Thus, the max-weight algorithm (developed by minimizing a bound on the Lyapunov drift) ensures the queueing network is strongly stable whenever the rate vector $\boldsymbol{\lambda}$ is interior to the capacity region Λ, with an average queue congestion bound that is inversely proportional to the distance the rate vector is away from the capacity region boundary.

As an example, assume $\lambda_1 = 0.3$ and $\lambda_2 = 0.7$, illustrated by the point Y of Fig. 3.1(b). Then $\epsilon_{max} = 0.12$. Assuming arrivals are Bernoulli so that $\mathbb{E}\{A_i^2\} = \mathbb{E}\{A_i\} = \lambda_i$ and using the value of $B = 1.45$ obtained from (3.15), we have:

$$\overline{Q_1 + Q_2} \leq \frac{1.45}{0.12} = 12.083 \text{ packets}$$

where $\overline{Q_1 + Q_2}$ represents the lim sup time average expected queue backlog in the network. By Little's Theorem (129), average delay satisfies:

$$\overline{W} = \frac{\overline{Q_1 + Q_2}}{\lambda_1 + \lambda_2} \leq 12.083 \text{ slots}$$

A simulation of the algorithm over 10^6 slots yields an empirical average queue backlog of $\overline{Q}_1^{empirical} + \overline{Q}_2^{empirical} = 3.058$ packets, and hence in this example, our upper bound overestimates backlog by roughly a factor of 4.

Thus, the actual max-weight algorithm performs much better than the bound would suggest. There are three reasons for this gap: (i) A simple upper bound was used when computing the Lyapunov drift in (3.12), (ii) The value B used an upper bound on the second moments of service, (iii) The drift inequality compares to a queue-unaware $\boldsymbol{S}$-only algorithm, whereas the actual drift is much better because our algorithm considers queue backlog. The third reason often dominates in networks with many queues. For example, in (100) it is shown that average congestion and delay in an N-queue wireless system with one server and ON/OFF channels is at least proportional to N if a queue-unaware algorithm is used (a related result is derived for $N \times N$ packet switches in (99)). However, a more sophisticated *queue grouping* analysis in (101) shows that the max-weight algorithm on the ON/OFF downlink system gives average backlog and delay that is $O(1)$, *independent of the number of queues*. For brevity, we do not include queue grouping concepts in this text. The interested reader is referred to the above references, see also queue grouping results in (102)(103)(104)(105).

3.1.5 SIMULATION OF THE MAX-WEIGHT ALGORITHM

Fig. 3.2 shows simulation results over 10^6 slots when the rate vector (λ_1, λ_2) is pushed up the line segment from X to Z in the figure, again assuming independent Bernoulli arrivals. The point Z is $(\lambda_1, \lambda_2) = (0.372, 0.868)$. In the figure, the x-axis is a normalization factor ρ that specifies the distance along the segment (so that $\rho = 0$ is the point X, $\rho = 1$ is the point Z, and $\rho = 0.806$ is the point Y). It can be seen that the network is strongly stable for all rates with $\rho < 1$, and it has average backlog that increases to infinity at the vertical asymptote defined by the capacity region boundary (i.e., at $\rho = 1$).

Also plotted in Fig. 3.2 is the upper-bound $B/\epsilon_{max}(\boldsymbol{\lambda})$ (where we have computed $\epsilon_{max}(\boldsymbol{\lambda})$ for each input rate vector $\boldsymbol{\lambda}$ simulated). This bound shows the same qualitative behavior, but it is roughly a factor of 4 larger than the empirically observed backlog.

3.2 STABILITY AND AVERAGE POWER MINIMIZATION

Now consider the same system, but define $p(t)$ as the *power expenditure* incurred by the transmission decision $\alpha(t)$ on slot t. To emphasize that power is a function of $\alpha(t)$, we write $p(t) = \hat{p}(\alpha(t))$ and

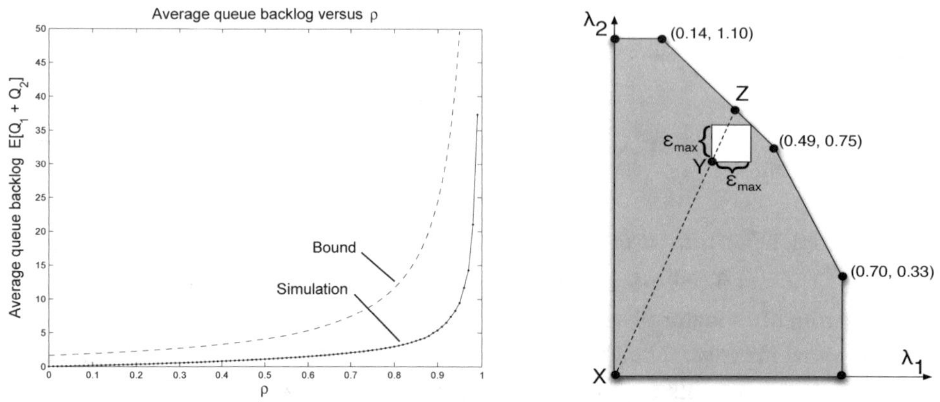

Figure 3.2: Average sum queue backlog (in units of packets) under the max-weight algorithm, as loading is pushed from point X (i.e., $\rho = 0$) to point Z (i.e., $\rho = 1$). Each simulated data point is an average over 10^6 slots.

assume the following simple power function:

$$\hat{p}(\alpha(t)) = \begin{cases} 1 & \text{if } \alpha(t) \in \{\text{“Transmit over channel 1,” “Transmit over channel 2”}\} \\ 0 & \text{if } \alpha(t) = \text{“Idle”} \end{cases}$$

That is, we spend 1 unit of power if we transmit over either channel, and no power is spent if we remain idle. Our goal is now to make transmission decisions to jointly stabilize the system while also striving to minimize average power expenditure.

For a given rate vector (λ_1, λ_2) in the capacity region Λ, define $\Psi(\lambda_1, \lambda_2)$ as the minimum average power that can be achieved by any $\boldsymbol{S}$-only algorithm that makes all queues rate stable. The value $\Psi(\lambda_1, \lambda_2)$ can be computed by solving the following linear program (compare with (3.3)-(3.7)):

$$\begin{array}{ll} \text{Minimize:} & \Psi \triangleq \sum_{(S_1,S_2)\in\mathcal{S}} Pr[S_1, S_2](q_1(S_1, S_2) + q_2(S_1, S_2)) \\ \text{Subject to:} & \lambda_1 \leq \sum_{(S_1,S_2)\in\mathcal{S}} Pr[S_1, S_2] S_1 q_1(S_1, S_2) \\ & \lambda_2 \leq \sum_{(S_1,S_2)\in\mathcal{S}} Pr[S_1, S_2] S_2 q_2(S_1, S_2) \\ & q_1(S_1, S_2) + q_2(S_1, S_2) \leq 1 \ \forall (S_1, S_2) \in \mathcal{S} \\ & q_1(S_1, S_2) \geq 0 \ , \ q_2(S_1, S_2) \geq 0 \ \forall (S_1, S_2) \in \mathcal{S} \end{array}$$

Thus, for each $\boldsymbol{\lambda} \in \Lambda$, there is an $\boldsymbol{S}$-only algorithm $\alpha^*(t)$ such that:

$$\mathbb{E}\left\{\hat{b}_1(\alpha^*(t), \boldsymbol{S}(t))\right\} \geq \lambda_1 \ , \ \mathbb{E}\left\{\hat{b}_2(\alpha^*(t), \boldsymbol{S}(t))\right\} \geq \lambda_2 \ , \ \mathbb{E}\left\{\hat{p}(\alpha^*(t))\right\} = \Psi(\lambda_1, \lambda_2)$$

It can be shown that $\Psi(\lambda_1, \lambda_2)$ is the minimum time average expected power expenditure that can be achieved by *any* control policy that stabilizes the system (including policies that are not $\boldsymbol{S}$-only) (21). Further, $\Psi(\lambda_1, \lambda_2)$ is continuous, convex, and entrywise non-decreasing.

Now assume that $\boldsymbol{\lambda} = (\lambda_1, \lambda_2)$ is interior to Λ, so that $(\lambda_1 + \epsilon, \lambda_2 + \epsilon) \in \Lambda$ for all ϵ such that $0 \leq \epsilon \leq \epsilon_{max}(\boldsymbol{\lambda})$. It follows that whenever $0 \leq \epsilon \leq \epsilon_{max}(\boldsymbol{\lambda})$, there exists an $\boldsymbol{S}$-only algorithm $\alpha^*(t)$ such that:

$$\mathbb{E}\left\{\hat{b}_1(\alpha^*(t), \boldsymbol{S}(t))\right\} \geq \lambda_1 + \epsilon \tag{3.22}$$

$$\mathbb{E}\left\{\hat{b}_2(\alpha^*(t), \boldsymbol{S}(t))\right\} \geq \lambda_2 + \epsilon \tag{3.23}$$

$$\mathbb{E}\left\{\hat{p}(\alpha^*(t))\right\} = \Psi(\lambda_1 + \epsilon, \lambda_2 + \epsilon) \tag{3.24}$$

3.2.1 DRIFT-PLUS-PENALTY

Define the same Lyapunov function $L(\boldsymbol{Q}(t))$ as in (3.11), and let $\Delta(\boldsymbol{Q}(t))$ represent the conditional Lyapunov drift for slot t. While taking actions to minimize a bound on $\Delta(\boldsymbol{Q}(t))$ every slot t would stabilize the system, the resulting average power expenditure might be unnecessarily large. For example, suppose the rate vector is $(\lambda_1, \lambda_2) = (0, 0.4)$, and recall that $Pr[S_2(t) = 2] = 0.3$. Then the drift-minimizing algorithm of the previous section would transmit over channel 2 whenever the queue is not empty and $S_2(t) \in \{1, 2\}$. In particular, it would sometimes use "inefficient" transmissions when $S_2(t) = 1$, which spend one unit of power but only deliver 1 packet. However, if we only transmit when $S_2(t) = 2$ and when the number of packets in the queue is at least 2, it can be shown that the system is still stable, but power expenditure is reduced to its minimum of $\lambda_2/2 = 0.2$ units/slot.

Instead of taking a control action to minimize a bound on $\Delta(\boldsymbol{Q}(t))$, we minimize a bound on the following *drift-plus-penalty* expression:

$$\Delta(\boldsymbol{Q}(t)) + V\mathbb{E}\left\{p(t)|\boldsymbol{Q}(t)\right\}$$

where $V \geq 0$ is a parameter that represents an "importance weight" on how much we emphasize power minimization. Such a control decision can be motivated as follows: We want to make $\Delta(\boldsymbol{Q}(t))$ small to push queue backlog towards a lower congestion state, but we also want to make $\mathbb{E}\{p(t)|\boldsymbol{Q}(t)\}$ small so that we do not incur a large power expenditure. We thus decide according to the above weighted sum. We now show that this intuitive algorithm leads to a provable power-backlog tradeoff: Average power can be pushed arbitrarily close to $\Psi(\lambda_1, \lambda_2)$ by using a large value of V, at the expense of incurring an average queue backlog that is $O(V)$.

We have already computed a bound on $\Delta(\boldsymbol{Q}(t))$ in (3.16), and so adding $V\mathbb{E}\{p(t)|\boldsymbol{Q}(t)\}$ to both sides of (3.16) yields a bound on the drift-plus-penalty:

$$\begin{aligned}\Delta(\boldsymbol{Q}(t)) + V\mathbb{E}\left\{p(t)|\boldsymbol{Q}(t)\right\} \leq\ & B + V\mathbb{E}\left\{\hat{p}(\alpha(t))|\boldsymbol{Q}(t)\right\} + \sum_{i=1}^{2} Q_i(t)\lambda_i \\ & -\mathbb{E}\left\{\sum_{i=1}^{2} Q_i(t)\hat{b}_i(\alpha(t), \boldsymbol{S}(t))|\boldsymbol{Q}(t)\right\}\end{aligned} \tag{3.25}$$

where we have used the fact that $p(t) = \hat{p}(\alpha(t))$. The *drift-plus-penalty* algorithm then observes $(Q_1(t), Q_2(t))$ and $(S_1(t), S_2(t))$ every slot t and chooses an action $\alpha(t)$ to minimize the right-hand-side of the above inequality. Again, using the concept of opportunistically minimizing an expectation, this is accomplished by greedily minimizing:

$$value = V\hat{p}(\alpha(t)) - \sum_{i=1}^{2} Q_i(t)\hat{b}_i(\alpha(t), \boldsymbol{S}(t))$$

We thus compare the following values and choose the action corresponding to the smallest (breaking ties arbitrarily):

- $value[1] = V - Q_1(t)S_1(t)$ if $\alpha(t) =$ "Transmit over channel 1."
- $value[2] = V - Q_2(t)S_2(t)$ if $\alpha(t) =$ "Transmit over channel 2."
- $value[\text{Idle}] = 0$ if $\alpha(t) =$ "Idle."

3.2.2 ANALYSIS OF THE DRIFT-PLUS-PENALTY ALGORITHM

Because our decisions $\alpha(t)$ minimize the right-hand-side of the drift-plus-penalty inequality (3.25) on every slot t (given the observed $\boldsymbol{Q}(t)$), we have:

$$\begin{aligned}\Delta(\boldsymbol{Q}(t)) + V\mathbb{E}\left\{p(t)|\boldsymbol{Q}(t)\right\} \leq\ & B + V\mathbb{E}\left\{\hat{p}(\alpha^*(t))|\boldsymbol{Q}(t)\right\} + \sum_{i=1}^{2} Q_i(t)\lambda_i \\ & -\mathbb{E}\left\{\sum_{i=1}^{2} Q_i(t)\hat{b}_i(\alpha^*(t), \boldsymbol{S}(t))|\boldsymbol{Q}(t)\right\} \end{aligned} \tag{3.26}$$

where $\alpha^*(t)$ is any other (possibly randomized) transmission decision that can be made on slot t. Now assume that $\boldsymbol{\lambda}$ is interior to Λ, and fix any value ϵ such that $0 \leq \epsilon \leq \epsilon_{max}(\boldsymbol{\lambda})$. Plugging the $\boldsymbol{S}$-only algorithm (3.22)-(3.24) into the right-hand-side of the above inequality and noting that this policy makes decisions independent of queue backlog yields:

$$\begin{aligned}\Delta(\boldsymbol{Q}(t)) + V\mathbb{E}\left\{p(t)|\boldsymbol{Q}(t)\right\} \leq\ & B + V\Psi(\lambda_1 + \epsilon, \lambda_2 + \epsilon) + \sum_{i=1}^{2} Q_i(t)\lambda_i \\ & - \sum_{i=1}^{2} Q_i(t)(\lambda_i + \epsilon) \\ =\ & B + V\Psi(\lambda_1 + \epsilon, \lambda_2 + \epsilon) - \epsilon\sum_{i=1}^{2} Q_i(t) \end{aligned} \tag{3.27}$$

Taking expectations of the above inequality and using the law of iterated expectations as before yields:

$$\mathbb{E}\{L(\boldsymbol{Q}(t+1))\} - \mathbb{E}\{L(\boldsymbol{Q}(t))\} + V\mathbb{E}\{p(t)\} \leq B + V\Psi(\lambda_1+\epsilon, \lambda_2+\epsilon) - \epsilon\sum_{i=1}^{2}\mathbb{E}\{Q_i(t)\}$$

Summing the above over $t \in \{0, 1, \ldots, T-1\}$ for some positive integer T yields:

$$\begin{aligned}\mathbb{E}\{L(\boldsymbol{Q}(T))\} - \mathbb{E}\{L(\boldsymbol{Q}(0))\} + V\sum_{t=0}^{T-1}\mathbb{E}\{p(t)\} &\leq BT + VT\Psi(\lambda_1+\epsilon, \lambda_2+\epsilon) \\ &\quad -\epsilon\sum_{t=0}^{T-1}\sum_{i=1}^{2}\mathbb{E}\{Q_i(t)\} \end{aligned} \tag{3.28}$$

Rearranging terms in the above and neglecting non-negative quantities where appropriate yields the following two inequalities:

$$\begin{aligned}\frac{1}{T}\sum_{t=0}^{T-1}\mathbb{E}\{p(t)\} &\leq \Psi(\lambda_1+\epsilon, \lambda_2+\epsilon) + \frac{B}{V} + \frac{\mathbb{E}\{L(\boldsymbol{Q}(0))\}}{VT} \\ \frac{1}{T}\sum_{t=0}^{T-1}\sum_{i=1}^{2}\mathbb{E}\{Q_i(t)\} &\leq \frac{B + V[\Psi(\lambda_1+\epsilon, \lambda_2+\epsilon) - \frac{1}{T}\sum_{t=0}^{T-1}\mathbb{E}\{p(t)\}]}{\epsilon} + \frac{\mathbb{E}\{L(\boldsymbol{Q}(0))\}}{\epsilon T}\end{aligned}$$

where the first inequality follows by dividing (3.28) by VT and the second follows by dividing (3.28) by ϵT. Taking limits as $T \to \infty$ shows that:[1]

$$\overline{p} \triangleq \lim_{T\to\infty}\frac{1}{T}\sum_{t=0}^{T-1}\mathbb{E}\{p(t)\} \leq \Psi(\lambda_1+\epsilon, \lambda_2+\epsilon) + \frac{B}{V} \tag{3.29}$$

$$\overline{Q_1+Q_2} \triangleq \lim_{T\to\infty}\frac{1}{T}\sum_{t=0}^{T-1}\sum_{i=1}^{2}\mathbb{E}\{Q_i(t)\} \leq \frac{B}{\epsilon} + \frac{V[\Psi(\lambda_1+\epsilon, \lambda_2+\epsilon) - \overline{p}]}{\epsilon} \tag{3.30}$$

3.2.3 OPTIMIZING THE BOUNDS

The bounds (3.29) and (3.30) hold for any ϵ that satisfies $0 \leq \epsilon \leq \epsilon_{max}(\boldsymbol{\lambda})$, and hence they can be optimized separately. Plugging $\epsilon_{max}(\boldsymbol{\lambda})$ into (3.30) shows that both queues are *strongly stable*. Using $\epsilon = 0$ in (3.29) thus yields:

$$\Psi(\lambda_1, \lambda_2) \leq \overline{p} \leq \Psi(\lambda_1, \lambda_2) + \frac{B}{V} \tag{3.31}$$

[1] In this simple example, the system evolves according to a countably infinite state space Discrete Time Markov Chain (DTMC), and it can be shown that the limits in (3.29) and (3.30) are well defined.

where the first inequality follows because our algorithm stabilizes the network and thus cannot yield time average expected power lower than $\Psi(\lambda_1, \lambda_2)$, the infimum time average expected power required for stability of any algorithm.

Because $\overline{p} \geq \Psi(\lambda_1, \lambda_2)$, it can be shown that:

$$\Psi(\lambda_1 + \epsilon, \lambda_2 + \epsilon) - \overline{p} \leq \Psi(\lambda_1 + \epsilon, \lambda_2 + \epsilon) - \Psi(\lambda_1, \lambda_2) \leq 2\epsilon$$

where the final inequality holds because it requires at most one unit of energy to support each new packet, and so increasing the total input rate from $\lambda_1 + \lambda_2$ to $\lambda_1 + \lambda_2 + 2\epsilon$ increases the minimum required average power by at most 2ϵ. Plugging the above into (3.30) yields:

$$\overline{Q_1 + Q_2} \leq \frac{B}{\epsilon} + 2V$$

The above holds for all ϵ that satisfy $0 \leq \epsilon \leq \epsilon_{max}(\boldsymbol{\lambda})$, and so plugging $\epsilon = \epsilon_{max}(\boldsymbol{\lambda})$ yields:

$$\overline{Q_1 + Q_2} \leq \frac{B}{\epsilon_{max}(\boldsymbol{\lambda})} + 2V \tag{3.32}$$

The performance bounds (3.31) and (3.32) demonstrate an $[O(1/V), O(V)]$ power-backlog tradeoff: We can use an arbitrarily large V to make B/V arbitrarily small, so that (3.31) implies the time average power $\overline{p}$ is arbitrarily close to the optimum $\Psi(\lambda_1, \lambda_2)$. This comes with a tradeoff: The average queue backlog bound in (3.32) is $O(V)$.

3.2.4 SIMULATIONS OF THE DRIFT-PLUS-PENALTY ALGORITHM

Consider the previous example of Bernoulli arrivals with $\lambda_1 = 0.3, \lambda_2 = 0.7, \epsilon_{max}(\boldsymbol{\lambda}) = 0.12, B = 1.45$, which corresponds to point Y in Fig. 3.1(b). Then the bounds (3.31)-(3.32) become:

$$\overline{p} \leq \Psi(\lambda_1, \lambda_2) + \frac{1.45}{V} \tag{3.33}$$

$$\overline{Q_1 + Q_2} \leq \frac{1.45}{0.12} + 2V \tag{3.34}$$

Figs. 3.3 and 3.4 plot simulations for this system together with the above power and backlog bounds. Each simulated data point represents a simulation over 2×10^6 slots using a particular value of V. Values of V in the range 0 to 100 are shown. It is clear from the figures that average power converges to the optimal $p^* = 0.7$ as V increases, while average backlog increases linearly in V.

Performance can be significantly improved by noting that the drift-plus-penalty algorithm given in Section 3.2.1 never transmits from queue 1 unless $Q_1(t) \geq V$ (else, $value[1]$ would be positive). Hence, $Q_1(t) \geq Q_1^{place} \triangleq \max[V - 1, 0]$ for all slots $t \geq 0$, provided that this holds at $t = 0$. Similarly, the algorithm never transmits from queue 2 unless $Q_2(t) \geq V/2$, and so $Q_2(t) \geq Q_2^{place} \triangleq \max[V/2 - 2, 0]$ for all slots $t \geq 0$, provided this holds at $t = 0$. It follows that we can stack the queues with *fake packets* (called *place-holder packets*) that never get transmitted, as described

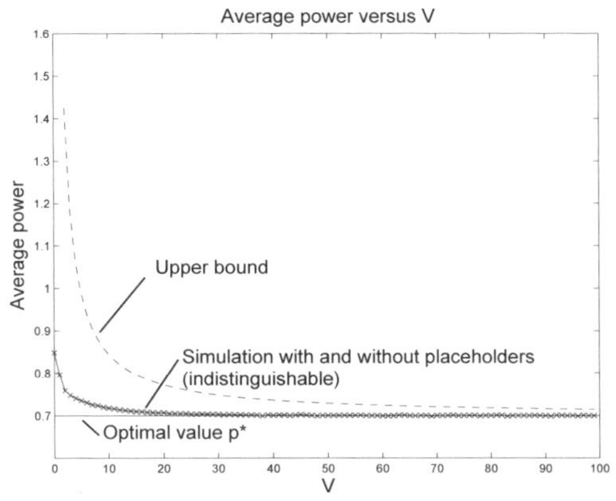

Figure 3.3: Average power versus V with $(\lambda_1, \lambda_2) = (0.3, 0.7)$.

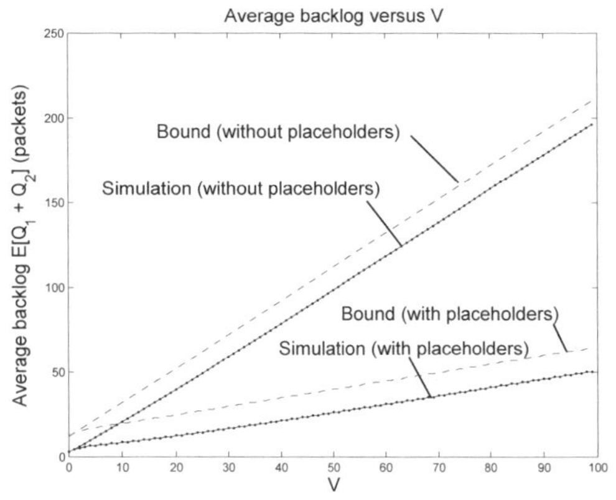

Figure 3.4: Average backlog versus V with $(\lambda_1, \lambda_2) = (0.3, 0.7)$.

in more detail in Section 4.8 of the next chapter. This *place-holder technique* yields the same power guarantee (3.33), but it has a significantly improved queue backlog bound given by:

$$\text{(with place-holders)} \quad \overline{Q_1 + Q_2} \leq \frac{1.45}{0.12} + 2V - \max[V - 1, 0] - \max[V/2 - 2, 0]$$

Thus, the average queue bound under the place-holder technique grows like $0.5V$, rather than $2V$ as suggested in (3.34), a dramatic savings when V is large. Simulations of the place-holder technique are also shown in Figs. 3.3 and 3.4. The queue backlog improvements due to placeholders are quite significant (Fig. 3.4), with no noticeable difference in power expenditure (Fig. 3.3). Indeed, the simulated power expenditure curves for the cases with and without place-holders are indistinguishable in Fig. 3.3. A plot of queue values over the first 3000 slots is given in Chapter 4, Fig. 4.2.

3.3 GENERALIZATIONS

The reader can easily see that the analysis in this chapter, which considers an example system of 2 queues, can be repeated for a larger system of K queues. Indeed, in that case the "min drift-plus-penalty" algorithm generalizes to choosing $\alpha(t)$ to maximize $\sum_{k=1}^{K} Q_k(t)\hat{b}_k(\alpha(t), \boldsymbol{S}(t)) - V\hat{p}(\alpha(t))$. This holds for systems with more general channel states $\boldsymbol{S}(t)$, more general resource allocation decisions $\alpha(t)$, and for arbitrary rate functions $\hat{b}_k(\alpha(t), \boldsymbol{S}(t))$ and "penalty functions" $\hat{p}(\alpha(t))$. In particular:

- The vector $\boldsymbol{S}(t)$ might have an infinite number of possible outcomes (rather than just 6 outcomes).
- The decision $\alpha(t)$ might represent one of an infinite number of possible power allocation options (rather than just one of three options). Alternatively, $\alpha(t)$ might represent one of an

infinite number of more sophisticated physical layer actions that can take place on slot t (such as modulation, coding, beamforming, etc.).

- The rate function $\hat{b}_k(\alpha(t), \boldsymbol{S}(t))$ can be any function that maps a resource allocation decision $\alpha(t)$ and a channel state vector $\boldsymbol{S}(t)$ into a transmission rate (and does not need to have the structure (3.2)).
- The "penalty" function $\hat{p}(\alpha(t))$ does not have to represent power, and it can be any general function of $\alpha(t)$.

The next chapter presents the general theory. It develops an important concept of *virtual queues* to ensure general time average equality and inequality constraints are satisfied. It also considers *variable* V algorithms that achieve the exact minimum average penalty subject to mean rate stability (which typically incurs infinite average backlog). Finally, it shows how to analyze systems with non-i.i.d. and non-ergodic arrival and channel processes.

CHAPTER 4

Optimizing Time Averages

This chapter considers the problem (1.1)-(1.5), which seeks to minimize the time average of a network attribute subject to additional time average constraints. We first develop the main results of Lyapunov drift and Lyapunov optimization theory.

4.1 LYAPUNOV DRIFT AND LYAPUNOV OPTIMIZATION

Consider a system of N queues, and let $\boldsymbol{\Theta}(t) = (\Theta_1(t), \ldots, \Theta_N(t))$ be the queue backlog vector. The reason we use notation $\boldsymbol{\Theta}(t)$ to represent a queue vector, instead of $\boldsymbol{Q}(t)$, is that in later sections we define $\boldsymbol{\Theta}(t) \triangleq [\boldsymbol{Q}(t), \boldsymbol{Z}(t), \boldsymbol{H}(t)]$, where $\boldsymbol{Q}(t)$ is a vector of *actual queues* in the network and $\boldsymbol{Z}(t)$, $\boldsymbol{H}(t)$ are suitably chosen *virtual queues*. Assume the $\boldsymbol{\Theta}(t)$ vector evolves over slots $t \in \{0, 1, 2, \ldots\}$ according to some probability law. The components $\Theta_n(t)$ are real numbers and can possibly be negative. Allowing $\Theta_n(t)$ to take negative values is often useful for the virtual queues that are defined later.

As a scalar measure of the "size" of the vector $\boldsymbol{\Theta}(t)$, define a *quadratic Lyapunov function* $L(\boldsymbol{\Theta}(t))$ as follows:

$$L(\boldsymbol{\Theta}(t)) \triangleq \frac{1}{2} \sum_{n=1}^{N} w_n \Theta_n(t)^2 \tag{4.1}$$

where $\{w_n\}_{n=1}^N$ are a collection of positive weights. We typically use $w_n = 1$ for all n, as in (3.11) of Chapter 3, although different weights are often useful to allow queues to be treated differently. This function $L(\boldsymbol{\Theta}(t))$ is always non-negative, and it is equal to zero if and only if all components of $\boldsymbol{\Theta}(t)$ are zero. Define the *one-slot conditional Lyapunov drift* $\Delta(\boldsymbol{\Theta}(t))$ as follows:[1]

$$\Delta(\boldsymbol{\Theta}(t)) \triangleq \mathbb{E}\{L(\boldsymbol{\Theta}(t+1)) - L(\boldsymbol{\Theta}(t)) | \boldsymbol{\Theta}(t)\} \tag{4.2}$$

This drift is the expected change in the Lyapunov function over one slot, given that the current state in slot t is $\boldsymbol{\Theta}(t)$.

4.1.1 LYAPUNOV DRIFT THEOREM

Theorem 4.1 *(Lyapunov Drift) Consider the quadratic Lyapunov function (4.1), and assume* $\mathbb{E}\{L(\boldsymbol{\Theta}(0))\} < \infty$. *Suppose there are constants* $B > 0$, $\epsilon \geq 0$ *such that the following drift condition*

[1] Strictly speaking, better notation would be $\Delta(\boldsymbol{\Theta}(t), t)$, as the drift may be due to a non-stationary policy. However, we use the simpler notation $\Delta(\boldsymbol{\Theta}(t))$ as a formal representation of the right-hand-side of (4.2).

holds for all slots $\tau \in \{0, 1, 2, \ldots\}$ *and all possible* $\boldsymbol{\Theta}(\tau)$*:*

$$\Delta(\boldsymbol{\Theta}(\tau)) \leq B - \epsilon \sum_{n=1}^{N} |\Theta_n(\tau)| \tag{4.3}$$

Then:

a) If $\epsilon \geq 0$ *then all queues* $\Theta_n(t)$ *are mean rate stable.*

b) If $\epsilon > 0$*, then all queues are strongly stable and:*

$$\limsup_{t\to\infty} \frac{1}{t} \sum_{\tau=0}^{t-1} \sum_{n=1}^{N} \mathbb{E}\left\{|\Theta_n(\tau)|\right\} \leq \frac{B}{\epsilon} \tag{4.4}$$

Proof. We first prove part (b). Taking expectations of (4.3) and using the law of iterated expectations yields:

$$\mathbb{E}\left\{L(\boldsymbol{\Theta}(\tau+1))\right\} - \mathbb{E}\left\{L(\boldsymbol{\Theta}(\tau))\right\} \leq B - \epsilon \sum_{n=1}^{N} \mathbb{E}\left\{|\Theta_n(\tau)|\right\}$$

Summing the above over $\tau \in \{0, 1, \ldots, t-1\}$ for some slot $t > 0$ and using the law of telescoping sums yields:

$$\mathbb{E}\left\{L(\boldsymbol{\Theta}(t))\right\} - \mathbb{E}\left\{L(\boldsymbol{\Theta}(0))\right\} \leq Bt - \epsilon \sum_{\tau=0}^{t-1} \sum_{n=1}^{N} \mathbb{E}\left\{|\Theta_n(\tau)|\right\} \tag{4.5}$$

Now assume that $\epsilon > 0$. Dividing by $t\epsilon$, rearranging terms, and using the fact that $\mathbb{E}\left\{L(\boldsymbol{\Theta}(t))\right\} \geq 0$ yields:

$$\frac{1}{t} \sum_{\tau=0}^{t-1} \sum_{n=1}^{N} \mathbb{E}\left\{|\Theta_n(\tau)|\right\} \leq \frac{B}{\epsilon} + \frac{\mathbb{E}\left\{L(\boldsymbol{\Theta}(0))\right\}}{\epsilon t} \tag{4.6}$$

The above holds for all slots $t > 0$. Taking a limit as $t \to \infty$ proves part (b).

To prove part (a), we have from (4.5) that for all slots $t > 0$:

$$\mathbb{E}\left\{L(\boldsymbol{\Theta}(t))\right\} - \mathbb{E}\left\{L(\boldsymbol{\Theta}(0))\right\} \leq Bt$$

Using the definition of $L(\boldsymbol{\Theta}(t))$ yields:

$$\frac{1}{2} \sum_{n=1}^{N} w_n \mathbb{E}\left\{\Theta_n(t)^2\right\} \leq \mathbb{E}\left\{L(\boldsymbol{\Theta}(0))\right\} + Bt$$

Therefore, for all $n \in \{1, \ldots, N\}$, we have:

$$\mathbb{E}\left\{\Theta_n(t)^2\right\} \leq \frac{2\mathbb{E}\left\{L(\boldsymbol{\Theta}(0))\right\}}{w_n} + \frac{2Bt}{w_n}$$

However, because the variance of $|\Theta_n(t)|$ cannot be negative, we have $\mathbb{E}\left\{\Theta_n(t)^2\right\} \geq \mathbb{E}\{|\Theta_n(t)|\}^2$. Thus, for all slots $t > 0$, we have:

$$\mathbb{E}\{|\Theta_n(t)|\} \leq \sqrt{\frac{2\mathbb{E}\{L(\boldsymbol{\Theta}(0))\}}{w_n} + \frac{2Bt}{w_n}} \tag{4.7}$$

Dividing by t and taking a limit as $t \to \infty$ proves that:

$$\lim_{t\to\infty} \frac{\mathbb{E}\{|\Theta_n(t)|\}}{t} \leq \lim_{t\to\infty} \sqrt{\frac{2\mathbb{E}\{L(\boldsymbol{\Theta}(0))\}}{t^2 w_n} + \frac{2B}{t w_n}} = 0$$

Thus, all queues $\Theta_n(t)$ are mean rate stable, proving part (a). □

The above theorem shows that if the drift condition (4.3) holds with $\epsilon \geq 0$, so that $\Delta(\boldsymbol{\Theta}(t)) \leq B$, then all queues are mean rate stable. Further, if $\epsilon > 0$, then all queues are strongly stable with time average expected queue backlog bounded by B/ϵ. We note that the proof reveals further detailed information concerning expected queue backlog for all slots $t > 0$, showing how the affect of the initial condition $\boldsymbol{\Theta}(0)$ decays over time (see (4.6) and (4.7)).

4.1.2 LYAPUNOV OPTIMIZATION THEOREM

Suppose that, in addition to the queues $\boldsymbol{\Theta}(t)$ that we want to stabilize, we have an associated stochastic "penalty" process $y(t)$ whose time average we want to make less than (or close to) some target value y^*. The process $y(t)$ can represent penalties incurred by control actions on slot t, such as power expenditures, packet drops, etc. Assume the expected penalty is lower bounded by a finite (possibly negative) value y_{min}, so that for all t and all possible control actions, we have:

$$\mathbb{E}\{y(t)\} \geq y_{min} \tag{4.8}$$

Theorem 4.2 *(Lyapunov Optimization) Suppose $L(\boldsymbol{\Theta}(t))$ and y_{min} are defined by (4.1) and (4.8), and that $\mathbb{E}\{L(\boldsymbol{\Theta}(0))\} < \infty$. Suppose there are constants $B \geq 0$, $V \geq 0$, $\epsilon \geq 0$, and y^* such that for all slots $\tau \in \{0, 1, 2, \ldots\}$ and all possible values of $\boldsymbol{\Theta}(\tau)$, we have:*

$$\Delta(\boldsymbol{\Theta}(\tau)) + V\mathbb{E}\{y(\tau)|\boldsymbol{\Theta}(\tau)\} \leq B + Vy^* - \epsilon \sum_{n=1}^{N} |\Theta_n(\tau)| \tag{4.9}$$

Then all queues $\Theta_n(t)$ are mean rate stable. Further, if $V > 0$ and $\epsilon > 0$ then time average expected penalty and queue backlog satisfy:

$$\limsup_{t\to\infty} \frac{1}{t}\sum_{\tau=0}^{t-1} \mathbb{E}\{y(\tau)\} \leq y^* + \frac{B}{V} \tag{4.10}$$

$$\limsup_{t\to\infty} \frac{1}{t}\sum_{\tau=0}^{t-1}\sum_{n=1}^{N} \mathbb{E}\{|\Theta_n(\tau)|\} \leq \frac{B + V(y^* - y_{min})}{\epsilon} \tag{4.11}$$

Finally, if $V = 0$ *then (4.11) still holds, and if* $\epsilon = 0$ *then (4.10) still holds.*

Proof. Fix any slot τ. Because (4.9) holds for this slot, we can take expectations of both sides and use the law of iterated expectations to yield:

$$\mathbb{E}\{L(\boldsymbol{\Theta}(\tau+1))\} - \mathbb{E}\{L(\boldsymbol{\Theta}(\tau))\} + V\mathbb{E}\{y(\tau)\} \leq B + Vy^* - \epsilon\sum_{n=1}^{N}\mathbb{E}\{|\Theta_n(\tau)|\}$$

Summing over $\tau \in \{0, 1, \ldots, t-1\}$ for some $t > 0$ and using the law of telescoping sums yields:

$$\mathbb{E}\{L(\boldsymbol{\Theta}(t))\} - \mathbb{E}\{L(\boldsymbol{\Theta}(0))\} + V\sum_{\tau=0}^{t-1}\mathbb{E}\{y(\tau)\} \leq (B + Vy^*)t - \epsilon\sum_{\tau=0}^{t-1}\sum_{n=1}^{N}\mathbb{E}\{|\Theta_n(\tau)|\} \tag{4.12}$$

Rearranging terms and neglecting non-negative terms when appropriate, it is easy to show that the above inequality directly implies the following two inequalities for all $t > 0$:

$$\frac{1}{t}\sum_{\tau=0}^{t-1}\mathbb{E}\{y(\tau)\} \leq y^* + \frac{B}{V} + \frac{\mathbb{E}\{L(\boldsymbol{\Theta}(0))\}}{Vt} \tag{4.13}$$

$$\frac{1}{t}\sum_{\tau=0}^{t-1}\sum_{n=1}^{N}\mathbb{E}\{|\Theta_n(\tau)|\} \leq \frac{B + V(y^* - y_{min})}{\epsilon} + \frac{\mathbb{E}\{L(\boldsymbol{\Theta}(0))\}}{\epsilon t} \tag{4.14}$$

where (4.13) follows by dividing (4.12) by Vt, and (4.14) follows by dividing (4.12) by ϵt. Taking limits of the above as $t \to \infty$ proves (4.10) and (4.11).

Rearranging (4.12) also yields:

$$\mathbb{E}\{L(\boldsymbol{\Theta}(t))\} \leq \mathbb{E}\{L(\boldsymbol{\Theta}(0))\} + (B + V(y^* - y_{min}))t$$

from which mean rate stability follows by an argument similar to that given in the proof of Theorem 4.1. □

Theorem 4.2 can be understood as follows: If for any parameter $V > 0$, we can design a control algorithm to ensure the drift condition (4.9) is satisfied on every slot τ, then the time average expected penalty satisfies (4.10) and hence is either less than the target value y^*, or differs from y^* by no more than a "fudge factor" B/V, which can be made arbitrarily small as V is increased. However, the time average queue backlog bound increases linearly in the V parameter, as shown by (4.11). This presents a performance-backlog tradeoff of $[O(1/V), O(V)]$. Because Little's Theorem tells us that average queue backlog is proportional to average delay (129), we often call this a performance-delay tradeoff. The proof reveals further details concerning the affect of the initial condition $\boldsymbol{\Theta}(0)$ on time average expectations at any slot t (see (4.13) and (4.14)).

This result suggests the following control strategy: Every slot τ, observe the current $\boldsymbol{\Theta}(\tau)$ values and take a control action that, subject to the known $\boldsymbol{\Theta}(\tau)$, greedily minimizes the drift-plus-penalty expression on the left-hand-side of the desired drift inequality (4.9):

$$\Delta(\boldsymbol{\Theta}(\tau)) + V\mathbb{E}\{y(\tau)|\boldsymbol{\Theta}(\tau)\} \tag{4.15}$$

It follows that if on every slot τ, there exists a particular control action that satisfies the drift requirement (4.9), then the drift-plus-penalty minimizing policy must also satisfy this drift requirement.

For intuition, note that taking an action on slot τ to minimize the drift $\Delta(\boldsymbol{\Theta}(\tau))$ alone would tend to push queues towards a lower congestion state, but it may incur a large penalty $y(\tau)$. Thus, we minimize a weighted sum of drift and penalty, where the penalty is scaled by an "importance" weight V, representing how much we emphasize penalty minimization. Using $V = 0$ corresponds to minimizing the drift $\Delta(\boldsymbol{\Theta}(\tau))$ alone, which reduces to the Tassiulas-Ephremides technique for network stability in (7)(8). While this does not provide any guarantees on the resulting time average penalty $y(t)$ (as the bound (4.10) becomes infinity for $V = 0$), it still ensures strong stability by (4.11). The case for $V > 0$ includes a weighted penalty term in the greedy minimization, and corresponds to our technique for joint stability and performance optimization, developed for utility optimal flow control in (17)(18) and used for average power optimization in (20)(21) and for problems similar to the type (1.1)-(1.5) and (1.6)-(1.11) in (22).

4.1.3 PROBABILITY 1 CONVERGENCE

Here we present a version of the Lyapunov optimization theorem that treats probability 1 convergence of sample path time averages, rather than time average expectations. We have the following preliminary lemma, related to the Kolmogorov law of large numbers:

Lemma 4.3 *Let $X(t)$ be a random process defined over $t \in \{0, 1, 2, \ldots\}$, and suppose that the following hold:*

- $\mathbb{E}\left\{X(t)^2\right\}$ *is finite for all $t \in \{0, 1, 2, \ldots\}$ and satisfies:*

$$\sum_{t=1}^{\infty} \frac{\mathbb{E}\left\{X(t)^2\right\}}{t^2} < \infty$$

- *There is a real-valued constant β such that for all $t \in \{1, 2, 3, \ldots\}$ and all possible $X(0), \ldots, X(t-1)$, the conditional expectation satisfies:*

$$\mathbb{E}\{X(t)|X(t-1), X(t-2), \ldots, X(0)\} \leq \beta$$

Then:

$$\limsup_{t\to\infty} \frac{1}{t}\sum_{\tau=0}^{t-1} X(\tau) \leq \beta \quad (w.p.1)$$

where "(w.p.1)" stands for "with probability 1."

A proof of this lemma is given in (138) as a simple application of the Kolmogorov law of large numbers for *martingale differences*. See (139)(140)(130)(141) for background on martingales and a statement and proof of the Kolmogorov law of large numbers. The lemma is used in (138) to prove the probability 1 version of the Lyapunov optimization theorem given below.

Let $\boldsymbol{\Theta}(t)$ be a vector of queues and $y(t)$ a penalty process, as before. Rather than defining a drift that conditions on $\boldsymbol{\Theta}(t)$, we must condition on the full *history* $\mathcal{H}(t)$, which includes values of $\boldsymbol{\Theta}(\tau)$ for $\tau \in \{0, \ldots, t\}$ and values of $y(\tau)$ for $\tau \in \{0, \ldots, t-1\}$. Specifically, for integers $t \geq 0$ define:

$$\mathcal{H}(t) \triangleq \{\boldsymbol{\Theta}(0), \boldsymbol{\Theta}(1), \ldots, \boldsymbol{\Theta}(t), y(0), y(1), \ldots, y(t-1)\}$$

Define $\Delta(t, \mathcal{H}(t))$ by:

$$\Delta(t, \mathcal{H}(t)) \triangleq \mathbb{E}\left\{L(\boldsymbol{\Theta}(t+1)) - L(\boldsymbol{\Theta}(t)) | \mathcal{H}(t)\right\}$$

Assume that:

- The penalty process $y(t)$ is deterministically lower bounded by a (possibly negative) constant y_{min}, so that:

$$y(t) \geq y_{min} \ \ \forall t \quad (w.p.1) \tag{4.16}$$

- The second moments $\mathbb{E}\left\{y(t)^2\right\}$ are finite for all $t \in \{0, 1, 2, \ldots\}$, and:

$$\sum_{t=1}^{\infty} \frac{\mathbb{E}\left\{y(t)^2\right\}}{t^2} < \infty \tag{4.17}$$

- There is a finite constant $D > 0$ such that for all $n \in \{1, \ldots, N\}$, all t, and all possible $\mathcal{H}(t)$, we have:

$$\mathbb{E}\left\{(\Theta_n(t+1) - \Theta_n(t))^4 | \mathcal{H}(t)\right\} \leq D \tag{4.18}$$

  so that conditional fourth moments of queue changes are uniformly bounded.

Theorem 4.4 *(Lyapunov Optimization with Probability 1 Convergence) Define $L(\boldsymbol{\Theta}(t))$ by (4.1), assume that $\boldsymbol{\Theta}(0)$ is finite with probability 1, and suppose that assumptions (4.16)-(4.18) hold. Suppose there are constants $B \geq 0$, $V > 0$, $\epsilon > 0$, and y^* such that for all slots $\tau \in \{0, 1, 2, \ldots\}$ and all possible $\mathcal{H}(\tau)$, we have:*

$$\Delta(\tau, \mathcal{H}(\tau)) + V\mathbb{E}\left\{y(\tau) | \mathcal{H}(\tau)\right\} \leq B + Vy^* - \epsilon \sum_{n=1}^{N} |\Theta_n(\tau)|$$

Then all queues $\Theta_n(t)$ are rate stable, and:

$$\limsup_{t\rightarrow\infty} \frac{1}{t}\sum_{\tau=0}^{t-1} y(\tau) \leq y^* + \frac{B}{V} \quad (w.p.1) \tag{4.19}$$

$$\limsup_{t\rightarrow\infty} \frac{1}{t}\sum_{\tau=0}^{t-1}\sum_{n=1}^{N} |\Theta_n(\tau)| \leq \frac{B + V(y^* - y_{min})}{\epsilon} \quad (w.p.1) \tag{4.20}$$

Further, if these same assumptions hold, and *if there is a value y' such that the following additional inequality also holds for all τ and all possible $\boldsymbol{\Theta}(\tau)$:*

$$\Delta(\tau, \mathcal{H}(\tau)) + V\mathbb{E}\{y(\tau)|\mathcal{H}(\tau)\} \leq B + Vy'$$

Then:

$$\limsup_{t\rightarrow\infty} \frac{1}{t}\sum_{\tau=0}^{t-1} y(\tau) \leq y' + B/V \quad (w.p.1) \tag{4.21}$$

Proof. Fix $\boldsymbol{\Theta}(0)$ as a given finite initial condition. Define the process $X(t)$ for $t \in \{0, 1, 2, \ldots\}$ as follows:

$$X(t) \triangleq L(\boldsymbol{\Theta}(t+1)) - L(\boldsymbol{\Theta}(t)) + Vy(t) - B - Vy^* + \epsilon\sum_{n=1}^{N}|\Theta_n(t)|$$

The conditions on $y(t)$ and $\boldsymbol{\Theta}(t)$ are shown in (138) to ensure that the queues $\Theta_n(t)$ are rate stable, that $\mathbb{E}\left\{X(t)^2\right\}$ is finite for all t, and that for all $t > 0$ and all possible values of $X(t-1), \ldots, X(0)$:

$$\sum_{t=1}^{\infty}\frac{\mathbb{E}\left\{X(t)^2\right\}}{t^2} < \infty \quad , \quad \mathbb{E}\{X(t)|X(t-1), X(t-2), \ldots, X(0)\} \leq 0$$

Thus, we can apply Lemma 4.3 to $X(t)$ to yield:

$$\limsup_{t\rightarrow\infty} \frac{1}{t}\sum_{\tau=0}^{t-1} X(\tau) \leq 0 \quad (w.p.1) \tag{4.22}$$

However, by definition of $X(t)$, we have for all $t > 0$:

$$\frac{1}{t}\sum_{\tau=0}^{t-1} X(\tau) = \frac{L(\boldsymbol{\Theta}(t)) - L(\boldsymbol{\Theta}(0))}{t} + \frac{1}{t}\sum_{\tau=0}^{t-1}\left[Vy(\tau) + \epsilon\sum_{n=1}^{N}|\Theta_n(\tau)|\right] - B - Vy^*$$

Rearranging terms in the above inequality and neglecting non-negative terms where appropriate directly leads to the following two inequalities that hold for all $t > 0$:

$$\frac{1}{Vt}\sum_{\tau=0}^{t-1} X(\tau) \geq \frac{-L(\boldsymbol{\Theta}(0))}{Vt} + \frac{1}{t}\sum_{\tau=0}^{t-1} y(\tau) - [B/V + y^*]$$
$$\frac{1}{\epsilon t}\sum_{\tau=0}^{t-1} X(\tau) \geq \frac{-L(\boldsymbol{\Theta}(0))}{\epsilon t} + \frac{1}{t}\sum_{\tau=0}^{t-1}\sum_{n=1}^{N} |\Theta_n(\tau)| - \frac{[B + V(y^* - y_{min})]}{\epsilon}$$

Taking limits of the above two inequalities and using (4.22) proves the results (4.19)-(4.20). A similar argument proves (4.21). □

Conditioning on the history $\mathcal{H}(t)$ is needed to prove Theorem 4.4 via Lemma 4.3. A policy that greedily minimizes $\Delta(t, \mathcal{H}(t)) + V\mathbb{E}\{y(t)|\mathcal{H}(t)\}$ every slot will *also* greedily minimize $\Delta(\boldsymbol{\Theta}(t)) + V\mathbb{E}\{y(t)|\boldsymbol{\Theta}(t)\}$. In this text, we focus primarily on time average expectations of the type (4.10) and (4.11), with the understanding that the same bounds can be shown to hold for time averages (with probability 1) if the additional assumptions (4.16)-(4.18) hold.

4.2 GENERAL SYSTEM MODEL

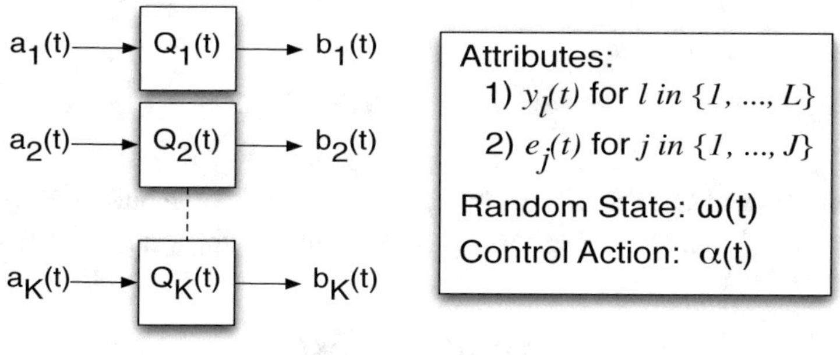

Figure 4.1: An illustration of a general K-queue network with attributes $y_l(t), e_j(t)$.

Consider now a system with queue backlog vector $\boldsymbol{Q}(t) = (Q_1(t), \ldots, Q_K(t))$, as shown in Fig. 4.1. Queue dynamics are given by:

$$Q_k(t+1) = \max[Q_k(t) - b_k(t), 0] + a_k(t) \tag{4.23}$$

where $\boldsymbol{a}(t) = (a_1(t), \ldots, a_K(t))$ and $\boldsymbol{b}(t) = (b_1(t), \ldots, b_K(t))$ are general functions of a *random event* $\omega(t)$ and a *control action* $\alpha(t)$:

$$a_k(t) = \hat{a}_k(\alpha(t), \omega(t)) \quad , \quad b_k(t) = \hat{b}_k(\alpha(t), \omega(t))$$

Every slot t the network controller observes $\omega(t)$ and chooses an action $\alpha(t) \in \mathcal{A}_{\omega(t)}$. The set $\mathcal{A}_{\omega(t)}$ is the *action space* associated with event $\omega(t)$. In addition to affecting these arrival and service variables, $\alpha(t)$ and $\omega(t)$ also determine the attribute vectors $\boldsymbol{x}(t)$, $\boldsymbol{y}(t)$, $\boldsymbol{e}(t)$ according to general functions $\hat{x}_m(\alpha, \omega)$, $\hat{y}_l(\alpha, \omega)$, $\hat{e}_j(\alpha, \omega)$, as described in Section 1.2.

We assume that $\omega(t)$ is a stationary process with a stationary probability distribution $\pi(\omega)$. Assume that $\omega(t)$ takes values in some sample space Ω. If Ω is a finite or countably infinite set, then for each $\omega \in \Omega$, $\pi(\omega)$ represents a probability mass function associated with the stationary distribution, and:

$$Pr[\omega(t) = \omega] = \pi(\omega) \;\; \forall t \in \{0, 1, 2, \ldots\} \tag{4.24}$$

If Ω is uncountably infinite, then we assume $\omega(t)$ is a random vector, and that $\pi(\omega)$ represents a probability density associated with the stationary distribution. The simplest model, which we mainly consider in this text, is the case when $\omega(t)$ is i.i.d. over slots t with stationary probabilities $\pi(\omega)$.

4.2.1 BOUNDEDNESS ASSUMPTIONS

The arrival function $\hat{a}_k(\alpha, \omega)$ is assumed to be non-negative for all $\omega \in \Omega$ and all $\alpha \in \mathcal{A}_\omega$. The service function $\hat{b}_k(\cdot)$ and the attribute functions $\hat{x}_m(\cdot)$, $\hat{y}_l(\cdot)$, $\hat{e}_j(\cdot)$ can possibly take negative values. All of these functions are general (possibly non-convex and discontinuous). However, we assume that these functions, together with the stationary probabilities $\pi(\omega)$, satisfy the following boundedness properties: For all t and all (possibly randomized) control decisions $\alpha(t) \in \mathcal{A}_{\omega(t)}$, we have:

$$\mathbb{E}\left\{\hat{a}_k(\alpha(t), \omega(t))^2\right\} \leq \sigma^2 \;\; \forall k \in \{1, \ldots, K\} \tag{4.25}$$

$$\mathbb{E}\left\{\hat{b}_k(\alpha(t), \omega(t))^2\right\} \leq \sigma^2 \;\; \forall k \in \{1, \ldots, K\} \tag{4.26}$$

$$\mathbb{E}\left\{\hat{x}_m(\alpha(t), \omega(t))^2\right\} \leq \sigma^2 \;\; \forall m \in \{1, \ldots, M\} \tag{4.27}$$

$$\mathbb{E}\left\{\hat{y}_l(\alpha(t), \omega(t))^2\right\} \leq \sigma^2 \;\; \forall l \in \{1, \ldots, L\} \tag{4.28}$$

$$\mathbb{E}\left\{\hat{e}_j(\alpha(t), \omega(t))^2\right\} \leq \sigma^2 \;\; \forall j \in \{1, \ldots, J\} \tag{4.29}$$

for some finite constant $\sigma^2 > 0$. Further, for all t and all actions $\alpha(t) \in \mathcal{A}_{\omega(t)}$, we require the expectation of $y_0(t)$ to be bounded by some finite constants $y_{0,min}$, $y_{0,max}$:

$$y_{0,min} \leq \mathbb{E}\left\{\hat{y}_0(\alpha(t), \omega(t))\right\} \leq y_{0,max} \tag{4.30}$$

4.3 OPTIMALITY VIA ω-ONLY POLICIES

For each $l \in \{0, 1, \ldots, L\}$, define $\overline{y}_l(t)$ as the time average expectation of $y_l(t)$ over the first t slots under a particular control strategy:

$$\overline{y}_l(t) \triangleq \frac{1}{t} \sum_{\tau=0}^{t-1} \mathbb{E}\left\{y_l(\tau)\right\}$$

where the expectation is over the randomness of the $\omega(\tau)$ values and the random control actions. Define time average expectations $\overline{a}_k(t), \overline{b}_k(t), \overline{e}_j(t)$ similarly. Define $\overline{y}_l$ and $\overline{e}_j$ as the limiting values of $\overline{y}_l(t)$ and $\overline{e}_j(t)$, assuming temporarily that these limits are well defined. We desire a control policy that solves the following problem:

$$\begin{aligned}
&\text{Minimize:} && \overline{y}_0 \\
&\text{Subject to:} \quad 1) && \overline{y}_l \leq 0 \ \forall l \in \{1, \ldots, L\} \\
&\qquad\qquad\quad 2) && \overline{e}_j = 0 \ \forall j \in \{1, \ldots, J\} \\
&\qquad\qquad\quad 3) && \text{Queues } Q_k(t) \text{ are mean rate stable } \forall k \in \{1, \ldots, K\} \\
&\qquad\qquad\quad 4) && \alpha(t) \in \mathcal{A}_{\omega(t)} \ \forall t
\end{aligned}$$

The above description of the problem is convenient, although we can state the problem more precisely without assuming limits are well defined as follows:

$$\text{Minimize:} \quad \limsup_{t\to\infty} \overline{y}_0(t) \tag{4.31}$$

$$\text{Subject to:} \quad 1) \ \limsup_{t\to\infty} \overline{y}_l(t) \leq 0 \ \forall l \in \{1, \ldots, L\} \tag{4.32}$$

$$2) \ \lim_{t\to\infty} \overline{e}_j(t) = 0 \ \forall j \in \{1, \ldots, J\} \tag{4.33}$$

$$3) \ \text{Queues } Q_k(t) \text{ are mean rate stable } \forall k \in \{1, \ldots, K\} \tag{4.34}$$

$$4) \ \alpha(t) \in \mathcal{A}_{\omega(t)} \ \forall t \tag{4.35}$$

An example of such a problem is when we have a K-queue wireless network that must be stabilized subject to average power constraints $\overline{P}_l \leq P_l^{av}$ for each node $l \in \{1, \ldots, L\}$, where $\overline{P}_l$ represents the time average power of node l, and P_l^{av} represents a pre-specified average power constraint. Suppose the goal is to maximize the time average of the total admitted traffic. Then $y_0(t)$ is -1 times the admitted traffic on slot t. We also define $y_l(t) = P_l(t) - P_l^{av}$, being the difference between the average power expenditure of node l and its time average constraint, so that $\overline{y}_l \leq 0$ corresponds to $\overline{P}_l \leq P_l^{av}$. In this example, there are no time average equality constraints, and so $J = 0$. See also Section 4.6 and Exercises 2.11, 4.7-4.14 for more examples.

Consider now the special class of stationary and randomized policies that we call *ω-only* policies, which observe $\omega(t)$ for each slot t and independently choose a control action $\alpha(t) \in \mathcal{A}_{\omega(t)}$ as a pure (possibly randomized) function of the observed $\omega(t)$. Let $\alpha^*(t)$ represent the decisions under such an ω-only policy over time $t \in \{0, 1, 2, \ldots\}$. Because $\omega(t)$ has the stationary distribution $\pi(\omega)$ for all t, the expectation of the arrival, service, and attribute values are the same for all t:

$$\begin{aligned}
\mathbb{E}\left\{\hat{y}_l(\alpha^*(t), \omega(t))\right\} &= \overline{y}_l \ \forall l \in \{0, 1, \ldots, L\} \\
\mathbb{E}\left\{\hat{e}_j(\alpha^*(t), \omega(t))\right\} &= \overline{e}_j \ \forall j \in \{1, \ldots, J\} \\
\mathbb{E}\left\{\hat{a}_k(\alpha^*(t), \omega(t))\right\} &= \overline{a}_k \ \forall k \in \{1, \ldots, K\} \\
\mathbb{E}\left\{\hat{b}_k(\alpha^*(t), \omega(t))\right\} &= \overline{b}_k \ \forall k \in \{1, \ldots, K\}
\end{aligned}$$

for some quantities $\overline{y}_l, \overline{e}_j, \overline{a}_k, \overline{b}_k$. In the case when Ω is finite or countably infinite, the expectations above can be understood as weighted sums over all ω values, weighted by the stationary distribution

$\pi(\omega)$. Specifically:

$$\mathbb{E}\left\{\hat{y}_l(\alpha^*(t), \omega(t))\right\} = \sum_{\omega\in\Omega} \pi(\omega)\mathbb{E}\left\{\hat{y}_l(\alpha^*(t), \omega)|\omega(t) = \omega\right\}$$

The above expectations $\overline{y}_l, \overline{e}_j, \overline{a}_k, \overline{b}_k$ are finite under any ω-only policy because of the boundedness assumptions (4.25)-(4.30). In addition to assuming $\omega(t)$ is a stationary process, we make the following mild "law of large numbers" assumption concerning time averages (not time average expectations): Under any ω-only policy $\alpha^*(t)$ that yields expectations $\overline{y}_l, \overline{e}_j, \overline{a}_k, \overline{b}_k$ on every slot t, the infinite horizon time averages of $\hat{y}_l(\alpha^*(t), \omega(t))$, $\hat{e}_j(\alpha^*(t), \omega(t))$, $\hat{a}_k(\alpha^*(t), \omega(t))$, $\hat{b}_k(\alpha^*(t), \omega(t))$ are equal to $\overline{y}_l, \overline{e}_j, \overline{a}_k, \overline{b}_k$ with probability 1. For example:

$$\lim_{t\to\infty} \frac{1}{t}\sum_{\tau=0}^{t-1} \hat{y}_l(\alpha^*(\tau), \omega(\tau)) = \overline{y}_l \quad (w.p.1)$$

where "(w.p.1)" means "with probability 1." This is a mild assumption that holds whenever $\omega(t)$ is i.i.d. over slots. This is because, by the law of large numbers, the resulting $\hat{y}_l(\alpha^*(t), \omega(t))$ process is i.i.d. over slots with finite mean $\overline{y}_l$. However, this also holds for a large class of other stationary processes, including stationary processes defined over finite state irreducible Discrete Time Markov Chains (as considered in Section 4.9). It does *not* hold, for example, for degenerate stationary processes where $\omega(0)$ can take different values according to some probability distribution, but is then held fixed for all slots thereafter so that $\omega(t) = \omega(0)$ for all t.

Under these assumptions, we say that the problem (4.31)-(4.35) is *feasible* if there exists a control policy that satisfies the constraints (4.32)-(4.35). Assuming feasibility, define y_0^{opt} as the infimum value of the cost metric (4.31) over all control policies that satisfy the constraints (4.32)-(4.35). This infimum is finite by (4.30). We emphasize that y_0^{opt} considers *all* possible control policies that choose $\alpha(t) \in \mathcal{A}_{\omega(t)}$ over slots t, not just ω-only policies. However, in Appendix 4.A, it is shown that y_0^{opt} can be computed in terms of ω-only policies. Specifically, it is shown that the set of all possible limiting time average expectations of the variables $[(y_l(t)), (e_j(t)), (a_k(t)), (b_k(t))]$, considering all possible algorithms, is equal to the closure of the set of all one-slot averages $[(\overline{y}_l), (\overline{e}_j), (\overline{a}_k), (\overline{b}_k)]$ achievable under ω-only policies. Further, the next theorem shows that if the problem (4.31)-(4.35) is feasible, then the utility y_0^{opt} and the constraints $\overline{y}_l \leq 0, \overline{e}_j \leq 0, \overline{a}_k \leq \overline{b}_k$ can be achieved arbitrarily closely by ω-only policies.

Theorem 4.5 *(Optimality over ω-only Policies) Suppose the $\omega(t)$ process is stationary with distribution $\pi(\omega)$, and that the system satisfies the boundedness assumptions (4.25)-(4.30) and the law of large numbers assumption specified above. If the problem (4.31)-(4.35) is feasible, then for any $\delta > 0$ there is an ω-only* *policy $\alpha^*(t)$ that satisfies $\alpha^*(t) \in \mathcal{A}_{\omega(t)}$ for all t, and:*

$$\mathbb{E}\left\{\hat{y}_0(\alpha^*(t), \omega(t))\right\} \leq y_0^{opt} + \delta \tag{4.36}$$
$$\mathbb{E}\left\{\hat{y}_l(\alpha^*(t), \omega(t))\right\} \leq \delta \quad \forall l \in \{1, \ldots, L\} \tag{4.37}$$
$$|\mathbb{E}\left\{\hat{e}_j(\alpha^*(t), \omega(t))\right\}| \leq \delta \quad \forall j \in \{1, \ldots, J\} \tag{4.38}$$
$$\mathbb{E}\left\{\hat{a}_k(\alpha^*(t), \omega(t))\right\} \leq \mathbb{E}\left\{\hat{b}_k(\alpha^*(t), \omega(t))\right\} + \delta \quad \forall k \in \{1, \ldots, K\} \tag{4.39}$$

Proof. See Appendix 4.A. □

The inequalities (4.36)-(4.39) are similar to those seen in Chapter 3, which related the existence of such randomized policies to the existence of linear programs that yield the desired time averages. The stationarity of $\omega(t)$ simplifies the proof of Theorem 4.5 but is not crucial to its result. Similar results are derived in (15)(21)(136) without the stationary assumption but under the additional assumption that $\omega(t)$ can take at most a finite (but arbitrarily large) number of values and has well defined time averages.

We have stated Theorem 4.5 in terms of arbitrarily small values $\delta > 0$. It may be of interest to note that for most practical systems, there exists an ω-only policy that satisfies all inequalities (4.36)-(4.39) with $\delta = 0$. Appendix 4.A shows that this holds whenever the set Γ, defined as the set of all one-slot expectations achievable under ω-only policies, is closed. Thus, one may prefer a more "aesthetically pleasing" version of Theorem 4.5 that assumes the additional mild closure property in order to remove the appearance of "δ" in the theorem statement. We have presented the theorem in the above form because it is sufficient for our purposes. In particular, we do not require the closure property in order to apply the Lyapunov optimization techniques developed next.

4.4 VIRTUAL QUEUES

To solve the problem (4.31)-(4.35), we first transform all inequality and equality constraints into queue stability problems. Specifically, define *virtual queues* $Z_l(t)$ and $H_j(t)$ for each $l \in \{1, \ldots, L\}$ and $j \in \{1, \ldots, J\}$, with update equations:

$$Z_l(t+1) = \max[Z_l(t) + y_l(t), 0] \tag{4.40}$$
$$H_j(t+1) = H_j(t) + e_j(t) \tag{4.41}$$

The virtual queue $Z_l(t)$ is used to enforce the $\overline{y}_l \leq 0$ constraint. Indeed, recall that if $Z_l(t)$ satisfies (4.40) then by our basic sample path properties in Chapter 2, we have for all $t > 0$:

$$\frac{Z_l(t)}{t} - \frac{Z_l(0)}{t} \geq \frac{1}{t}\sum_{\tau=0}^{t-1} y_l(\tau)$$

Taking expectations of the above and taking $t \to \infty$ shows:

$$\limsup_{t\to\infty} \frac{\mathbb{E}\{Z_l(t)\}}{t} \geq \limsup_{t\to\infty} \overline{y}_l(t)$$

where we recall that $\overline{y}_l(t)$ is the time average expectation of $y_l(\tau)$ over $\tau \in \{0, \ldots, t-1\}$. Thus, if $Z_l(t)$ is mean rate stable, the left-hand-side of the above inequality is 0 and so:

$$\limsup_{t\to\infty} \overline{y}_l(t) \leq 0$$

This means our desired time average constraint for $y_l(t)$ is satisfied. This turns the problem of satisfying a time average inequality constraint into a pure queue stability problem! This discussion is of course just a repeated derivation of Theorem 2.5 (as well as Exercise 2.11).

The virtual queue $H_j(t)$ is designed to turn the time average *equality constraint* $\overline{e}_j = 0$ into a pure queue stability problem. The $H_j(t)$ queue has a different structure, and can possibly be negative, because it enforces an equality constraint rather than an inequality constraint. It is easy to see by summing (4.41) that for any $t > 0$:

$$H_j(t) - H_j(0) = \sum_{\tau=0}^{t-1} e_j(\tau)$$

Taking expectations and dividing by t yields:

$$\frac{\mathbb{E}\left\{H_j(t)\right\} - \mathbb{E}\left\{H_j(0)\right\}}{t} = \overline{e}_j(t) \tag{4.42}$$

Therefore, if $H_j(t)$ is mean rate stable then:[2]

$$\lim_{t\to\infty} \overline{e}_j(t) = 0$$

so that the desired equality constraint for $e_j(t)$ is satisfied.

It follows that if we can design a control algorithm that chooses $\alpha(t) \in \mathcal{A}_{\omega(t)}$ for all t, makes all actual queues $Q_k(t)$ and virtual queues $Z_l(t)$, $H_j(t)$ mean rate stable, and yields a time average expectation of $y_0(t)$ that is equal to our target y_0^{opt}, then we have solved the problem (4.31)-(4.35). This transforms the original problem into a problem of minimizing the time average of a cost function subject to queue stability. We assume throughout that initial conditions satisfy $Z_l(0) \geq 0$ for all $l \in \{1, \ldots, L\}$, $H_j(0) \in \mathbb{R}$ for all $j \in \{1, \ldots, J\}$, and that $\mathbb{E}\left\{Z_l(0)^2\right\} < \infty$ and $\mathbb{E}\left\{H_j(0)^2\right\} < \infty$ for all l and j.

[2]Note by Jensen's inequality that $0 \leq |\mathbb{E}\{H(t)\}| \leq \mathbb{E}\{|H(t)|\}$, and so if $\mathbb{E}\{|H(t)|\}/t \to 0$, then $\mathbb{E}\{H(t)\}/t \to 0$.

4.5 THE MIN DRIFT-PLUS-PENALTY ALGORITHM

Let $\boldsymbol{\Theta}(t) = [\boldsymbol{Q}(t), \boldsymbol{Z}(t), \boldsymbol{H}(t)]$ be a concatenated vector of all actual and virtual queues, with update equations (4.23), (4.40), (4.41). Define the Lyapunov function:

$$L(\boldsymbol{\Theta}(t)) \triangleq \frac{1}{2}\sum_{k=1}^{K} Q_k(t)^2 + \frac{1}{2}\sum_{l=1}^{L} Z_l(t)^2 + \frac{1}{2}\sum_{j=1}^{J} H_j(t)^2 \tag{4.43}$$

If there are no equality constraints, we have $J = 0$ and we remove the $H_j(t)$ queues. If there are no inequality constraints, then $L = 0$ and we remove the $Z_l(t)$ queues.

Lemma 4.6 *Suppose $\omega(t)$ is i.i.d. over slots. Under any control algorithm, the drift-plus-penalty expression has the following upper bound for all t, all possible values of $\boldsymbol{\Theta}(t)$, and all parameters $V \geq 0$:*

$$\begin{aligned}\Delta(\boldsymbol{\Theta}(t)) + V\mathbb{E}\{y_0(t)|\boldsymbol{\Theta}(t)\} \;\leq\; & B + V\mathbb{E}\{y_0(t)|\boldsymbol{\Theta}(t)\} + \sum_{k=1}^{K} Q_k(t)\mathbb{E}\{a_k(t) - b_k(t) \mid \boldsymbol{\Theta}(t)\} \\ & + \sum_{l=1}^{L} Z_l(t)\mathbb{E}\{y_l(t)|\boldsymbol{\Theta}(t)\} + \sum_{j=1}^{J} H_j(t)\mathbb{E}\left\{e_j(t)|\boldsymbol{\Theta}(t)\right\}\end{aligned} \tag{4.44}$$

where B is a positive constant that satisfies the following for all t:

$$\begin{aligned}B \;\geq\; & \frac{1}{2}\sum_{k=1}^{K}\mathbb{E}\left\{a_k(t)^2 + b_k(t)^2 \mid \boldsymbol{\Theta}(t)\right\} + \frac{1}{2}\sum_{l=1}^{L}\mathbb{E}\left\{y_l(t)^2|\boldsymbol{\Theta}(t)\right\} \\ & + \frac{1}{2}\sum_{j=1}^{J}\mathbb{E}\left\{e_j(t)^2|\boldsymbol{\Theta}(t)\right\} - \sum_{k=1}^{K}\mathbb{E}\left\{\tilde{b}_k(t)a_k(t)|\boldsymbol{\Theta}(t)\right\}\end{aligned} \tag{4.45}$$

where we recall that $\tilde{b}_k(t) = \min[Q_k(t), b_k(t)]$. Such a constant B exists because $\omega(t)$ is i.i.d. and the boundedness assumptions in Section 4.2.1 hold.

Proof. Squaring the queue update equation (4.23) and using the fact that $\max[q-b, 0]^2 \leq (q-b)^2$ yields:

$$\begin{aligned}Q_k(t+1)^2 \;&\leq\; (Q_k(t) - b_k(t))^2 + a_k(t)^2 + 2\max[Q_k(t) - b_k(t), 0]a_k(t) \\ &=\; (Q_k(t) - b_k(t))^2 + a_k(t)^2 + 2(Q_k(t) - \tilde{b}_k(t))a_k(t)\end{aligned} \tag{4.46}$$

Therefore:

$$\frac{Q_k(t+1)^2 - Q_k(t)^2}{2} \leq \frac{a_k(t)^2 + b_k(t)^2}{2} - \tilde{b}_k(t)a_k(t) + Q_k(t)[a_k(t) - b_k(t)]$$

Similarly,

$$\frac{Z_l(t+1)^2 - Z_l(t)^2}{2} \leq \frac{y_l(t)^2}{2} + Z_l(t)y_l(t) \tag{4.47}$$
$$\frac{H_j(t+1)^2 - H_j(t)^2}{2} = \frac{e_j(t)^2}{2} + H_j(t)e_j(t)$$

Taking conditional expectations of the above three equations and summing over $k \in \{1, \ldots, K\}$, $l \in \{1, \ldots, L\}$, $j \in \{1, \ldots, J\}$ gives a bound on $\Delta(\boldsymbol{\Theta}(t))$. Adding $V\mathbb{E}\{y_0(t)|\boldsymbol{\Theta}(t)\}$ to both sides proves the result. □

Rather than directly minimize the expression $\Delta(\boldsymbol{\Theta}(t)) + V\mathbb{E}\{y_0(t)|\boldsymbol{\Theta}(t)\}$ every slot t, our strategy actually seeks to minimize the bound given in the right-hand-side of (4.44). This is done via the framework of opportunistically minimizing a (conditional) expectation as described in Section 1.8 (see also Exercise 4.5), and the resulting algorithm is given below.

Min Drift-Plus-Penalty Algorithm for solving (4.31)-(4.35): Every slot t, observe the current queue states $\boldsymbol{\Theta}(t)$ and the random event $\omega(t)$, and make a control decision $\alpha(t) \in \mathcal{A}_{\omega(t)}$ as follows:

$$\begin{aligned} \text{Minimize:} \quad & V\hat{y}_0(\alpha(t), \omega(t)) + \textstyle\sum_{k=1}^{K} Q_k(t)[\hat{a}_k(\alpha(t), \omega(t)) - \hat{b}_k(\alpha(t), \omega(t))] \\ & + \textstyle\sum_{l=1}^{L} Z_l(t)\hat{y}_l(\alpha(t), \omega(t)) + \sum_{j=1}^{J} H_j(t)\hat{e}_j(\alpha(t), \omega(t)) && (4.48) \\ \text{Subject to:} \quad & \alpha(t) \in \mathcal{A}_{\omega(t)} && (4.49) \end{aligned}$$

Then update the virtual queues $Z_l(t)$ and $H_j(t)$ according to (4.40) and (4.41), and the actual queues $Q_k(t)$ according to (4.23).

A remarkable property of this algorithm is that it does not need to know the probabilities $\pi(\omega)$. After observing $\omega(t)$, it seeks to minimize a (possibly non-linear, non-convex, and discontinuous) function of α over all $\alpha \in \mathcal{A}_{\omega(t)}$. Its complexity depends on the structure of the functions $\hat{a}_k(\cdot)$, $\hat{b}_k(\cdot)$, $\hat{y}_l(\cdot)$, $\hat{e}_j(\cdot)$. However, in the case when the set $\mathcal{A}_{\omega(t)}$ contains a finite (and small) number of possible control actions, the policy simply evaluates the function over each option and chooses the best one.

Before presenting the analysis, we note that the problem (4.48)-(4.49) may not have a well defined minimum when the set $\mathcal{A}_{\omega(t)}$ is infinite. However, rather than assuming our decisions obtain the exact minimum every slot (or come close to the infimum), we analyze the performance when our implementation comes within an additive constant of the infimum in the right-hand-side of (4.44).

Definition 4.7 For a given constant $C \geq 0$, a *C-additive approximation* of the drift-plus-penalty algorithm is one that, every slot t and given the current $\boldsymbol{\Theta}(t)$, chooses a (possibly randomized) action $\alpha(t) \in \mathcal{A}_{\omega(t)}$ that yields a conditional expected value on the right-hand-side of the drift expression (4.44) (given $\boldsymbol{\Theta}(t)$) that is within a constant C from the infimum over all possible control actions.

Definition 4.7 allows the deviation from the infimum to be in an expected sense, rather than a deterministic sense, which is useful in some applications. These C-additive approximations are also useful for implementations with out-of-date queue backlog information, as shown in Exercise 4.10, and for achieving maximum throughput in interference networks via approximation algorithms, as shown in Chapter 6.

Theorem 4.8 *(Performance of Min Drift-Plus-Penalty Algorithm) Suppose that $\omega(t)$ is i.i.d. over slots with probabilities $\pi(\omega)$, the problem (4.31)-(4.35) is feasible, and that $\mathbb{E}\{L(\mathbf{\Theta}(0))\} < \infty$. Fix a value $C \geq 0$. If we use a C-additive approximation of the algorithm every slot t, then:*

a) Time average expected cost satisfies:

$$\limsup_{t\to\infty} \frac{1}{t}\sum_{\tau=0}^{t-1} \mathbb{E}\{y_0(\tau)\} \leq y_0^{opt} + \frac{B+C}{V} \tag{4.50}$$

where y_0^{opt} is the infimum time average cost achievable by any policy that meets the required constraints, and B is defined in (4.45).

b) All queues $Q_k(t)$, $Z_l(t)$, $H_j(t)$ are mean rate stable, and all required constraints (4.32)-(4.35) are satisfied.

c) Suppose there are constants $\epsilon > 0$ and $\Psi(\epsilon)$ for which the Slater condition of Assumption A1 holds, stated below in (4.61)-(4.64). Then:

$$\limsup_{t\to\infty} \frac{1}{t}\sum_{\tau=0}^{t-1}\sum_{k=1}^{K} \mathbb{E}\{Q_k(\tau)\} \leq \frac{B+C+V[\Psi(\epsilon)-y_0^{opt}]}{\epsilon} \tag{4.51}$$

where $[\Psi(\epsilon) - y_0^{opt}] \leq y_{0,max} - y_{0,min}$, and $y_{0,min}$, $y_{0,max}$ are defined in (4.30).

We note that the bounds given in (4.50) and (4.51) are not just infinite horizon bounds: Inequalities (4.58) and (4.59) in the below proof show that these bounds hold for all time $t > 0$ in the case when all initial queue backlogs are zero, and that a "fudge factor" that decays like $O(1/t)$ must be included if initial queue backlogs are non-zero. The above theorem is for the case when $\omega(t)$ is i.i.d. over slots. The same algorithm can be shown to offer similar performance under more general ergodic $\omega(t)$ processes as well as for non-ergodic processes, as discussed in Section 4.9.

Proof. (Theorem 4.8) Because, every slot t, our implementation comes within an additive constant C of minimizing the right-hand-side of the drift expression (4.44) over all $\alpha(t) \in \mathcal{A}_{\omega(t)}$, we have for each slot t:

$$\begin{aligned}\Delta(\mathbf{\Theta}(t)) + V\mathbb{E}\{y_0(t)|\mathbf{\Theta}(t)\} \leq\; & B + C + V\mathbb{E}\left\{y_0^*(t)|\mathbf{\Theta}(t)\right\} \\ & + \sum_{l=1}^{L} Z_l(t)\mathbb{E}\left\{y_l^*(t)|\mathbf{\Theta}(t)\right\} + \sum_{j=1}^{J} H_j(t)\mathbb{E}\left\{e_j^*(t)|\mathbf{\Theta}(t)\right\} \\ & + \sum_{k=1}^{K} Q_k(t)\mathbb{E}\left\{a_k^*(t) - b_k^*(t) \mid \mathbf{\Theta}(t)\right\}\end{aligned} \tag{4.52}$$

where $a_k^*(t)$, $b_k^*(t)$, $y_l^*(t)$, $e_j^*(t)$ are the resulting arrival, service, and attribute values under any alternative (possibly randomized) decision $\alpha^*(t) \in \mathcal{A}_{\omega(t)}$. Specifically, $a_k^*(t)\triangleq\hat{a}_k(\alpha^*(t), \omega(t))$, $b_k^*(t)\triangleq\hat{b}_k(\alpha^*(t), \omega(t))$, $y_l^*(t)\triangleq\hat{y}_l(\alpha^*(t), \omega(t))$, $e_j^*(t)\triangleq\hat{e}_j(\alpha^*(t), \omega(t))$.

Now fix $\delta > 0$, and consider the ω-only policy $\alpha^*(t)$ that yields (4.36)-(4.39). Because this is an ω-only policy, and $\omega(t)$ is i.i.d. over slots, the resulting values of $y_0^*(t)$, $a_k^*(t)$, $b_k^*(t)$, $e_j^*(t)$ are independent of the current queue backlogs $\boldsymbol{\Theta}(t)$, and we have from (4.36)-(4.39):

$$\mathbb{E}\left\{y_0^*(t)|\boldsymbol{\Theta}(t)\right\} = \mathbb{E}\left\{y_0^*(t)\right\} \leq y_0^{opt} + \delta \tag{4.53}$$

$$\mathbb{E}\left\{y_l^*(t)|\boldsymbol{\Theta}(t)\right\} = \mathbb{E}\left\{y_l^*(t)\right\} \leq \delta \quad \forall l \in \{1, \ldots, L\} \tag{4.54}$$

$$|\mathbb{E}\left\{e_j^*(t)|\boldsymbol{\Theta}(t)\right\}| = |\mathbb{E}\left\{e_j^*(t)\right\}| \leq \delta \quad \forall j \in \{1, \ldots, J\} \tag{4.55}$$

$$\mathbb{E}\left\{a_k^*(t) - b_k^*(t)|\boldsymbol{\Theta}(t)\right\} = \mathbb{E}\left\{a_k^*(t) - b_k^*(t)\right\} \leq \delta \quad \forall k \in \{1, \ldots, K\} \tag{4.56}$$

Plugging these into the right-hand-side of (4.52) and taking $\delta \to 0$ yields:

$$\Delta(\boldsymbol{\Theta}(t)) + V\mathbb{E}\left\{y_0(t)|\boldsymbol{\Theta}(t)\right\} \leq B + C + Vy_0^{opt} \tag{4.57}$$

This is in the exact form for application of the Lyapunov Optimization Theorem (Theorem 4.2). Hence, all queues are mean rate stable, and so all required time average constraints are satisfied, which proves part (b). Further, from the above drift expression, we have for any $t > 0$ (from (4.13) of Theorem 4.2, or simply from taking iterated expectations and telescoping sums):

$$\frac{1}{t}\sum_{\tau=0}^{t-1}\mathbb{E}\left\{y_0(\tau)\right\} \leq y_0^{opt} + \frac{B+C}{V} + \frac{\mathbb{E}\left\{L(\boldsymbol{\Theta}(0))\right\}}{Vt} \tag{4.58}$$

which proves part (a) by taking a lim sup as $t \to \infty$.

To prove part (c), assume Assumption A1 holds (stated below). Plugging the ω-only policy that yields (4.61)-(4.64) into the right-hand-side of the drift bound (4.52) yields:

$$\Delta(\boldsymbol{\Theta}(t)) + V\mathbb{E}\left\{y_0(t)|\boldsymbol{\Theta}(t)\right\} \leq B + C + V\Psi(\epsilon) - \epsilon\sum_{k=1}^{K}Q_k(t)$$

Taking iterated expectations, summing the telescoping series, and rearranging terms as usual yields:

$$\frac{1}{t}\sum_{\tau=0}^{t-1}\sum_{k=1}^{K}\mathbb{E}\left\{Q_k(\tau)\right\} \leq \frac{B + C + V[\Psi(\epsilon) - \frac{1}{t}\sum_{\tau=0}^{t-1}\mathbb{E}\left\{y_0(\tau)\right\}]}{\epsilon} + \frac{\mathbb{E}\left\{L(\boldsymbol{\Theta}(0))\right\}}{\epsilon t} \tag{4.59}$$

However, because our algorithm satisfies all of the desired constraints of the optimization problem (4.31)-(4.35), its limiting time average expectation for $y_0(t)$ cannot be better than y_0^{opt}:

$$\liminf_{t\to\infty}\frac{1}{t}\sum_{\tau=0}^{t-1}\mathbb{E}\left\{y_0(\tau)\right\} \geq y_0^{opt} \tag{4.60}$$

Indeed, this fact is shown in Appendix 4.A (equation (4.96)). Taking a lim sup of (4.59) as $t \to \infty$ and using (4.60) yields:

$$\limsup_{t\to\infty} \frac{1}{t}\sum_{\tau=0}^{t-1}\sum_{k=1}^{K}\mathbb{E}\{Q_k(\tau)\} \leq \frac{B + C + V[\Psi(\epsilon) - y_0^{opt}]}{\epsilon}$$

□

The following is the Assumption A1 needed in part (c) of Theorem 4.8.

Assumption A1 (Slater Condition): There are values $\epsilon > 0$ and $\Psi(\epsilon)$ (where $y_0^{min} \leq \Psi(\epsilon) \leq y_0^{max}$) and an ω-only policy $\alpha^*(t)$ that satisfies:

$$\mathbb{E}\left\{\hat{y}_0(\alpha^*(t), \omega(t))\right\} = \Psi(\epsilon) \quad (4.61)$$

$$\mathbb{E}\left\{\hat{y}_l(\alpha^*(t), \omega(t))\right\} \leq 0 \quad \forall l \in \{1, \dots, L\} \quad (4.62)$$

$$\mathbb{E}\left\{\hat{e}_j(\alpha^*(t), \omega(t))\right\} = 0 \quad \forall j \in \{1, \dots, J\} \quad (4.63)$$

$$\mathbb{E}\left\{\hat{a}_k(\alpha^*(t), \omega(t))\right\} \leq \mathbb{E}\left\{\hat{b}_k(\alpha^*(t), \omega(t))\right\} - \epsilon \quad \forall k \in \{1, \dots, K\} \quad (4.64)$$

Assumption A1 ensures strong stability of the $Q_k(t)$ queues. However, often the structure of a particular problem allows stronger *deterministic* queue bounds, even without Assumption A1 (see Exercise 4.9). A variation on the above proof that considers probability 1 convergence is treated in Exercise 4.6.

4.5.1 WHERE ARE WE USING THE I.I.D. ASSUMPTIONS?

In (4.53)-(4.56) of the above proof, we used equalities of the form $\mathbb{E}\left\{y_l^*(t)|\boldsymbol{\Theta}(t)\right\} = \mathbb{E}\left\{y_l^*(t)\right\}$, which hold for any ω-only policy $\alpha^*(t)$ when $\omega(t)$ is i.i.d. over slots. Because past values of $\omega(\tau)$ for $\tau < t$ have influenced the current queue states $\boldsymbol{\Theta}(t)$, this influence might skew the conditional distribution of $\omega(t)$ (given $\boldsymbol{\Theta}(t)$) unless $\omega(t)$ is independent of the past. However, while the i.i.d. assumption is crucial for the above proof, it is *not* crucial for efficient performance of the algorithm, as shown in Section 4.9.

4.6 EXAMPLES

Here we provide examples of using the drift-plus-penalty algorithm for the same systems considered in Sections 2.3.1 and 2.3.2. More examples are given in Exercises 4.7-4.15.

4.6.1 DYNAMIC SERVER SCHEDULING

Example Problem: Consider the 3-queue, 2-server system described in Section 2.3.1 (see Fig. 2.1). Define $\omega(t) \triangleq (a_1(t), a_2(t), a_3(t))$ as the random arrivals on slot t, and assume $\omega(t)$ is i.i.d. over slots with $\mathbb{E}\{a_i(t)\} = \lambda_i, \mathbb{E}\left\{a_i(t)^2\right\} = \mathbb{E}\left\{a_i^2\right\}$ for $i \in \{1, 2, 3\}$.

a) Suppose $(\lambda_1, \lambda_2, \lambda_3) \in \Lambda$, where we recall that Λ is defined by the constraints $0 \leq \lambda_i \leq 1$ for all $i \in \{1, 2, 3\}$, and $\lambda_1 + \lambda_2 + \lambda_3 \leq 2$. State the drift-plus-penalty algorithm (with $V = 0$ and $C = 0$) for stabilizing all three queues.

b) Suppose the Slater condition (Assumption A1) holds for a value $\epsilon > 0$. Using the drift-plus-penalty algorithm with $V = 0$, $C = 0$, derive a value B such that time average queue backlog satisfies $\overline{Q_1 + Q_2 + Q_3} \leq B/\epsilon$, where $\overline{Q_1 + Q_2 + Q_3}$ is the lim sup time average expected backlog in the system.

c) Suppose we must choose $\boldsymbol{b}(t) \in \{(1, 1, 0), (1, 0, 1), (0, 1, 1)\}$ every slot t. Suppose that choosing $\boldsymbol{b}(t) = (1, 1, 0)$ or $\boldsymbol{b}(t) = (1, 0, 1)$ consumes one unit of power per slot, but using the vector $\boldsymbol{b}(t) = (0, 1, 1)$ uses two units of power per slot. State the drift-plus-penalty algorithm (with $V > 0$ and $C = 0$) that seeks to minimize time average power subject to queue stability. Conclude that $\overline{p} \leq p^{opt} + B/V$, where $\overline{p}$ is the lim sup time average expected power expenditure of the algorithm, and p^{opt} is the minimum possible time average power expenditure required for queue stability. Assuming the Slater condition of part (b), conclude that $\overline{Q_1 + Q_2 + Q_3} \leq (B + V)/\epsilon$.

Solution:

a) We have $K = 3$ with queues $Q_1(t), Q_2(t), Q_3(t)$. There is no penalty to minimize, so $y_0(t) = 0$ (and so we also choose $V = 0$). There are no additional $y_l(t)$ or $e_j(t)$ attributes, and so $L = J = 0$. The control action $\alpha(t)$ determines the server allocations, so that $\alpha(t) = (b_1(t), b_2(t), b_3(t))$, and the set of possible action vectors is $\mathcal{A} = \{(1, 1, 0), (1, 0, 1), (0, 1, 1)\}$ (so that we choose which two queues to serve on each slot). The control action does not affect the arrivals, and so $\hat{a}_k(\alpha(t), \omega(t)) = a_k(t)$. The algorithm (4.48)-(4.49) with $V = 0$ reduces to observing the queue backlogs every slot t and choosing $(b_1(t), b_2(t), b_3(t))$ as follows:

$$\text{Minimize:} \quad -\sum_{k=1}^{3} Q_k(t) b_k(t) \tag{4.65}$$
$$\text{Subject to:} \quad (b_1(t), b_2(t), b_3(t)) \in \{(1, 1, 0), (1, 0, 1), (0, 1, 1)\} \tag{4.66}$$

Then update the queues $Q_k(t)$ according to (4.23). Note that the problem (4.65)-(4.66) is equivalent to minimizing $\sum_{k=1}^{3} Q_k(t)[a_k(t) - b_k(t)]$ subject to the same constraints, but to minimize this, it suffices to minimize only the terms we can control (so we can remove the $\sum_{k=1}^{3} Q_k(t)a_k(t)$ term that is the same regardless of our control decision). It is easy to see that the problem (4.65)-(4.66) reduces to choosing the two largest queues to serve every slot, breaking ties arbitrarily. This simple policy does not require any knowledge of $(\lambda_1, \lambda_2, \lambda_3)$, yet ensures all queues are mean rate stable whenever possible!

b) From (4.45) and using the fact that $L = J = 0$ and $\tilde{b}_k(t)a_k(t) \geq 0$, we want to find a value B that satisfies:

$$B \geq \frac{1}{2}\sum_{k=1}^{3} \mathbb{E}\left\{a_k^2(t)|\boldsymbol{\Theta}(t)\right\} + \frac{1}{2}\mathbb{E}\left\{\sum_{k=1}^{3} b_k(t)^2|\boldsymbol{\Theta}(t)\right\}$$

Because $a_k(t)$ is i.i.d. over slots, it is independent of $\boldsymbol{\Theta}(t)$ and so $\mathbb{E}\left\{a_k(t)^2|\boldsymbol{\Theta}(t)\right\} = \mathbb{E}\left\{a_k^2\right\}$. Further, $b_k(t)^2 = b_k(t)$ (because $b_k(t) \in \{0, 1\}$). Thus, it suffices to find a value B that satisfies:

$$B \geq \frac{1}{2}\sum_{k=1}^{3} \mathbb{E}\left\{a_k^2\right\} + \frac{1}{2}\mathbb{E}\left\{\sum_{k=1}^{3} b_k(t)|\boldsymbol{\Theta}(t)\right\}$$

However, since $b_1(t) + b_2(t) + b_3(t) \leq 2$ for all t (regardless of $\boldsymbol{\Theta}(t)$), we can choose:

$$B = \frac{1}{2}\sum_{k=1}^{3}\mathbb{E}\left\{a_k^2\right\} + 1$$

Because Assumption A1 is satisfied and $V = C = 0$, we have from (4.51) that:

$$\overline{Q_1 + Q_2 + Q_3} \leq B/\epsilon$$

c) We now define penalty $y_0(t) = \hat{y}_0(b_1(t), b_2(t), b_3(t))$, where:

$$\hat{y}_0(b_1(t), b_2(t), b_3(t)) = \begin{cases} 1 & \text{if } (b_1(t), b_2(t), b_3(t)) \in \{(1,1,0) \cup (1,0,1)\} \\ 2 & \text{if } (b_1(t), b_2(t), b_3(t)) = (0,1,1) \end{cases}$$

Then the drift-plus-penalty algorithm (with $V > 0$) now observes $(Q_1(t), Q_2(t), Q_3(t))$ every slot t and chooses a server allocation to solve:

$$\text{Minimize:} \quad V\hat{y}_0(b_1(t), b_2(t), b_3(t)) - \sum_{k=1}^{2} Q_k(t)b_k(t) \quad (4.67)$$
$$\text{Subject to:} \quad (b_1(t), b_2(t), b_3(t)) \in \{(1,1,0), (1,0,1), (0,1,1)\} \quad (4.68)$$

This can be solved easily by comparing the value of (4.67) associated with each option:

- Option $(1, 1, 0)$: value = $V - Q_1(t) - Q_2(t)$.
- Option $(1, 0, 1)$: value = $V - Q_1(t) - Q_3(t)$.
- Option $(0, 1, 1)$: value = $2V - Q_2(t) - Q_3(t)$.

Thus, every slot t we pick the option with the smallest of the above three values, breaking ties arbitrarily. This is again a simple dynamic algorithm that does not require knowledge of the rates $(\lambda_1, \lambda_2, \lambda_3)$. By (4.50), we know that the achieved time average power $\overline{p}$ (where $\overline{p}\triangleq\overline{y}_0$) satisfies $\overline{p} \leq p^{opt} + B/V$, where B is defined in part (b). Because $y_{0,max} = 2$ and $y_{0,min} = 1$, by (4.51), we know the resulting average backlog satisfies $\overline{Q_1 + Q_2 + Q_3} \leq (B + (2-1)V)/\epsilon$, where ϵ is defined in (b). This illustrates the $[O(1/V), O(V)]$ tradeoff between average power and average backlog.

The above problem assumes we must allocate exactly two servers on every slot. The problem can of course be modified if we allow the option of serving only 1 queue, or 0 queues, at some reduced power expenditure.

4.6.2 OPPORTUNISTIC SCHEDULING

Example Problem: Consider the 2-queue wireless system with ON/OFF channels described in Section 2.3.2 (see Fig. 2.2). Suppose channel vectors $(S_1(t), S_2(t))$ are i.i.d. over slots with $S_i(t) \in \{ON, OFF\}$, as before. However, suppose that new arrivals are not immediately sent into the queue, but are only admitted via a *flow control decision*. Specifically, suppose that $(A_1(t), A_2(t))$ represents the random vector of new packet arrivals on slot t, where $A_1(t)$ is i.i.d. over slots and Bernoulli with $Pr[A_1(t) = 1] = \lambda_1$, and $A_2(t)$ is i.i.d. over slots and Bernoulli with $Pr[A_2(t) = 1] = \lambda_2$. Every slot a flow controller observes $(A_1(t), A_2(t))$ and makes an *admission decision* $a_1(t), a_2(t)$, subject to the constraints:

$$a_1(t) \in \{0, A_1(t)\}, a_2(t) \in \{0, A_2(t)\}$$

Packets that are not admitted are dropped. We thus have $\omega(t) = [(S_1(t), S_2(t)), (A_1(t), A_2(t))]$. The control action is given by $\alpha(t) = [(\alpha_1(t), \alpha_2(t)); (\beta_1(t), \beta_2(t))]$ where $\alpha_k(t)$ is a binary value that is 1 if we choose to admit the packet (if any) arriving to queue k on slot t, and $\beta_k(t)$ is a binary value that is 1 if we choose serve queue k on slot t, with the constraint $\beta_1(t) + \beta_2(t) \leq 1$.

a) Use the drift-plus-penalty method (with $V > 0$ and $C = 0$) to stabilize the queues while seeking to maximize the linear utility function of throughput $w_1\overline{a}_1 + w_2\overline{a}_2$, where w_1 and w_2 are given positive weights and $\overline{a}_k$ represents the time average rate of data admitted to queue k.

b) Assuming the Slater condition of Assumption A1 holds for some value $\epsilon > 0$, state the resulting utility and average backlog performance.

c) Redo parts (a) and (b) with the additional constraint that $\overline{a}_1 \geq 0.1$ (assuming this constraint, is feasible).

Solution:

a) We have $K = 2$ queues to stabilize. We have penalty function $y_0(t) = -w_1a_1(t) - w_2a_2(t)$ (so that minimizing the time average of this penalty maximizes $w_1\overline{a}_1 + w_2\overline{a}_2$). There are no other attributes $y_l(t)$ or $e_j(t)$, so $L = J = 0$. The arrival and service variables are given by $a_k(t) = \hat{a}_k(\alpha_k(t), A_k(t))$ and $b_k(t) = \hat{b}_k(\beta_k(t), S_k(t))$ for $k \in \{1, 2\}$, where:

$$\hat{a}_k(\alpha_k(t), A_k(t)) = \alpha_k(t)A_k(t) \quad , \quad \hat{b}_k(\beta_k(t), S_k(t)) = \beta_k(t)1_{\{S_k(t)=ON\}}$$

where $1_{\{S_k(t)=ON\}}$ is an indicator function that is 1 if $S_k(t) = ON$, and 0 else. The drift-plus-penalty algorithm of (4.48) thus reduces to observing the queue backlogs $(Q_1(t), Q_2(t))$ and the current network state $\omega(t) = [(S_1(t), S_2(t)), (A_1(t), A_2(t))]$ and making flow control and transmission actions $\alpha_k(t)$ and $\beta_k(t)$ to solve:

$$\text{Min: } -V[w_1\alpha_1(t)A_1(t) + w_2\alpha_2(t)A_2(t)] + \sum_{k=1}^{2} Q_k(t)[\alpha_k(t)A_k(t) - \beta_k(t)1_{\{S_k(t)=ON\}}]$$
$$\text{Subj. to: } \alpha_k(t) \in \{0, 1\} \ \forall k \in \{1, 2\} \ , \ \beta_k(t) \in \{0, 1\} \ \forall k \in \{1, 2\}, \ \beta_1(t) + \beta_2(t) \leq 1$$

The flow control and transmission decisions appear in separate terms in the above problem, and so they can be chosen to minimize their respective terms separately. This reduces to the following simple algorithm:

- (Flow Control) For each $k \in \{1, 2\}$, choose $\alpha_k(t) = 1$ (so that we admit $A_k(t)$ to queue k) whenever $Vw_k \geq Q_k(t)$, and choose $\alpha_k(t) = 0$ else.

- (Transmission) Choose $(\beta_1(t), \beta_2(t))$ subject to the constraints to maximize $Q_1(t)\beta_1(t)1_{\{S_1(t)=ON\}} + Q_2(t)\beta_2(t)1_{\{S_2(t)=ON\}}$. This reduces to the "Longest Connected Queue" algorithm of (8). Specifically, we place the server to the queue that is ON and that has the largest value of queue backlog, breaking ties arbitrarily.

b) We compute B from (4.45). Because $L = J = 0$, we choose B to satisfy:

$$B \geq \frac{1}{2}\sum_{k=1}^{2}\mathbb{E}\left\{a_k(t)^2|\Theta(t)\right\} + \frac{1}{2}\sum_{k=1}^{2}\mathbb{E}\left\{b_k(t)^2|\Theta(t)\right\}$$

Because arrivals are i.i.d. Bernoulli, they are independent of queue backlog and so $\mathbb{E}\left\{a_k(t)^2|\Theta(t)\right\} = \mathbb{E}\left\{a_k(t)^2\right\} = \mathbb{E}\left\{a_k(t)\right\} = \lambda_k$. Further, $b_k(t)^2 = b_k(t)$, and $b_1(t) + b_2(t) \leq 1$. Thus we can choose: $B = (\lambda_1 + \lambda_2 + 1)/2$. It follows from (4.50) that:

$$w_1\overline{a}_1 + w_2\overline{a}_2 \geq utility^{opt} - B/V$$

where $utility^{opt}$ is the maximum possible utility value subject to stability. Further, because $y_{0,min} = -(w_1 + w_2)$ and $y_{0,max} = 0$, we have from (4.51):

$$\overline{Q}_1 + \overline{Q}_2 \leq (B + V(w_1 + w_2))/\epsilon$$

c) The constraint $\overline{a}_1 \geq 0.1$ is equivalent to $0.1 - \overline{a}_1 \leq 0$. To enforce this constraint, we simply introduce a virtual queue $Z_1(t)$ as follows:

$$Z_1(t+1) = \max[Z_1(t) + 0.1 - a_1(t), 0] \tag{4.69}$$

This can be viewed as introducing an additional penalty $y_1(t) = 0.1 - a_1(t)$. The drift-plus-penalty algorithm (4.48) reduces to observing the queue backlogs and network state $\omega(t)$ every slot t and making actions to solve

$$\begin{aligned}\text{Min:} \quad & -V[w_1\alpha_1(t)A_1(t) + w_2\alpha_2(t)A_2(t)] + \textstyle\sum_{k=1}^{2} Q_k(t)[\alpha_k(t)A_k(t) - \beta_k(t)1_{\{S_k(t)=ON\}}] \\ & +Z_1(t)[0.1 - \alpha_1(t)A_1(t)] \\ \text{Subj. to:} \quad & \alpha_k(t) \in \{0,1\}\ \forall k \in \{1,2\}\ ,\ \beta_k(t) \in \{0,1\}\ \forall k \in \{1,2\},\ \beta_1(t) + \beta_2(t) \leq 1\end{aligned}$$

Then update virtual queue $Z_1(t)$ according to (4.69) at the end of the slot, and update the queues $Q_k(t)$ according to (4.23). This reduces to:

- (Flow Control) Choose $\alpha_1(t) = 1$ whenever $Vw_1 + Z_1(t) \geq Q_1(t)$, and choose $\alpha_1(t) = 0$ else. Choose $\alpha_2(t) = 1$ whenever $Vw_2 \geq Q_2(t)$, and choose $\alpha_2(t) = 0$ else.

- (Transmission) Choose $(\beta_1(t), \beta_2(t))$ the same as in part (a).

4.7 VARIABLE V ALGORITHMS

The $[O(1/V), O(V)]$ performance-delay tradeoff suggests that if we use a *variable* parameter $V(t)$ that gradually increases with time, then we can maintain mean rate stability while driving the time average penalty to its exact optimum value y_0^{opt}. This is shown below, and is analogous to diminishing stepsize methods for static convex optimization problems (133)(134).

Theorem 4.9 *Suppose that $\omega(t)$ is i.i.d. over slots with probabilities $\pi(\omega)$, the problem (4.31)-(4.35) is feasible, and $\mathbb{E}\{L(\boldsymbol{\Theta}(0))\} < \infty$. Suppose that every slot t, we implement a C-additive approximation that comes within $C \geq 0$ of the infimum of a* modified *right-hand-side of (4.44), where the V parameter is replaced with $V(t)$, defined:*

$$V(t) \triangleq V_0(t+1)^{\beta} \;\; \forall t \in \{0, 1, 2, \ldots\} \tag{4.70}$$

for some constants $V_0 > 0$ and β such that $0 < \beta < 1$. Then all queues are mean rate stable, all required constraints (4.32)-(4.35) are satisfied, and:

$$\lim_{t\to\infty} \frac{1}{t}\sum_{\tau=0}^{t-1} \mathbb{E}\{y_0(\tau)\} = y_0^{opt}$$

The manner in which the V_0 and β parameters affect convergence is described in the proof, specifically in (4.72) and (4.73).

While this variable V approach yields the exact optimum y_0^{opt}, its disadvantage is that we achieve only mean rate stability and not strong stability, so that there is no finite bound on average queue size and average delay. In fact, it is known that for typical problems (except for those with a trivial structure), average backlog and delay *necessarily* grow to infinity as we push performance closer and closer to optimal, becoming infinity at the optimal point (50)(51)(52)(53). The very large queue sizes incurred by this variable V algorithm also make it more difficult to adapt to changes in system parameters, whereas fixed V algorithms can easily adapt.

Proof. (Theorem 4.9) Repeating the proof of Theorem 4.8 by replacing V with $V(t)$ for a given slot t, the equation (4.57) becomes:

$$\Delta(\boldsymbol{\Theta}(t)) + V(t)\mathbb{E}\{y_0(t)|\boldsymbol{\Theta}(t)\} \leq B + C + V(t)y_0^{opt}$$

Taking expectations of both sides of the above and using iterated expectations yields:

$$\mathbb{E}\{L(\boldsymbol{\Theta}(t+1))\} - \mathbb{E}\{L(\boldsymbol{\Theta}(t))\} + V(t)\mathbb{E}\{y_0(t)\} \leq B + C + V(t)y_0^{opt} \tag{4.71}$$

Noting that $\mathbb{E}\{y_0(t)\} \geq y_{0,min}$ yields:

$$\mathbb{E}\{L(\boldsymbol{\Theta}(t+1))\} - \mathbb{E}\{L(\boldsymbol{\Theta}(t))\} \leq B + C + V(t)(y_0^{opt} - y_{0,min})$$

The above holds for all $t \geq 0$. Summing over $\tau \in \{0, \ldots, t-1\}$ yields:

$$\mathbb{E}\{L(\boldsymbol{\Theta}(t))\} - \mathbb{E}\{L(\boldsymbol{\Theta}(0))\} \leq (B+C)t + (y_0^{opt} - y_{0,min})\sum_{\tau=0}^{t-1} V(\tau)$$

Using the definition of the Lyapunov function in (4.43) yields the following for all $t > 0$:

$$\sum_{k=1}^{K} \mathbb{E}\left\{Q_k(t)^2\right\} + \sum_{l=1}^{L} \mathbb{E}\left\{Z_l(t)^2\right\} + \sum_{j=1}^{J} \mathbb{E}\left\{H_j(t)^2\right\} \leq$$

$$2(B+C)t + 2\mathbb{E}\{L(\boldsymbol{\Theta}(0))\} + 2(y_0^{opt} - y_{0,min})\sum_{\tau=0}^{t-1} V(\tau)$$

Take any queue $Q_k(t)$. Because $\mathbb{E}\{Q_k(t)\}^2 \leq \mathbb{E}\left\{Q_k(t)^2\right\}$, we have for all queues $Q_k(t)$:

$$\mathbb{E}\{Q_k(t)\} \leq \sqrt{2(B+C)t + 2\mathbb{E}\{L(\boldsymbol{\Theta}(0))\} + 2(y_0^{opt} - y_{0,min})\sum_{\tau=0}^{t-1} V(\tau)}$$

and the same bound holds for $\mathbb{E}\{Z_l(t)\}$ and $\mathbb{E}\left\{|H_j(t)|\right\}$ for all $l \in \{1, \ldots, L\}$, $j \in \{1, \ldots, J\}$. Dividing both sides of the above inequality by t yields the following for all $t > 0$:

$$\frac{\mathbb{E}\{Q_k(t)\}}{t} \leq \sqrt{\frac{2(B+C)}{t} + \frac{2\mathbb{E}\{L(\boldsymbol{\Theta}(0))\}}{t^2} + 2(y_0^{opt} - y_{0,min})\frac{1}{t^2}\sum_{\tau=0}^{t-1} V(\tau)} \tag{4.72}$$

and the same bound holds for all $\mathbb{E}\{Z_l(t)\}/t$ and $\mathbb{E}\left\{|H_j(t)|\right\}/t$. However, we have:

$$0 \leq \frac{1}{t^2}\sum_{\tau=0}^{t-1} V(\tau) = \frac{V_0}{t^2}\sum_{\tau=0}^{t-1}(1+\tau)^\beta \leq \frac{V_0}{t^2}\int_0^t (1+v)^\beta dv = \frac{V_0}{t^2}\left[\frac{(1+t)^{1+\beta} - 1}{1+\beta}\right]$$

Because $0 < \beta < 1$, taking a limit of the above as $t \to \infty$ shows that $\frac{1}{t^2}\sum_{\tau=0}^{t-1} V(\tau) \to 0$. Using this and taking a limit of (4.72) shows that all queues are mean rate stable, and hence (by Section 4.4)) all required constraints (4.32)-(4.35) are satisfied.

To prove that the time average expectation of $y_0(t)$ converges to y_0^{opt}, consider again the inequality (4.71), which holds for all t. Dividing both sides of (4.71) by $V(t)$ yields:

$$\frac{\mathbb{E}\{L(\boldsymbol{\Theta}(t+1))\} - \mathbb{E}\{L(\boldsymbol{\Theta}(t))\}}{V(t)} + \mathbb{E}\{y_0(t)\} \leq \frac{B+C}{V(t)} + y_0^{opt}$$

Summing the above over $\tau \in \{0, 1, \ldots, t-1\}$ and collecting terms yields:

$$\frac{\mathbb{E}\{L(\boldsymbol{\Theta}(t))\}}{V(t-1)} - \frac{\mathbb{E}\{L(\boldsymbol{\Theta}(0))\}}{V(0)} + \sum_{\tau=1}^{t-1}\mathbb{E}\{L(\boldsymbol{\Theta}(\tau))\}\left[\frac{1}{V(\tau-1)} - \frac{1}{V(\tau)}\right] + \sum_{\tau=0}^{t-1}\mathbb{E}\{y_0(\tau)\} \leq$$

$$t y_0^{opt} + (B+C)\sum_{\tau=0}^{t-1}\frac{1}{V(\tau)}$$

Because $V(t)$ is non-decreasing, we have for all $\tau \geq 1$:

$$\left[\frac{1}{V(\tau-1)} - \frac{1}{V(\tau)}\right] \geq 0$$

Using this in the above inequality and dividing by t yields:

$$\frac{1}{t}\sum_{\tau=0}^{t-1}\mathbb{E}\left\{y_0(\tau)\right\} \leq y_0^{opt} + (B+C)\frac{1}{t}\sum_{\tau=0}^{t-1}\frac{1}{V(\tau)} + \frac{\mathbb{E}\left\{L(\boldsymbol{\Theta}(0))\right\}}{V(0)t} \tag{4.73}$$

However:

$$0 \leq \frac{1}{t}\sum_{\tau=0}^{t-1}\frac{1}{V(\tau)} \leq \frac{1}{tV(0)} + \frac{1}{V_0 t}\int_0^{t-1}\frac{1}{(1+v)^{\beta}}dv = \frac{1}{tV(0)} + \frac{1}{V_0 t}\left[\frac{t^{1-\beta}-1}{1-\beta}\right]$$

Taking a limit as $t \to \infty$ shows that this term vanishes, and so the lim sup of the left-hand-side in (4.73) is less than or equal to y_0^{opt}. However, the policy satisfies all constraints (4.32)-(4.35) and so the lim inf must be greater than or equal to y_0^{opt} (by the Appendix 4.A result (4.96)), so the limit exists and is equal to y_0^{opt}. □

4.8 PLACE-HOLDER BACKLOG

Here we present a simple delay improvement for the fixed-V drift-plus-penalty algorithm. The queue backlogs under this algorithm can be viewed as a stochastic version of a Lagrange multiplier for classical static convex optimization problems (see (45)(37) for more intuition on this), and they need to be large to appropriately inform the stochastic optimizer about good decisions to take. However, for many such problems, we can *trick* the stochastic optimizer by making it think actual queue backlog is larger than it really is. This allows the same performance with reduced queue backlog. To develop the technique, we make the following three preliminary observations:

- The infinite horizon time average expected penalty and backlog bounds of Theorem 4.8 are insensitive to the initial condition $\boldsymbol{\Theta}(0)$.
- All sample paths of backlog and penalty are the same under any service order for the $Q_k(t)$ queues, provided that queueing dynamics satisfy (4.23). In particular, the results are the same if service is First-In-First-Out (FIFO) or Last-In-First-Out (LIFO).
- It is often the case that, under the drift-plus-penalty algorithm (or a particular C-additive approximation of it), some queues are never served until they have at least a certain minimum amount of backlog.

The third observation motivates the following definition.

Definition 4.10 (Place-Holder Values) A non-negative value Q_k^{place} is a *place-holder value* for network queue $Q_k(t)$ with respect to a given algorithm if for all possible sample paths, we have

$Q_k(t) \geq Q_k^{place}$ for all slots $t \geq 0$ whenever $Q_k(0) \geq Q_k^{place}$. Likewise, a non-negative value Z_l^{place} is a place-holder value for queue $Z_l(t)$ if for all possible sample paths, we have $Z_l(t) \geq Z_l^{place}$ for all $t \geq 0$ whenever $Z_l(0) \geq Z_l^{place}$.

Clearly 0 is a place-holder value for all queues $Q_k(t)$ and $Z_l(t)$, but the idea is to compute the largest possible place-holder values. It is often easy to pre-compute positive place-holder values without knowing anything about the system probabilities. This is done in the Chapter 3 example for minimizing average power expenditure subject to stability (see Section 3.2.4), and Exercises 4.8 and 4.11 provide further examples. Suppose now we run the algorithm with initial queue backlog $Q_k(0) = Q_k^{place}$ for all $k \in \{1, \ldots, K\}$. Then we achieve exactly the same backlog and penalty sample paths under either FIFO or LIFO. However, none of the initial backlog Q_k^{place} would ever exit the system under LIFO! Thus, we can achieve the same performance by replacing this initial backlog Q_k^{place} with *fake backlog*, called *place-holder backlog* (142)(143). Whenever a transmission opportunity arises, we transmit only actual data whenever possible, serving the actual data in any order we like (such as FIFO or LIFO). Because queue backlog never dips below Q_k^{place}, we never have to serve any fake data. Thus, the *actual* queue backlog under this implementation is equal to $Q_k^{actual}(t) = Q_k(t) - Q_k^{place}$ for all t, which reduces the actual backlog by an amount exactly equal to Q_k^{place}. This does not affect the sample path and hence does not affect the time average penalty.

Specifically, for all $k \in \{1, \ldots, K\}$ and $l \in \{1, \ldots, L\}$, we initialize the *actual* backlog $Q_k^{actual}(0) = Z_l^{actual}(0) = 0$, but we use place-holder backlogs Q_k^{place}, Z_l^{place} so that:

$$Q_k(0) = Q_k^{place} \quad , \quad Z_l(0) = Z_l^{place} \;\; \forall k \in \{1, \ldots, K\}, l \in \{1, \ldots, L\}$$

We then operate the algorithm using the $Q_k(t)$ and $Z_l(t)$ values (not the actual values $Q_k^{actual}(t)$ and $Z_l^{actual}(t)$). The above discussion ensures that for all time t, we have:

$$Q_k^{actual}(t) = Q_k(t) - Q_k^{place} \quad , \quad Z_l^{actual}(t) = Z_l(t) - Z_l^{place} \;\; \forall t \geq 0$$

Because the bounds in Theorem 4.8 are independent of the initial condition, the same penalty and backlog bounds are achieved. However, the *actual* backlog is reduced by exactly Q_k^{place} and Z_l^{place} at every instant of time. This is a "free" reduction in the queue backlog, with no impact on the limiting time average penalty. This has already been illustrated in the example minimum average power problem of the previous chapter (Section 3.2.4, Figs. 3.3-3.4). The Fig. 4.2 below provides further insight: Fig. 4.2 shows a sample path of $Q_2(t)$ for the same example system of Section 3.2.4 (using $V = 100$ and $(\lambda_1, \lambda_2) = (0.3, 0.7)$). We use $Q_2^{place} = \min[V - 2, 0] = 48$ as the initial backlog, and the figure illustrates that $Q_2(t)$ indeed never drops below 48. The place-holder savings is illustrated in the figure.

We developed this method of place-holder bits in (143) for use in dynamic data compression problems and in (142) for general constrained cost minimization problems (including multi-hop wireless networks with unreliable channels). The reader is referred to the examples and simulations

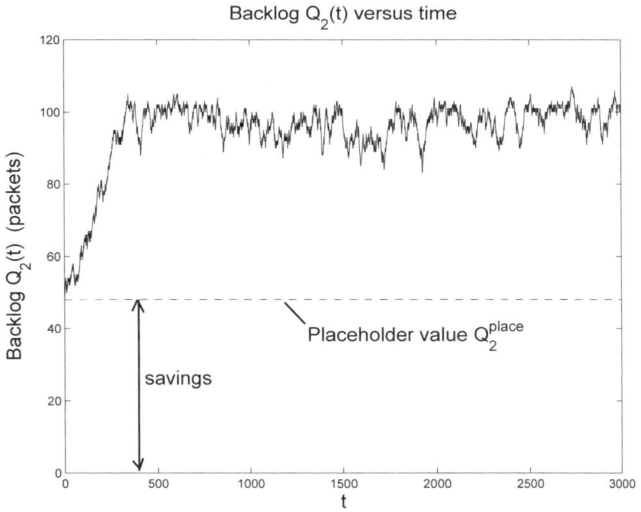

Figure 4.2: A sample path of $Q_2(t)$ over 3000 slots for the example system of Section 3.2.4.

given in (143)(142). A more aggressive place-holder technique is developed in (37). The idea of (37) can be illustrated easily from Fig. 4.2: While the figure illustrates that $Q_2(t)$ never drops below Q_2^{place}, the backlog actually increases until it reaches a "plateau" around 100 packets, and then oscillates with some noise about this value. Intuitively, we can almost double the place-holder value in the figure, raising the horizontal line up to a level that is close to the minimum backlog value seen in the plateau. While we cannot guarantee that backlog will never drop below this new line, the idea is to show that such events occur *rarely*. Work in (45) shows that scaled queue backlog converges to a *Lagrange multiplier* of a related static optimization problem, and work in (37) shows that actual queue backlog oscillates very closely about this Lagrange multiplier. Specifically, it is shown in (37) that, under mild assumptions, the steady state backlog distribution decays exponentially in distance from the Lagrange multiplier value. It then develops an algorithm that uses a place-holder that is a distance of $O(\log^2(V))$ from the Lagrange multiplier, showing that deviations by more than this amount are rare and can be handled separately by dropping a small amount of packets. The result fundamentally changes the performance-backlog tradeoff from $[O(1/V), O(V)]$ to $[O(1/V), O(\log^2(V))]$ (within a logarithmic factor of the optimal tradeoff shown in (52)(51)(53)).

A disadvantage of this aggressive approach is that Lagrange multipliers must be known in advance, which is difficult as they may depend on system statistics and they may be different for each queue in the system. This is handled elegantly in a Last-In-First-Out (LIFO) implementation of the drift-plus-penalty method, developed in (54). That LIFO can improve delay can be understood by Fig. 4.2: First, a LIFO implementation would achieve all of the savings of the original place-holder value of $Q_2^{place} = 48$ (at the cost of never serving the first 48 packets). Next, a LIFO implementation would intuitively lead to delays of "most" packets that are on the order of the magnitude of noise variations in the plateau area. That is, LIFO can achieve the more aggressive place-holder gains without computing the Lagrange multipliers! This is formally proven in (55). Experiments with the LIFO drift-plus-penalty method on an actual multi-hop wireless network deployment in (54) show a dramatic improvement in delay (by more than an order of magnitude) for all but 2% of the packets.

4.9 NON-I.I.D. MODELS AND UNIVERSAL SCHEDULING

Here we show that the same drift-plus-penalty algorithm provides similar $[O(1/V), O(V)]$ performance guarantees when $\omega(t)$ varies according to a more general ergodic (possibly non-i.i.d.) process. We then show it also provides efficient performance for arbitrary (possibly non-ergodic) sample paths. The main proof techniques are the same as those we have already developed, with the exception that we use a multi-slot drift analysis rather than a 1-slot drift analysis.

We consider the same system as in Section 4.2.1, with K queues with dynamics (4.23), and attributes $y_l(t) = \hat{y}_l(\alpha(t), \omega(t))$ for $l \in \{1, \ldots, L\}$. For simplicity, we eliminate the attributes $e_j(t)$ associated with equality constraints (so that $J = 0$). We seek an algorithm for choosing $\alpha(t) \in \mathcal{A}_{\omega(t)}$ every slot to minimize $\overline{y}_0$ subject to mean rate stability of all queues $Q_k(t)$ and subject to $\overline{y}_l \leq 0$ for all $l \in \{1, \ldots, L\}$. The virtual queues $Z_l(t)$ for $l \in \{1, \ldots, L\}$ are the same as before, defined in (4.40). For simplicity of exposition, we assume:

- The exact drift-plus-penalty algorithm of (4.48)-(4.49) is used, rather than a C-additive approximation (so that $C = 0$).
- The functions $\hat{a}_k(\cdot), \hat{b}_k(\cdot), \hat{y}_l(\cdot)$ are *deterministically bounded*, so that:

$$0 \leq \hat{a}_k(\alpha(t), \omega(t)) \leq a_k^{max} \quad \forall k \in \{1, \ldots, K\}, \forall \omega(t), \alpha(t) \in \mathcal{A}_{\omega(t)} \tag{4.74}$$

$$0 \leq \hat{b}_k(\alpha(t), \omega(t)) \leq b_k^{max} \quad \forall k \in \{1, \ldots, K\}, \forall \omega(t), \alpha(t) \in \mathcal{A}_{\omega(t)} \tag{4.75}$$

$$y_l^{min} \leq \hat{y}_l(\alpha(t), \omega(t)) \leq y_l^{max} \quad \forall l \in \{0, 1, \ldots, L\}, \forall \omega(t), \alpha(t) \in \mathcal{A}_{\omega(t)} \tag{4.76}$$

Define $\boldsymbol{\Theta}(t) \triangleq [\boldsymbol{Q}(t), \boldsymbol{Z}(t)]$, and define the Lyapunov function $L(\boldsymbol{\Theta}(t))$ as follows:

$$L(\boldsymbol{\Theta}(t)) \triangleq \frac{1}{2}\sum_{k=1}^{K} Q_k(t)^2 + \frac{1}{2}\sum_{l=1}^{L} Z_l(t)^2 \tag{4.77}$$

We have the following preliminary lemma.

Lemma 4.11 *(T-slot Drift) Assume (4.74)-(4.76) hold. For any slot t, any queue backlogs $\boldsymbol{\Theta}(t)$, and any integer $T > 0$, the drift-plus-penalty algorithm ensures that:*

$$\begin{aligned} L(\boldsymbol{\Theta}(t+T)) - L(\boldsymbol{\Theta}(t)) + V\sum_{\tau=t}^{t+T-1}\hat{y}_0(\alpha(\tau),\omega(\tau)) \leq DT^2 + V\sum_{\tau=t}^{t+T-1}\hat{y}_0(\alpha^*(\tau),\omega(\tau)) \\ + \sum_{l=1}^{L} Z_l(t)\sum_{\tau=t}^{t+T-1}[\hat{y}_l(\alpha^*(\tau),\omega(\tau))] \\ + \sum_{k=1}^{K} Q_k(t)\sum_{\tau=t}^{t+T-1}[\hat{a}_k(\alpha^*(\tau),\omega(\tau)) - \hat{b}_k(\alpha^*(\tau),\omega(\tau))] \end{aligned}$$

where $L(\boldsymbol{\Theta}(t))$ is defined in (4.77), $\alpha^*(\tau)$ for $\tau \in \{t, \ldots, t+T-1\}$ is any sequence of alternative decisions that satisfy $\alpha^*(\tau) \in \mathcal{A}_{\omega(\tau)}$, and the constant D is defined:

$$D \triangleq \frac{1}{2}\sum_{k=1}^{K}[(a_k^{max})^2 + (b_k^{max})^2] + \frac{1}{2}\sum_{l=1}^{L}\max[(y_l^{min})^2, (y_l^{max})^2] \tag{4.78}$$

Proof. From (4.46)-(4.47), we have for any slot τ:

$$\begin{aligned} L(\boldsymbol{\Theta}(\tau+1)) - L(\boldsymbol{\Theta}(\tau)) \leq D + \sum_{k=1}^{K} Q_k(\tau)[\hat{a}_k(\alpha(\tau),\omega(\tau)) - \hat{b}_k(\alpha(\tau),\omega(\tau))] \\ + \sum_{l=1}^{L} Z_l(\tau)\hat{y}_l(\alpha(\tau),\omega(\tau)) \end{aligned}$$

where D is defined in (4.78). We then add $V\hat{y}_0(\alpha(\tau),\omega(\tau))$ to both sides. Because the drift-plus-penalty algorithm is designed to choose $\alpha(\tau)$ to deterministically minimize the right-hand-side of the resulting inequality when this term is added, it follows that:

$$\begin{aligned} L(\boldsymbol{\Theta}(\tau+1)) - L(\boldsymbol{\Theta}(\tau)) + V\hat{y}_0(\alpha(\tau),\omega(\tau)) \leq D + V\hat{y}_0(\alpha^*(\tau),\omega(\tau)) \\ + \sum_{k=1}^{K} Q_k(\tau)[\hat{a}_k(\alpha^*(\tau),\omega(\tau)) - \hat{b}_k(\alpha^*(\tau),\omega(\tau))] \\ + \sum_{l=1}^{L} Z_l(\tau)\hat{y}_l(\alpha^*(\tau),\omega(\tau)) \end{aligned}$$

where $\alpha^*(\tau)$ is any other decision that satisfies $\alpha^*(\tau) \in \mathcal{A}_{\omega(\tau)}$. However, we now note that for all $\tau \in \{t, \ldots, t+T-1\}$:

$$|Q_k(\tau) - Q_k(t)| \leq (\tau - t)\max[a_k^{max}, b_k^{max}]$$
$$|Z_l(\tau) - Z_l(t)| \leq (\tau - t)\max[|y_l^{max}|, |y_l^{min}|]$$

Plugging these in, it can be shown that:

$$\begin{aligned} L(\boldsymbol{\Theta}(\tau+1)) - L(\boldsymbol{\Theta}(\tau)) + V\hat{y}_0(\alpha(\tau), \omega(\tau)) \leq D + 2D \times (\tau - t) + V\hat{y}_0(\alpha^*(\tau), \omega(\tau)) \\ + \sum_{l=1}^{L} Z_l(t)\hat{y}_l(\alpha^*(\tau), \omega(\tau)) \\ + \sum_{k=1}^{K} Q_k(t)[\hat{a}_k(\alpha^*(\tau), \omega(\tau)) - \hat{b}_k(\alpha^*(\tau), \omega(\tau))] \end{aligned}$$

Summing the above over $\tau \in \{t, \ldots, t+T-1\}$ and using the fact that $\sum_{\tau=t}^{t+T-1}(\tau - t) = (T-1)T/2$ yields the result. □

4.9.1 MARKOV MODULATED PROCESSES

Here we present a method developed in (144) for proving that the $[O(1/V), O(V)]$ behavior of the drift-plus-penalty algorithm is preserved in ergodic (but non-i.i.d.) contexts. Let $\Phi(t)$ be an irreducible (possibly not aperiodic) Discrete Time Markov Chain (DTMC) with a finite state space $\mathcal{S}$.[3] Let π_i represent the stationary distribution over states $i \in \mathcal{S}$. Such a distribution always exists (and is unique) for irreducible finite state Markov chains. It is well known that all π_i probabilities are positive, and the time average fraction of time being in state i is π_i with probability 1. Further, $1/\pi_i$ represents the (finite) mean recurrence time to state i, which is the average number of slots required to get back to state i, given that we start in state i. Finally, it is known that second moments of recurrence time are also finite (see (132)(130) for more details on DTMCs).

The random network event process $\omega(t)$ is modulated by the DTMC $\Phi(t)$ as follows: Whenever $\Phi(t) = i$, the value of $\omega(t)$ is chosen independently with some distribution $p_i(\omega)$. Then the stationary distribution of $\omega(t)$ is given by:

$$Pr[\omega(t) = \omega] = \sum_{i \in \mathcal{S}} \pi_i p_i(\omega)$$

Assume the state space $\mathcal{S}$ has a state "0" that we designate as a "renewal" state. Assume for simplicity that $\Phi(0) = 0$, and let the sequence $\{T_0, T_1, T_2, \ldots\}$ represent the recurrence times to state 0. Clearly $\{T_r\}_{r=0}^{\infty}$ is an i.i.d. sequence with $\mathbb{E}\{T_r\} = 1/\pi_0$ for all r. Define $\mathbb{E}\{T\}$ and $\mathbb{E}\{T^2\}$ as the first and second moments of these recurrence times (so that $\mathbb{E}\{T\} = 1/\pi_0$). Define $t_0 = 0$, and for integers

[3] This subsection (Subsection 4.9.1) assumes familiarity with DTMC theory and can be skipped without loss of continuity.

$r > 0$ define t_r as the time of the rth revisitation to state 0, so that $t_r = \sum_{j=1}^{r} T_j$. We now define the *variable slot drift* $\Delta(\boldsymbol{\Theta}(t_r))$ as follows:

$$\Delta(\boldsymbol{\Theta}(t_r)) \triangleq \mathbb{E}\left\{L(\boldsymbol{\Theta}(t_{r+1})) - \dot{L}(\boldsymbol{\Theta}(t_r))|\boldsymbol{\Theta}(t_r)\right\}$$

This drift represents the expected change in the Lyapunov function from renewal time t_r to renewal time t_{r+1}, where the expectation is over the random duration of the renewal period and the random events on each slot of this period. By plugging $t = t_r$ and $T = T_r$ into Lemma 4.11 and taking conditional expectations given $\boldsymbol{\Theta}(t_r)$, we have the following variable-slot drift-plus-penalty expression:

$$\begin{aligned}
\Delta(\boldsymbol{\Theta}(t_r)) + V\mathbb{E}\left\{\sum_{\tau=t_r}^{t_r+T_r-1} \hat{y}_0(\alpha(\tau), \omega(\tau))|\boldsymbol{\Theta}(t_r)\right\} &\leq D\mathbb{E}\left\{T_r^2|\boldsymbol{\Theta}(t_r)\right\} \\
&+V\mathbb{E}\left\{\sum_{\tau=t_r}^{t_r+T_r-1} \hat{y}_0(\alpha^*(\tau), \omega(\tau))|\boldsymbol{\Theta}(t_r)\right\} \\
&+\sum_{l=1}^{L} Z_l(t_r)\mathbb{E}\left\{\sum_{\tau=t_r}^{t_r+T_r-1} \hat{y}_l(\alpha^*(\tau), \omega(\tau))|\boldsymbol{\Theta}(t_r)\right\} \\
+\sum_{k=1}^{K} Q_k(t_r)\mathbb{E}\left\{\sum_{\tau=t_r}^{t_r+T_r-1} [\hat{a}_k(\alpha^*(\tau), \omega(\tau)) - \hat{b}_k(\alpha^*(\tau), \omega(\tau))]|\boldsymbol{\Theta}(t_r)\right\}&
\end{aligned}$$

where $\alpha^*(\tau)$ are decisions from any other policy. First note that $\mathbb{E}\left\{T_r^2|\boldsymbol{\Theta}(t_r)\right\} = \mathbb{E}\left\{T^2\right\}$ because the renewal duration is independent of the queue state $\boldsymbol{\Theta}(t_r)$. Next, note that the conditional expectations in the next three terms on the right-hand-side of the above inequality can be changed into pure expectations (given that t_r is a renewal time) under the assumption that the policy $\alpha^*(\tau)$ is ω-only. Thus:

$$\begin{aligned}
\Delta(\boldsymbol{\Theta}(t_r)) + V\mathbb{E}\left\{\sum_{\tau=t_r}^{t_r+T_r-1} \hat{y}_0(\alpha(\tau), \omega(\tau))|\boldsymbol{\Theta}(t_r)\right\} &\leq D\mathbb{E}\left\{T^2\right\} \\
&+V\mathbb{E}\left\{\sum_{\tau=t_r}^{t_r+T_r-1} \hat{y}_0(\alpha^*(\tau), \omega(\tau))\right\} \\
&+\sum_{l=1}^{L} Z_l(t_r)\mathbb{E}\left\{\sum_{\tau=t_r}^{t_r+T_r-1} \hat{y}_l(\alpha^*(\tau), \omega(\tau))\right\} \\
+\sum_{k=1}^{K} Q_k(t_r)\mathbb{E}\left\{\sum_{\tau=t_r}^{t_r+T_r-1} [\hat{a}_k(\alpha^*(\tau), \omega(\tau)) - \hat{b}_k(\alpha^*(\tau), \omega(\tau))]\right\}&
\end{aligned} \tag{4.79}$$

The expectations in the final terms are expected rewards over a renewal period, and so by basic renewal theory (130)(66), we have for all $l \in \{0, 1, \ldots, L\}$ and all $k \in \{1, \ldots, K\}$:

$$\mathbb{E}\left\{\sum_{\tau=t_r}^{t_r+T_r-1} \hat{y}_l(\alpha(\tau), \omega(\tau))\right\} = \mathbb{E}\{T\} y_l^* \tag{4.80}$$

$$\mathbb{E}\left\{\sum_{\tau=t_r}^{t_r+T_r-1} [\hat{a}_k(\alpha^*(\tau), \omega(\tau)) - \hat{b}_k(\alpha^*(\tau), \omega(\tau))]\right\} = \mathbb{E}\{T\}(a_k^* - b_k^*) \tag{4.81}$$

where y_l^*, a_k^*, b_k^* are the infinite horizon time average values achieved for the $\hat{y}_l(\alpha^*(t), \omega(t))$, $\hat{a}_k(\alpha^*(t), \omega(t))$, and $\hat{b}_k(\alpha^*(t), \omega(t))$ processes under the ω-only policy $\alpha^*(t)$. This basic renewal theory fact can easily be understood as follows (with the below equalities holding with probability 1):[4]

$$\begin{aligned}
y_l^* &= \lim_{R\to\infty} \frac{1}{t_R} \sum_{\tau=0}^{t_R-1} \hat{y}_l(\alpha^*(\tau), \omega(\tau)) \\
&= \lim_{R\to\infty} \frac{\sum_{r=0}^{R-1} \sum_{\tau=t_r}^{t_r+T_r-1} \hat{y}_l(\alpha^*(\tau), \omega(\tau))}{\sum_{r=0}^{R-1} T_r} \\
&= \frac{\lim_{R\to\infty} \frac{1}{R} \sum_{r=0}^{R-1} \sum_{\tau=t_r}^{t_r+T_r-1} \hat{y}_l(\alpha^*(\tau), \omega(\tau))}{\lim_{R\to\infty} \frac{1}{R} \sum_{r=0}^{R-1} T_r} \\
&= \frac{\mathbb{E}\left\{\sum_{\tau=0}^{T_0-1} \hat{y}_l(\alpha^*(\tau), \omega(\tau))\right\}}{\mathbb{E}\{T\}}
\end{aligned}$$

where the final equality holds by the strong law of large numbers (noting that both the numerator and denominator are just a time average of i.i.d. quantities). In particular, the numerator is a sum of i.i.d. quantities because the policy $\alpha^*(t)$ is ω-only, and so the sum penalty over each renewal period is independent but identically distributed. Plugging (4.80)-(4.81) into (4.79) yields:

$$\begin{aligned}
\Delta(\boldsymbol{\Theta}(t_r)) + V\mathbb{E}\left\{\sum_{\tau=t_r}^{t_r+T_r-1} \hat{y}_0(\alpha(\tau), \omega(\tau))|\boldsymbol{\Theta}(t_r)\right\} &\leq D\mathbb{E}\left\{T^2\right\} + V\mathbb{E}\{T\} y_0^* \\
&\quad + \sum_{l=1}^{L} Z_l(t)\mathbb{E}\{T\} y_l^* + \sum_{k=1}^{K} Q_k(t)\mathbb{E}\{T\}(a_k^* - b_k^*)
\end{aligned}$$

The above holds for any time averages $\{y_l^*, a_k^*, b_k^*\}$ that can be achieved by ω-only policies. However, by Theorem 4.5, we know that if the problem is feasible, then either there is a single ω-only policy that achieves time averages $y_0^* = y_0^{opt}$, $y_l^* \leq 0$ for all $l \in \{1, \ldots, L\}$, $(a_k^* - b_k^*) \leq 0$ for all $k \in \{1, \ldots, K\}$,

[4]Because the processes are deterministically bounded and have time averages that converge with probability 1, the Lebesgue Dominated Convergence Theorem (145) ensures the time average expectations are the same as the pure time averages (see Exercise 7.9).

or there is an infinite sequence of ω-only policies that approach these averages. Plugging this into the above yields:

$$\Delta(\boldsymbol{\Theta}(t_r)) + V\mathbb{E}\left\{\sum_{\tau=t_r}^{t_r+T_r-1} \hat{y}_0(\alpha(\tau), \omega(\tau))|\boldsymbol{\Theta}(t_r)\right\} \leq D\mathbb{E}\left\{T^2\right\} + V\mathbb{E}\{T\}\, y_0^{opt}$$

Taking expectations of the above, summing the resulting telescoping series over $r \in \{0, \ldots, R-1\}$, and dividing by $VR\mathbb{E}\{T\}$ yields:

$$\frac{\mathbb{E}\{L(\boldsymbol{\Theta}(t_R))\} - \mathbb{E}\{L(\boldsymbol{\Theta}(0))\}}{V\mathbb{E}\{T\}R} + \frac{1}{\mathbb{E}\{T\}R}\mathbb{E}\left\{\sum_{\tau=0}^{t_R-1} \hat{y}_0(\alpha(\tau), \omega(\tau))\right\} \leq y_0^{opt} + \frac{D\mathbb{E}\left\{T^2\right\}}{V\mathbb{E}\{T\}}$$

Because $t_R/R \rightarrow \mathbb{E}\{T\}$ with probability 1 (by the law of large numbers), it can be shown that the middle term has a lim sup that is equal to the lim sup time average expected penalty. Thus, assuming $\mathbb{E}\{L(\boldsymbol{\Theta}(0))\} < \infty$, we have:

$$\overline{y}_0 \triangleq \limsup_{t\rightarrow\infty} \frac{1}{t}\sum_{\tau=0}^{t-1} \mathbb{E}\left\{\hat{y}_0(\alpha(\tau), \omega(\tau))\right\} \leq y_0^{opt} + \frac{D\mathbb{E}\left\{T^2\right\}}{V\mathbb{E}\{T\}} = y_0^{opt} + O(1/V) \tag{4.82}$$

where we note that the constants D, $\mathbb{E}\{T\}$, and $\mathbb{E}\left\{T^2\right\}$ do not depend on V. Similarly, it can be shown that if the problem is feasible then all queues are mean rate stable, and if the slackness condition of Assumption A1 holds, then sum average queue backlog is $O(V)$ (144). This leads to the following theorem.

Theorem 4.12 *(Markov Modulated Processes (144)) Assume the $\omega(t)$ process is modulated by the DTMC $\Phi(t)$ as described above, the boundedness assumptions (4.74)-(4.76) hold, $\mathbb{E}\{L(\boldsymbol{\Theta}(0))\} < \infty$, and that the drift-plus-penalty algorithm is used every slot t. If the problem is feasible, then:*

(a) The penalty satisfies (4.82), so that $\overline{y}_0 \leq y_0^{opt} + O(1/V)$.

(b) All queues are mean rate stable, and so $\overline{y}_l \leq 0$ for all $l \in \{1, \ldots, L\}$.

(c) If the Slackness Assumption A1 holds, then all queues $Q_k(t)$ are strongly stable with average backlog $O(V)$.

4.9.2 NON-ERGODIC MODELS AND ARBITRARY SAMPLE PATHS

Now assume that the $\omega(t)$ process follows an arbitrary sample path, possibly one with non-ergodic behavior. However, continue to assume that the deterministic bounds (4.74)-(4.76) hold, so that Lemma 4.11 applies. We present a technique developed in (41)(40) for stock market trading and modified in (39)(38) for use in wireless networks with arbitrary traffic, channels and mobility. Because $\omega(t)$ follows an arbitrary sample path, usual "equilibrium" notions of optimality are not relevant, and so we use a different metric for evaluation of the drift-plus-penalty algorithm, called the *T-slot lookahead metric*. Specifically, let T and R be positive integers, and consider the first RT slots

$\{0, 1, \ldots, RT-1\}$ being divided into R frames of size T. For the rth frame (for $r \in \{0, \ldots, R-1\}$), we define c_r^* as the optimal cost associated with the following static optimization problem, called the T-slot lookahead problem. This problem has variables $\alpha(\tau) \in \{rT, \ldots, (r+1)T-1\}$, and treats the $\omega(\tau)$ values in this interval as known quantities:

$$\begin{aligned}
\text{Minimize:} \quad & c_r \overset{\triangle}{=} \frac{1}{T} \sum_{\tau=rT}^{(r+1)T-1} \hat{y}_0(\alpha(\tau), \omega(\tau)) && (4.83)\\
\text{Subject to:} \quad & 1) \sum_{\tau=rT}^{(r+1)T-1} \hat{y}_l(\alpha(\tau), \omega(\tau)) \leq 0 \ \forall l \in \{1, \ldots, L\}\\
& 2) \sum_{\tau=rT}^{(r+1)T-1} [\hat{a}_k(\alpha(\tau), \omega(\tau)) - \hat{b}_k(\alpha(\tau), \omega(\tau))] \leq 0 \ \forall k \in \{1, \ldots, K\}\\
& 3) \ \alpha(\tau) \in \mathcal{A}_{\omega(\tau)} \ \forall \tau \in \{rT, \ldots, (r+1)T-1\}
\end{aligned}$$

The value c_r^* thus represents the optimal empirical average penalty for frame r over all policies that have full knowledge of the future $\omega(\tau)$ values over the frame and that satisfy the constraints.[5] We assume throughout that the constraints are feasible for the above problem. Feasibility is often guaranteed when there is an "idle" action, such as the action of admitting and transmitting no data, which can be used on all slots to trivially satisfy the constraints in the form $0 \leq 0$.

Frame r consists of slots $\tau \in \{rT, \ldots, (r+1)T-1\}$. Let $\alpha^*(\tau)$ represent the decisions that solve the T-slot lookahead problem (4.83) over this frame to achieve cost c_r^*.[6] It is generally impossible to solve for the $\alpha^*(\tau)$ decisions, as these would require knowledge of the $\omega(\tau)$ values up to T-slots into the future. However, the $\alpha^*(\tau)$ values exist, and can still be plugged into Lemma 4.11 to yield the following (using $t = rT$ and T as the frame size):

$$\begin{aligned}
&L(\boldsymbol{\Theta}(rT+T)) - L(\boldsymbol{\Theta}(rT)) + V \sum_{\tau=rT}^{rT+T-1} \hat{y}_0(\alpha(\tau), \omega(\tau))\\
&\leq \ DT^2 + V \sum_{\tau=rT}^{rT+T-1} \hat{y}_0(\alpha^*(\tau), \omega(\tau)) + \sum_{l=1}^{L} Z_l(rT) \sum_{\tau=rT}^{rT+T-1} [\hat{y}_l(\alpha^*(\tau), \omega(\tau))]\\
&\quad + \sum_{k=1}^{K} Q_k(rT) \sum_{\tau=rT}^{rT+T-1} [\hat{a}_k(\alpha^*(\tau), \omega(\tau)) - \hat{b}_k(\alpha^*(\tau), \omega(\tau))]\\
&\leq \ DT^2 + VTc_r^*
\end{aligned}$$

where the final inequality follows by noting that the $\alpha^*(\tau)$ policy satisfies the constraints of the T-slot lookahead problem (4.83) and yields cost c_r^*.

[5] Theorem 4.13 holds exactly as stated in the extended case when c_r^* is re-defined by a T-slot lookahead problem that allows actions $[(\tilde{y}_l^*(\tau)), (\tilde{a}_k^*(\tau)), (\tilde{b}_k^*(\tau))]$ every slot τ to be taken within the convex hull of the set of all possible values of $[(\hat{y}_l(\alpha, \omega(\tau))), (\hat{a}_k(\alpha, \omega(\tau))), (\hat{b}_k(\alpha, \omega(\tau)))]$ under $\alpha \in \mathcal{A}_{\omega(\tau)}$, but we skip this extension for simplicity of exposition.

[6] For simplicity, we assume the infimum cost is achievable. Else, we can derive the same result by taking a limit over policies that approach the infimum.

Summing the above over $r \in \{0, \ldots, R-1\}$ (for any integer $R > 0$) yields:

$$L(\boldsymbol{\Theta}(RT)) - L(\boldsymbol{\Theta}(0)) + V\sum_{\tau=0}^{RT-1}\hat{y}_0(\alpha(\tau), \omega(\tau)) \leq DT^2R + VT\sum_{r=0}^{R-1}c_r^* \tag{4.84}$$

Dividing by VTR, using the fact that $L(\boldsymbol{\Theta}(RT)) \geq 0$, and rearranging terms yields:

$$\frac{1}{RT}\sum_{\tau=0}^{RT-1}\hat{y}_0(\alpha(\tau), \omega(\tau)) \leq \frac{1}{R}\sum_{r=0}^{R-1}c_r^* + \frac{DT}{V} + \frac{L(\boldsymbol{\Theta}(0))}{VTR} \tag{4.85}$$

where we recall that $\alpha(\tau)$ represents the decisions under the drift-plus-penalty algorithm. The inequality (4.85) holds for all integers $R > 0$. When R is large, the final term on the right-hand-side above goes to zero (this term is exactly zero if $L(\boldsymbol{\Theta}(0)) = 0$). Thus, we have that the time average cost is within $O(1/V)$ of the time average of the c_r^* values. The above discussion proves part (a) of the following theorem:

Theorem 4.13 *(Universal Scheduling) Assume the $\omega(t)$ sample path satisfies the boundedness assumptions (4.74)-(4.76), and that initial queue backlog is finite. Fix any integers $R > 0$ and $T > 0$, and assume the T-slot lookahead problem (4.83) is feasible for every frame $r \in \{0, 1, \ldots, R-1\}$. If the drift-plus-penalty algorithm is implemented every slot t, then:*

(a) The time average cost over the first RT slots satisfies (4.85). In particular,[7]

$$\limsup_{t\to\infty}\frac{1}{t}\sum_{\tau=0}^{t-1}\hat{y}_0(\alpha(\tau), \omega(\tau)) \leq \limsup_{R\to\infty}\frac{1}{R}\sum_{r=0}^{R-1}c_r^* + DT/V$$

where c_r^* is the optimal cost in the T-slot lookahead problem (4.83) for frame r, and D is defined in (4.78).

(b) All actual and virtual queues are rate stable, and so we have:

$$\limsup_{t\to\infty}\frac{1}{t}\sum_{\tau=0}^{t-1}\hat{y}_l(\alpha(\tau), \omega(\tau)) \leq 0 \;\; \forall l \in \{1, \ldots, L\}$$

(c) Suppose there exists an $\epsilon > 0$ and a sequence of decisions $\tilde{\alpha}(\tau) \in \mathcal{A}_{\omega(\tau)}$ that satisfies the following slackness assumptions for all frames r:

$$\sum_{\tau=rT}^{rT+T-1}\hat{y}_l(\tilde{\alpha}(\tau), \omega(\tau)) \leq 0 \;\; \forall l \in \{1, \ldots, L\} \tag{4.86}$$

$$\frac{1}{T}\sum_{\tau=rT}^{rT+T-1}[\hat{a}_k(\tilde{\alpha}(\tau), \omega(\tau)) - \hat{b}_k(\tilde{\alpha}(\tau), \omega(\tau))] \leq -\epsilon \;\; \forall k \in \{1, \ldots, K\} \tag{4.87}$$

[7] It is clear that the lim sup over times sampled every T slots is the same as the regular lim sup because the $\hat{y}_0(\cdot)$ values are bounded. Indeed, we have $\sum_{\tau=0}^{\lfloor t/T\rfloor T}\hat{y}_0(\alpha(\tau), \omega(\tau)) + Ty_0^{min} \leq \sum_{\tau=0}^{t}\hat{y}_0(\alpha(\tau), \omega(\tau)) \leq \sum_{\tau=0}^{\lfloor t/T\rfloor T}\hat{y}_0(\alpha(\tau), \omega(\tau)) + Ty_0^{max}$. Dividing both sides by t and taking limits shows these limits are equal.

Then:

$$\limsup_{t\to\infty} \frac{1}{t}\sum_{\tau=0}^{t-1}\sum_{k=1}^{K} Q_k(\tau) \leq \frac{DT}{\epsilon} + \frac{V(y_0^{max} - y_0^{min})}{\epsilon} + \frac{T-1}{2}\sum_{k=1}^{K}\max[a_k^{max}, b_k^{max}]$$

Proof. Part (a) has already been shown in the above discussion. We provide a summary of parts (b) and (c): The inequality (4.84) plus the boundedness assumptions (4.74)-(4.76) imply that there is a finite constant $F > 0$ such that $L(\boldsymbol{\Theta}(RT)) \leq FR$ for all R. By an argument similar to part (a) of Theorem 4.1, it can then be shown that $\lim_{R\to\infty} Q_k(RT)/(RT) = 0$ for all $k \in \{1, \ldots, K\}$ and $\lim_{R\to\infty} Z_l(RT)/(RT) = 0$ for all $l \in \{1, \ldots, L\}$. Further, these limits that sample only on slots RT (as $R \to \infty$) are clearly the same when taken over all $t \to \infty$ because the queues can change by at most a constant proportional to T in between the sample times. This proves part (b).

Part (c) follows by plugging the policy $\tilde{\alpha}(\tau)$ for $\tau \in \{rT, \ldots, (r+1)T - 1\}$ into Lemma 4.11 and using (4.86)-(4.87) to yield:

$$L(\boldsymbol{\Theta}(rT+T)) - L(\boldsymbol{\Theta}(rT)) + V\sum_{\tau=rT}^{rT+T-1} \hat{y}_0(\alpha(\tau), \omega(\tau)) \leq DT^2 + VTy_0^{max} - T\epsilon\sum_{k=1}^{K} Q_k(rT)$$

and hence:

$$\begin{aligned}
L(\boldsymbol{\Theta}(rT+T)) - L(\boldsymbol{\Theta}(rT)) &\leq DT^2 + VT(y_0^{max} - y_0^{min}) - T\epsilon\sum_{k=1}^{K} Q_k(rT) \\
&\leq DT^2 + VT(y_0^{max} - y_0^{min}) - \epsilon\sum_{k=1}^{K}\sum_{j=0}^{T-1} Q_k(rT+j) \\
&\quad + \epsilon\sum_{k=1}^{K}\sum_{j=0}^{T-1} j\max[a_k^{max}, b_k^{max}] \\
&= DT^2 + VT(y_0^{max} - y_0^{min}) - \epsilon\sum_{k=1}^{K}\sum_{j=0}^{T-1} Q_k(rT+j) \\
&\quad + \frac{\epsilon(T-1)T}{2}\sum_{k=1}^{K}\max[a_k^{max}, b_k^{max}]
\end{aligned}$$

Summing the above over $r \in \{0, \ldots, R-1\}$ yields:

$$\begin{aligned}
L(\boldsymbol{\Theta}(RT)) - L(\boldsymbol{\Theta}(0)) + \epsilon\sum_{\tau=0}^{RT-1}\sum_{k=1}^{K} Q_k(\tau) \leq RDT^2 + RVT(y_0^{max} - y_0^{min}) \\
+ \frac{\epsilon R(T-1)T}{2}\sum_{k=1}^{K}\max[a_k^{max}, b_k^{max}]
\end{aligned}$$

Using $L(\boldsymbol{\Theta}(RT)) \geq 0$, dividing by ϵRT and taking a lim sup as $R \to \infty$ yields:

$$\limsup_{R\to\infty} \frac{1}{RT} \sum_{\tau=0}^{RT-1} \sum_{k=1}^{K} Q_k(\tau) \leq \frac{DT}{\epsilon} + \frac{V(y_0^{max} - y_0^{min})}{\epsilon} + \frac{T-1}{2} \sum_{k=1}^{K} \max[a_k^{max}, b_k^{max}]$$

□

Inequality (4.85) holds for all R and T, and hence it can be viewed as a family of bounds that apply to the same sample path under the drift-plus-penalty algorithm. Note also that increasing the value of T changes the frame size and typically improves the c_r^* values (as it allows these values to be achieved with a larger future lookahead). However, this affects the error term DT/V, requiring V to also be increased as T increases. Increasing V creates a larger queue backlog. We thus see a similar $[O(1/V), O(V)]$ cost-backlog tradeoff for this sample path context. If the slackness assumptions (4.86)-(4.87) are modified to also include slackness in the $y_l(\cdot)$ constraints, a modified argument can be used to show the worst case queue backlog is bounded for all time by a constant that is $O(V)$ (see also (146)(39)(38)).

The target value $\frac{1}{R}\sum_{r=0}^{R-1} c_r^*$ that we use for comparison does *not* represent the optimal cost that can be achieved over the full horizon RT if the entire future were known. However, when T is large it still represents a meaningful target that is not trivial to achieve, as it is one that is defined in terms of an ideal policy with T-slot lookahead. It is remarkable that the drift-plus-penalty algorithm can closely track such an "ideal" T-slot lookahead algorithm.

4.10 EXERCISES

Exercise 4.1. Let $\boldsymbol{Q} = (Q_1, \ldots, Q_K)$ and $L(\boldsymbol{Q}) = \frac{1}{2}\sum_{k=1}^{K} Q_k^2$.

a) If $L(\boldsymbol{Q}) \leq 25$, show that $Q_k \leq \sqrt{50}$ for all $k \in \{1, \ldots, K\}$.

b) If $L(\boldsymbol{Q}) > 25$, show that $Q_k > \sqrt{50/K}$ for at least one queue $k \in \{1, \ldots, K\}$.

c) Let $K = 2$. Plot the region of all non-negative vectors (Q_1, Q_2) such that $L(\boldsymbol{Q}) = 2$. Also plot for $L(\boldsymbol{Q}) = 2.5$. Give an example where $L(Q_1(t), Q_2(t)) = 2.5$, $L(Q_1(t+1), Q_2(t+1)) = 2$, but where $Q_1(t) < Q_1(t+1)$.

Exercise 4.2. For any constants $Q \geq 0, b \geq 0, a \geq 0$, show that:

$$(\max[Q - b, 0] + a)^2 \leq Q^2 + b^2 + a^2 + 2Q(a - b)$$

Exercise 4.3. Let $\boldsymbol{Q}(t)$ be a discrete time vector process with $\boldsymbol{Q}(0) = \boldsymbol{0}$, and let $f(t)$ and $g(t)$ be discrete time real valued processes. Suppose there is a non-negative function $L(\boldsymbol{Q}(t))$ such that

$L(\mathbf{0}) = 0$, and such that its conditional drift $\Delta(\mathbf{Q}(t))$ satisfies the following every slot τ and for all possible $\mathbf{Q}(\tau)$:

$$\Delta(\mathbf{Q}(\tau)) + \mathbb{E}\{f(\tau)|\mathbf{Q}(\tau)\} \leq \mathbb{E}\{g(\tau)|\mathbf{Q}(\tau)\}$$

a) Use the law of iterated expectations to prove that:

$$\mathbb{E}\{L(\mathbf{Q}(\tau+1))\} - \mathbb{E}\{L(\mathbf{Q}(\tau))\} + \mathbb{E}\{f(\tau)\} \leq \mathbb{E}\{g(\tau)\}$$

b) Use telescoping sums together with part (a) to prove that for any $t > 0$:

$$\tfrac{1}{t}\textstyle\sum_{\tau=0}^{t-1}\mathbb{E}\{f(\tau)\} \leq \tfrac{1}{t}\textstyle\sum_{\tau=0}^{t-1}\mathbb{E}\{g(\tau)\}$$

Exercise 4.4. (Opportunistically Minimizing an Expectation) Consider the game described in Section 1.8. Suppose that ω is a Gaussian random variable with mean m and variance σ^2. Define $c(\alpha, \omega) = \omega^2 + \omega(3 - 2\alpha) + \alpha^2$.

a) Compute the optimal choice of α (as a function of the observed ω) to minimize $\mathbb{E}\{c(\alpha, \omega)\}$. Compute $\mathbb{E}\{c(\alpha, \omega)\}$ under your optimal policy.

b) Suppose that ω is exponentially distributed with mean $1/\lambda$. Does the optimal policy change? Does $\mathbb{E}\{c(\alpha, \omega)\}$ change?

c) Let $\omega = (\omega_1, \ldots, \omega_K)$, $\alpha = (\alpha_1, \ldots, \alpha_K)$, $\mathbf{\Theta} = (\Theta_1, \ldots, \Theta_K)$ be non-negative vectors. Define $c(\alpha, \omega, \mathbf{\Theta}) = \sum_{k=1}^{K}\left[V\alpha_k - \Theta_k \log(1 + \alpha_k\omega_k)\right]$, where $\log(\cdot)$ denotes the natural logarithm and $V \geq 0$. We choose α subject to $0 \leq \alpha_k \leq 1$ for all k, and $\alpha_k\alpha_j = 0$ for $k \neq j$. Design a policy that observes ω and chooses α to minimize $\mathbb{E}\{c(\alpha, \omega, \mathbf{\Theta})|\mathbf{\Theta}\}$. Hint: First compute the solution assuming that $\alpha_k > 0$.

Exercise 4.5. (The Drift-Plus-Penalty Method) Explain, using the game of opportunistically minimizing an expectation described in Section 1.8, how choosing $\alpha(t) \in \mathcal{A}_{\omega(t)}$ according to (4.48)-(4.49) minimizes the right-hand-side of (4.44).

Exercise 4.6. (Probability 1 Convergence) Consider the fixed-V drift-plus-penalty algorithm (4.48)-(4.49), but assume the following modified Slater condition holds:

Assumption A2: There is an $\epsilon > 0$ such that for any J-dimensional vector $\mathbf{h} = (h_1, \ldots, h_J)$ that consists only of values 1 and -1, there is an ω-only policy $\alpha^*(t)$ (which depends on $\mathbf{h}$) that satisfies:

$$\mathbb{E}\left\{\hat{y}_0(\alpha^*(t), \omega(t))\right\} \leq y_{0,max} \tag{4.88}$$
$$\mathbb{E}\left\{\hat{y}_l(\alpha^*(t), \omega(t))\right\} \leq -\epsilon \quad \forall l \in \{1, \ldots, L\} \tag{4.89}$$
$$\mathbb{E}\left\{\hat{e}_j(\alpha^*(t), \omega(t))\right\} = \epsilon h_j \quad \forall j \in \{1, \ldots, J\} \tag{4.90}$$
$$\mathbb{E}\left\{\hat{a}_k(\alpha^*(t), \omega(t))\right\} \leq \mathbb{E}\left\{\hat{b}_k(\alpha^*(t), \omega(t))\right\} - \epsilon \quad \forall k \in \{1, \ldots, K\} \tag{4.91}$$

Using $\mathcal{H}(t)$ and $\Delta(t, \mathcal{H}(t))$ as defined in Section 4.1.3, it can be shown that for all t and all possible $\mathcal{H}(t)$, we have (compare with (4.52)):

$$\begin{aligned}\Delta(t, \mathcal{H}(t)) + V\mathbb{E}\{y_0(t)|\mathcal{H}(t)\} \leq\ & B + C + V\mathbb{E}\{y_0^*(t)|\mathcal{H}(t)\} \\ & + \sum_{l=1}^{L} Z_l(t)\mathbb{E}\{y_l^*(t)|\mathcal{H}(t)\} + \sum_{j=1}^{J} H_j(t)\mathbb{E}\left\{e_j^*(t)|\mathcal{H}(t)\right\} \\ & + \sum_{k=1}^{K} Q_k(t)\mathbb{E}\{a_k^*(t) - b_k^*(t) \mid \mathcal{H}(t)\} \qquad (4.92)\end{aligned}$$

where $y_l^*(t)$, $e_j^*(t)$, $a_k^*(t)$, $b_k^*(t)$ represent decisions under any other (possibly randomized) action $\alpha^*(t)$ that can be made on slot t (so that $y_l^*(t) = \hat{y}_l(\alpha^*(t), \omega(t))$, etc.).

a) Define $\boldsymbol{h} = (h_1, \ldots, h_J)$ by:

$$h_j = \begin{cases} -1 & \text{if } H_j(t) \geq 0 \\ 1 & \text{if } H_j(t) < 0 \end{cases}$$

Using this $\boldsymbol{h}$, plug the ω-only policy $\alpha^*(t)$ from (4.88)-(4.91) into the right-hand-side of (4.92) to obtain:

$$\begin{aligned}\Delta(t, \mathcal{H}(t)) + V\mathbb{E}\{y_0(t)|\mathcal{H}(t)\} \leq\ & B + C + Vy_{0,max} \\ & -\epsilon\left[\sum_{l=1}^{L} Z_l(t) + \sum_{k=1}^{K} Q_k(t) + \sum_{j=1}^{J} |H_j(t)|\right]\end{aligned}$$

b) Assume that (4.16)-(4.17) hold for $y_0(t)$, and that the fourth moment assumption (4.18) holds. Use this with part (a) to obtain probability 1 bounds on the lim sup time average queue backlog via Theorem 4.4.

c) Now consider the ω-only policy that yields (4.53)-(4.56), and plug this into the right-hand-side of (4.92) to yield a probability 1 bound on the lim sup time average of $y_0(t)$, again by Theorem 4.4.

Exercise 4.7. (Min Average Power (21)) Consider a wireless downlink with arriving data $\boldsymbol{a}(t) = (a_1(t), \ldots, a_K(t))$ every slot t. The data is stored in separate queues $\boldsymbol{Q}(t) = (Q_1(t), \ldots, Q_K(t))$ for transmission over K different channels. The update equation is (4.23). Service variables $b_k(t)$ are determined by a power allocation vector $\boldsymbol{P}(t) = (P_1(t), \ldots, P_K(t))$ according to $b_k(t) = \log(1 + S_k(t)P_k(t))$, where $\log(\cdot)$ denotes the natural logarithm, and $\boldsymbol{S}(t) = (S_1(t), \ldots, S_K(t))$ is a vector of channel attenuations. Assume that $\boldsymbol{S}(t)$ is known at the beginning of each slot t, and satisfies $0 \leq S_k(t) \leq 1$ for all k. Power is allocated subject to $\boldsymbol{P}(t) \in \mathcal{A}$, where $\mathcal{A}$ is the set of all power vectors with at most one non-zero element and such that $0 \leq P_k \leq P_{max}$ for all $k \in \{1, \ldots, K\}$, where P_{max} is a peak power constraint. Assume that the vectors $\boldsymbol{a}(t)$ and $\boldsymbol{S}(t)$ are i.i.d. over slots, and that $0 \leq a_k(t) \leq a_k^{max}$ for all t, for some finite constants a_k^{max}.

a) Using $\omega(t) \triangleq (\boldsymbol{a}(t), \boldsymbol{S}(t)), \alpha(t) = \boldsymbol{P}(t), J = 0, L = 0, y_0(t) = \sum_{k=1}^{K} P_k(t)$, state the drift-plus-penalty algorithm for a fixed V in this context.

b) Assume we use an exact implementation of the algorithm in part (a) (so that $C = 0$), and that the problem is feasible. Use Theorem 4.8 to conclude that all queues are mean rate stable, and compute a value B such that:

$$\limsup_{t\to\infty} \frac{1}{t} \sum_{\tau=0}^{t-1} \sum_{k=1}^{K} \mathbb{E}\{P_k(\tau)\} \leq P_{av}^{opt} + B/V$$

where P_{av}^{opt} is the minimum average power over any stabilizing algorithm.

c) Assume Assumption A1 holds for a given $\epsilon > 0$. Use Theorem 4.8c to give a bound on the time average sum of queue backlog in all queues.

Exercise 4.8. (Place-Holder Backlog)

a) Show that for any values V, p, s, q such that $V > 0, p \geq 0, q \geq 0, 0 \leq s \leq 1$, if $q < V$, we have $Vp - q\log(1 + sp) > 0$ whenever $p > 0$ (where $\log(\cdot)$ denotes the natural logarithm). Conclude that the algorithm from Exercise 4.7 chooses $P_k(t) = 0$ whenever $Q_k(t) < V$.

b) Use part (a) to conclude that $Q_k(t) \geq \max[V - \log(1 + P_{max}), 0]$ for all t greater than or equal to the time t^* for which this inequality first holds. By how much can place-holder bits reduce average backlog from the bound given in part (c) of Exercise 4.7? This exercise computes a simple place-holder Q_k^{place} that is not the largest possible. A more detailed analysis in (143) computes a larger place-holder value.

Exercise 4.9. (Maximum Throughput Subject to Peak and Average Power Constraints (21)) Consider the same system of Exercise 4.7, with the exception that it is now a wireless *uplink*, and queue backlogs now satisfy:

$$Q_k(t + 1) = \max[Q_k(t) - b_k(t), 0] + x_k(t)$$

where $x_k(t)$ is a *flow control decision* for slot t, made subject to the constraint $0 \leq x_k(t) \leq a_k(t)$ for all t. The control action is now a joint flow control and power allocation decision $\alpha(t) = [\boldsymbol{x}(t), \boldsymbol{P}(t)]$. We want the average power expenditure over each link k to be less than or equal to P_k^{av}, where P_k^{av} is a fixed constant for each $k \in \{1, \ldots, K\}$ (satisfying $P_k^{av} \leq P_{max}$). The new goal is to maximize a weighted sum of admission rates $\sum_{k=1}^{K} \theta_k \overline{x}_k$ subject to queue stability and to all average power constraints, where $\{\theta_1, \ldots, \theta_K\}$ are a given set of positive weights.

a) Using $J = 0, L = K, y_0(t) = -\sum_{k=1}^{K} \theta_k x_k(t)$, and a fixed V, state the drift-plus-penalty algorithm for this problem. Note that the constraints $\overline{P}_k \leq P_k^{av}$ should be enforced by virtual queues $Z_k(t)$ of the form (4.40) with a suitable definition of $y_k(t)$.

b) Use Theorem 4.8 to conclude that all queues are mean rate stable (and hence all average power constraints are met), and compute a value B such that:

$$\liminf_{t\to\infty} \frac{1}{t}\sum_{\tau=0}^{t-1}\sum_{k=1}^{K}\theta_k \mathbb{E}\{x_k(\tau)\} \geq util^{opt} - B/V$$

where $util^{opt}$ is the optimal weighted sum of admitted rates into the network under any algorithm that stabilizes the queues and satisfies all average power constraints.

c) Show that the algorithm is such that $x_k(t) = 0$ whenever $Q_k(t) > V\theta_k$. Assume that all queues are initially empty, and compute values Q_k^{max} such that $Q_k(t) \leq Q_k^{max}$ for all $t \geq 0$ and all $k \in \{1, \ldots, K\}$. This shows that queues are *deterministically bounded*, even without the Slater condition of Assumption A1.

d) Show that the algorithm is such that $P_k(t) = 0$ whenever $Z_k(t) > Q_k(t)$. Conclude that $Z_k(t) \leq Z_k^{max}$, where Z_k^{max} is defined $Z_k^{max} \triangleq Q_k^{max} + (P_{max} - P_k^{av})$.

e) Use part (d) and the sample path input-output inequality (2.3) to conclude that for any positive integer T, the total power expended by each link k over any T-slot interval is deterministically less than or equal to $TP_k^{av} + Z_k^{max}$. That is:

$$\sum_{\tau=t_0}^{t_0+T-1} P_k(\tau) \leq TP_k^{av} + Z_k^{max} \quad \forall t_0 \in \{0, 1, 2, \ldots\}, \forall T \in \{1, 2, 3, \ldots\}$$

f) Suppose link k is a wireless transmitter with a battery that has initial energy E_k. Use part (e) to provide a guarantee on the lifetime of the link.

Exercise 4.10. (Out-of-Date Queue Backlog Information) Consider the K-queue problem with $L = J = 0$, and $0 \leq a_k(t) \leq a_{max}$ and $0 \leq b_k(t) \leq b_{max}$ for all k and all t, for some finite constants a_{max} and b_{max}. The network controller attempts to perform the drift-plus-penalty algorithm (4.48)-(4.49) every slot. However, it does not have access to the current queue backlogs $Q_k(t)$, and only receives delayed information $Q_k(t-T)$ for some integer $T \geq 0$. It thus uses $Q_k(t-T)$ in place of $Q_k(t)$ in (4.48). Let $\alpha^{ideal}(t)$ be the optimal decision of (4.48)-(4.49) in the ideal case when current queue backlogs $Q_k(t)$ are used, and let $\alpha^{approx}(t)$ be the implemented decision that uses the out-of-date queue backlogs $Q_k(t-T)$. Show that $\alpha^{approx}(t)$ yields a C-additive approximation for some finite constant C. Specifically, compute a value C such that:

$$V\hat{y}_0(\alpha^{approx}(t), \omega(t)) + \sum_{k=1}^{K} Q_k(t)[\hat{a}_k(\alpha^{approx}(t), \omega(t)) - \hat{b}_k(\alpha^{approx}(t), \omega(t))] \leq$$
$$V\hat{y}_0(\alpha^{ideal}(t), \omega(t)) + \sum_{k=1}^{K} Q_k(t)[\hat{a}_k(\alpha^{ideal}(t), \omega(t)) - \hat{b}_k(\alpha^{ideal}(t), \omega(t))] + C$$

This shows that we can still optimize the system and provide stability with out-of-date queue backlog information. Treatment of delayed queue information for Lyapunov drift arguments was perhaps first used in (147), where random delays without a deterministic bound are also considered.

	t	0	1	2	3	4	5	6	7	8
Arrivals	$a_1(t)$	3	0	3	0	0	1	0	1	0
	$a_2(t)$	2	0	1	0	1	1	0	0	0
Channels	$S_1(t)$	G	G	M	M	G	G	M	M	G
	$S_2(t)$	M	M	B	M	B	M	B	G	B
Max $Q_i b_i$	$Q_1(t)$	0	$\boxed{3}$	0	$\boxed{3}$	$\boxed{1}$	0	1	1	$\boxed{2}$
Policy	$Q_2(t)$	0	2	$\boxed{2}$	2	2	$\boxed{3}$	$\boxed{2}$	$\boxed{1}$	0

Figure 4.3: Arrivals, channel conditions, and queue backlogs for a two queue wireless downlink.

Exercise 4.11. (Simulation) Consider a 2-queue system with time varying channels $(S_1(t), S_2(t))$, where $S_i(t) \in \{G, M, B\}$, representing "Good," "Medium," "Bad" channel conditions for $i \in \{1, 2\}$. Only one channel can be served per slot. All packets have fixed length, and 3 packets can be served when a channel is "Good," 2 when "Medium," and 1 when "Bad." Exactly one unit of power is expended when we serve any channel (regardless of its condition). A sample path example is given in Fig. 4.3, which expends 8 units of power over the first 9 slots under the policy that serves the queue that yields the largest $Q_i(t)b_i(t)$ value, which is a special case of the drift-plus-penalty algorithm for $K = 2$, $J = L = 0$, $V = 0$.

a) Given the full future arrival and channel events as shown in the table, and given $Q_1(0) = Q_2(0) = 0$, select a different set of channels to serve over slots $\{0, 1, \ldots, 8\}$ that also leaves the system empty on slot 9, but that minimizes the amount of power required to do so (so that more than 1 slot will be idle). How much power is used?

b) Assume these arrivals and channels are repeated periodically every 9 slots. Simulate the system using the drift-plus-penalty policy of choosing the queue i that maximizes $Q_i(t)b_i(t) - V$ whenever this quantity is non-negative, and remains idle if this is negative for both $i = 1$ and $i = 2$. Find the empirical average power expenditure and the empirical average queue backlog over 10^6 slots when $V = 0$. Repeat for $V = 1$, $V = 5$, $V = 10$, $V = 20$, $V = 50$, $V = 100$, $V = 200$.

c) Repeat part (b) in the case when arrival vectors $(a_1(t), a_2(t))$ and channel vectors $(S_1(t), S_2(t))$ are independent and i.i.d. over slots with the same empirical distribution as that achieved over 9 slots in the table, so that $Pr[(a_1, a_2) = (3, 2)] = 1/9$, $Pr[(S_1, S_2) = (G, M)] = 3/9$, $Pr[(S_1, S_2) = (M, B)] = 2/9$, etc. *Note: You should find that the resulting minimum power that is approached as V is increased is the same as part (b), and is strictly less than the empirical power expenditure of part (a).*

d) Show that queue i is only served if $Q_i(t) \geq \lceil V/3 \rceil$. Conclude that $Q_i(t) \geq \max[\lceil V/3 \rceil - 3, 0] \triangleq Q^{place}$ for all t, provided that this inequality holds for $Q_i(0)$. Hence, using Q^{place} place-holder packets would reduce average backlog by exactly this amount, with no loss of power performance.

Exercise 4.12. (Wireless Network Coding) Consider a system of 4 wireless users that communicate to each other through a base station (Fig. 4.4). User 1 desires to send data to user 2 and user 2 desires to send data to user 1. Likewise, user 3 desires to send data to user 4 and user 4 desires to send data to user 3.

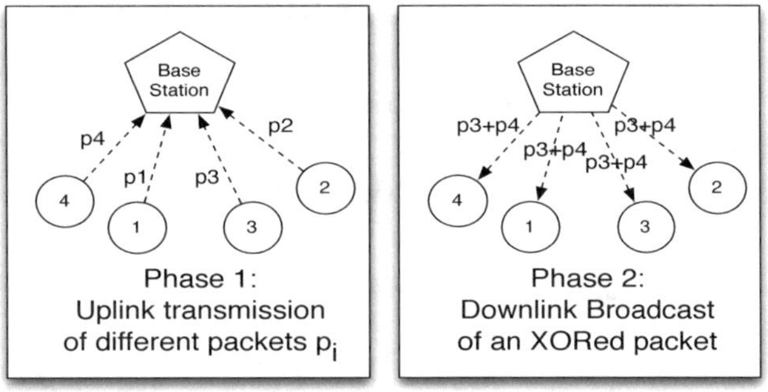

Figure 4.4: An illustration of the 2 phases forming a cycle.

Let $t \in \{0, 1, 2, \ldots\}$ index a *cycle*. Each cycle t is divided into 2 *phases*: In the first phase, users 1, 2, 3, and 4 all send a new packet (if any) to the base station (this can be accomplished, for example, using TDMA or FDMA in the first phase). In the second phase, the base station makes a transmission decision $\alpha(t) \in \{\{1, 2\}, \{3, 4\}\}$. If $\alpha(t) = \{1, 2\}$, the head-of-line packets for users 1 and 2 are XORed together, XORing with 0 if only one packet is available, and creating a null packet if no packets from users 1 or 2 are available. The XORed packet (or null packet) is then broadcast to all users. We assume all packets are labeled with sequence numbers, and the sequence numbers of both XORed packets are placed in a packet header. As in (148), users 1 and 2 can decode the new data if they keep copies of the previous packets they sent. If $\alpha(t) = \{3, 4\}$, a similar XOR operation is done for user 3 and 4 packets.

Assume that downlink channel conditions are time-varying and known at the beginning of each cycle, with channel state vector $S(t) = (S_1(t), S_2(t), S_3(t), S_4(t))$, where $S_i(t) \in \{ON, OFF\}$. Only users with ON channel states can receive the transmission. The queueing dynamics from one cycle to the next thus satisfy:

$$Q_1(t+1) = \max[Q_1(t) - b_1(t), 0] + a_2(t) \ , \quad Q_2(t+1) = \max[Q_2(t) - b_2(t), 0] + a_1(t)$$
$$Q_3(t+1) = \max[Q_3(t) - b_3(t), 0] + a_4(t) \ , \quad Q_4(t+1) = \max[Q_4(t) - b_4(t), 0] + a_3(t)$$

where $Q_k(t)$ is the integer number of packets waiting in the base station for transmission to destination k, $b_k(t) \in \{0, 1\}$ is the number of packets transmitted over the downlink to node k during cycle t, satisfying:

$$b_k(t) = \hat{b}_k(\alpha(t), S(t)) = \begin{cases} 1 & \text{if } S_k(t) = ON \text{ and } k \in \alpha(t) \\ 0 & \text{otherwise} \end{cases}$$

and $a_k(t)$ is the number of packets arriving over the uplink from node k during cycle t (notice that data destined for node 1 arrives as the process $a_2(t)$, etc.). Suppose that $S(t)$ is i.i.d. over cycles, with probabilities $\pi_s = Pr[S(t) = s]$, where $s = (S_1, S_2, S_3, S_4)$. Arrivals $a_k(t)$ are i.i.d. over cycles with rate $\lambda_k = \mathbb{E}\{a_k(t)\}$, for $k \in \{1, \ldots, 4\}$, and with bounded second moments.

a) Suppose that $S(t) = (ON, ON, OFF, ON)$ and that $Q_k(t) > 0$ for all queues $k \in \{1, 2, 3, 4\}$. It is tempting to assume that mode $\alpha(t) = \{1, 2\}$ is the best choice in this case, although this is not always true. Give an example where it is *impossible* to stabilize the system if the controller always chooses $\alpha(t) = \{1, 2\}$ whenever $S(t) = (ON, ON, OFF, ON)$ or $S(t) = (ON, ON, ON, OFF)$, but where a more intelligent control choice *would* stabilize the system.[8]

b) Define $L(\boldsymbol{Q}(t)) = \frac{1}{2}\sum_{k=1}^{4} Q_k(t)^2$. Compute $\Delta(\boldsymbol{Q}(t))$ and show it has the form:

$$\Delta(\boldsymbol{Q}(t)) \leq B - \mathbb{E}\left\{\sum_{k=1}^{4} Q_k(t)[b_k(t) - \lambda_{m(k)}] \middle| \boldsymbol{Q}(t)\right\} \tag{4.93}$$

where $m(1) = 2$, $m(2) = 1$, $m(3) = 4$, $m(4) = 3$, and where $B < \infty$. Design a control policy that observes $S(t)$ and chooses actions $\alpha(t)$ to minimize the right-hand-side of (4.93) over all feasible control policies.

c) Consider all possible *S-only algorithms* that choose a transmission mode as a stationary and random function of the observed $S(t)$ (and independent of queue backlog). Define the S-only throughput region Λ as the set of all $(\lambda_1, \lambda_2, \lambda_3, \lambda_4)$ vectors for which there exists an S-only policy $\alpha^*(t)$ such that:

$$\mathbb{E}\left\{\hat{b}_1(\alpha^*(t), S(t)), \hat{b}_2(\alpha^*(t), S(t)), \hat{b}_3(\alpha^*(t), S(t)), \hat{b}_4(\alpha^*(t), S(t))\right\} \geq (\lambda_2, \lambda_1, \lambda_4, \lambda_3)$$

Suppose that $(\lambda_1, \lambda_2, \lambda_3, \lambda_4)$ is *interior* to Λ, so that $(\lambda_1 + \epsilon, \lambda_2 + \epsilon, \lambda_3 + \epsilon, \lambda_4 + \epsilon) \in \Lambda$ for some value $\epsilon > 0$. Conclude that the drift-minimizing policy of part (b) makes all queues strongly stable, and provide an upper bound on time average expected backlog.

Exercise 4.13. (A modified algorithm) Suppose the conditions of Theorem 4.8 hold. However, suppose that every slot t we observe $\boldsymbol{\Theta}(t), \omega(t)$ and choose an action $\alpha(t) \in \mathcal{A}_{\omega(t)}$ that minimizes the

[8]It can also be shown that an algorithm that always chooses $\alpha(t) = \{1, 2\}$ under states (ON, ON, OFF, ON) or (ON, ON, ON, OFF) *and* when there are indeed two packets to serve will not necessarily work—we need to take queue length into account. See (10) for related examples in the context of a 3×3 packet switch.

exact drift-plus-penalty expression $\Delta(\boldsymbol{\Theta}(t)) + V\mathbb{E}\left\{\hat{y}_0(\alpha(t), \omega(t))|\boldsymbol{\Theta}(t)\right\}$, rather than minimizing the upper bound on the right-hand-side of (4.44).

a) Show that the same performance guarantees of Theorem 4.8 hold.

b) Using (2.2), state this algorithm (for $C = 0$) in the special case when $L = J = 0$, $y_l(t) = e_j(t) = 0, \omega(t) = [(a_1(t), \ldots, a_K(t)), (S_1(t), \ldots, S_K(t))], \hat{a}_k(\alpha(t), \omega(t)) = a_k(t), \alpha(t) \in \{1, \ldots, K\}$ (representing a single queue that we serve every slot), and:

$$\hat{b}_k(\alpha(t), \omega(t)) = \begin{cases} S_k(t) & \text{if } \alpha(t) = k \\ 0 & \text{if } \alpha(t) \neq k \end{cases}$$

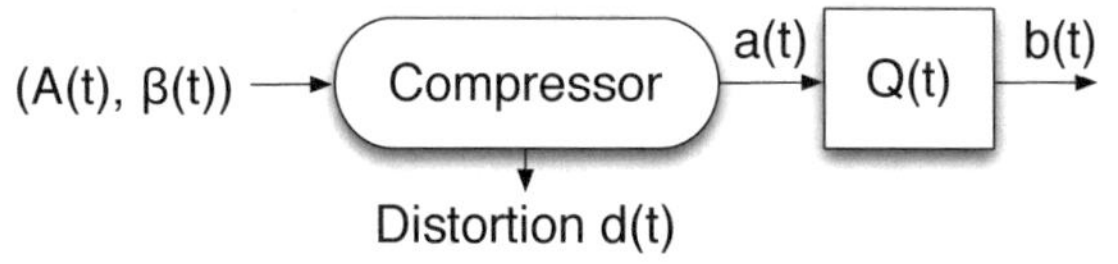

Figure 4.5: A dynamic data compression system for Exercise 4.14.

Exercise 4.14. (Distortion-Aware Data Compression (143)) Consider a single queue $Q(t)$ with dynamics (2.1), where $b(t)$ is an i.i.d. transmission rate process with bounded second moments. As shown in Fig. 4.5, the arrival process $a(t)$ is generated as the output of a *data compression operation*. Specifically, every slot t a new packet of size $A(t)$ bits arrives to the system (where $A(t) = 0$ if no packet arrives). This packet has *meta-data* $\beta(t)$, where $\beta(t) \in \mathcal{B}$, where $\mathcal{B}$ represents a set of different data types. Assume the pair $(A(t), \beta(t))$ is i.i.d. over slots. Every slot t, a network controller observes $(A(t), \beta(t))$ and chooses a *data compression option* $c(t) \in \{0, 1, \ldots, C\}$, where $c(t)$ indexes a collection of possible data compression algorithms. The output of the compressor is a *compressed packet* of random size $a(t) = \hat{a}(A(t), \beta(t), c(t))$, causing a random distortion $d(t) = \hat{d}(A(t), \beta(t), c(t))$. Note that $\hat{a}(\cdot)$ and $\hat{d}(\cdot)$ are *random functions*. Assume the pair $(a(t), d(t))$ is i.i.d. over all slots with the same $A(t), \beta(t), c(t)$. Define functions $m(A, \beta, c)$ and $\delta(A, \beta, c)$ as follows:

$$\begin{aligned} m(A, \beta, c) &\triangleq \mathbb{E}\left\{\hat{a}(A(t), \beta(t), c(t))|A(t) = A, \beta(t) = \beta, c(t) = c\right\} \\ \delta(A, \beta, c) &\triangleq \mathbb{E}\left\{\hat{d}(A(t), \beta(t), c(t))|A(t) = A, \beta(t) = \beta, c(t) = c\right\} \end{aligned}$$

Assume that $c(t) = 0$ corresponds to *no compression*, so that $m(A, \beta, 0) = A$, $\delta(A, \beta, 0) = 0$ for all (A, β). Further, assume that $c(t) = C$ corresponds to throwing the packet away, so that $m(A, \beta, C) = 0$ for all (A, β). Further assume there is a finite constant σ^2 such that for all (A, β, c), we have:

$$\begin{aligned} \mathbb{E}\left\{\hat{a}(A(t), \beta(t), c(t))^2|A(t) = A, \beta(t) = \beta, c(t) = c\right\} &\leq \sigma^2 \\ \mathbb{E}\left\{\hat{d}(A(t), \beta(t), c(t))^2|A(t) = A, \beta(t) = \beta, c(t) = c\right\} &\leq \sigma^2 \end{aligned}$$

Assume the functions $m(A, \beta, c)$ and $\delta(A, \beta, c)$ are known. We want to design an algorithm that minimizes the time average expected distortion $\overline{d}$ subject to queue stability. It is clear that this problem is *feasible*, as we can always choose $c(t) = C$ (although this would maximize distortion). Use the drift-plus-penalty framework (with fixed V) to design such an algorithm. Hint: Use iterated expectations to claim that:

$$\begin{aligned}\mathbb{E}\left\{\hat{a}(A(t), \beta(t), c(t))|Q(t)\right\} &= \mathbb{E}\left\{\mathbb{E}\left\{\hat{a}(A(t), \beta(t), c(t))|Q(t), A(t), \beta(t), c(t)\right\}|Q(t)\right\} \\ &= \mathbb{E}\left\{m(A(t), \beta(t), c(t))|Q(t)\right\}\end{aligned}$$

Exercise 4.15. (Weighted Lyapunov Functions) Recompute the drift-plus-penalty bound in Lemma 4.6 under the following modified Lyapunov function:

$$L(\boldsymbol{\Theta}(t)) = \frac{1}{2}\sum_{k=1}^{K} w_k Q_k(t)^2 + \frac{1}{2}\sum_{l=1}^{L} Z_l(t)^2 + \frac{1}{2}\sum_{j=1}^{J} H_j(t)^2$$

where $\{w_k\}_{k=1}^{K}$ are a positive weights. How does the drift-plus-penalty algorithm change?

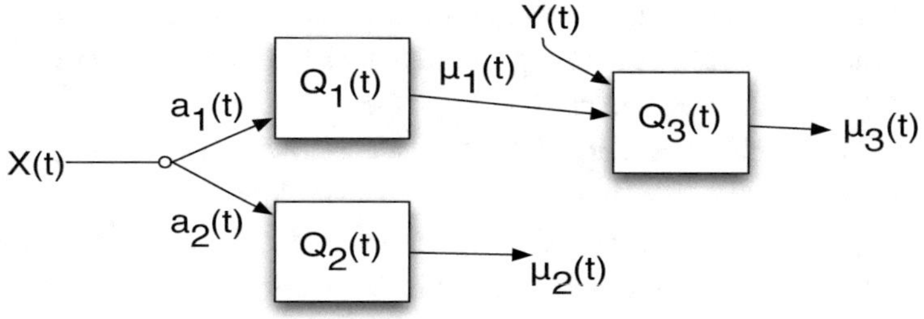

Figure 4.6: The 3-node multi-hop network for Exercise 4.16.

Exercise 4.16. (Multi-Hop with Orthogonal Channels) Consider the 3-node wireless network of Fig. 4.6. The network operates in discrete time with unit time slots $t \in \{0, 1, 2, \ldots\}$. It has orthogonal channels, so that node 3 can send and receive at the same time. The network controller makes *power allocation decisions* and *routing decisions*.

- (Power Allocation) Let $\mu_i(t)$ be the transmission rate at node i on slot t, for $i \in \{1, 2, 3\}$. This transmission rate depends on the channel state $S_i(t)$ and the power allocation decision $P_i(t)$ by the following function:

$$\mu_i(t) = \log(1 + P_i(t)S_i(t)) \quad \forall i \in \{1, 2, 3\}, \forall t$$

where $\log(\cdot)$ denotes the natural logarithm. Every time slot t, the network controller observes the channels $(S_1(t), S_2(t), S_3(t))$ and determines the *power allocation decisions* $(P_1(t), P_2(t), P_3(t))$, made subject to the following constraints:

$$0 \leq P_i(t) \leq 1 \ \forall i \in \{1, 2, 3\}, \forall t$$

- (Routing) There are two arrival processes $X(t)$ and $Y(t)$, taking units of bits. The $X(t)$ process can be routed to either queue 1 or 2. The $Y(t)$ process goes directly into queue 3. Let $a_1(t)$ and $a_2(t)$ represent the *routing decision variables*, where $a_1(t)$ is the amount of bits routed to queue 1, and $a_2(t)$ is the amount of bits routed to queue 2. The network controller observes $X(t)$ every slot and makes decisions for $(a_1(t), a_2(t))$ subject to the following constraints:

$$a_1(t) \geq 0 \ , \ a_2(t) \geq 0 \ , \ a_1(t) + a_2(t) = X(t) \ \forall t$$

It can be shown that the Lyapunov drift $\Delta(\boldsymbol{Q}(t))$ satisfies the following every slot t:

$$\Delta(\boldsymbol{Q}(t)) \leq B + Q_1(t)\mathbb{E}\{a_1(t) - \mu_1(t)|\boldsymbol{Q}(t)\} + Q_2(t)\mathbb{E}\{a_2(t) - \mu_2(t)|\boldsymbol{Q}(t)\}$$
$$+ Q_3(t)\mathbb{E}\{\mu_1(t) + Y(t) - \mu_3(t)|\boldsymbol{Q}(t)\}$$

where B is a positive constant. We want to design a dynamic algorithm that solves the following problem:

$$\begin{array}{lll} \text{Minimize:} & & \overline{P}_1 + \overline{P}_2 + \overline{P}_3 \\ \text{Subject to:} & 1) & Q_i(t) \text{ is mean rate stable } \forall i \in \{1, 2, 3\} \\ & 2) & a_1(t) \geq 0 \ , \ a_2(t) \geq 0 \ , \ a_1(t) + a_2(t) = X(t) \ \forall t \\ & 3) & 0 \leq P_i(t) \leq 1 \ \forall i \in \{1, 2, 3\}, \forall t \end{array}$$

a) Using a fixed parameter $V > 0$, state the drift-plus-penalty algorithm for this problem. The algorithm should have separable power allocation and routing decisions.

b) Suppose that $V = 20$, $Q_1(t) = 50$, $Q_2(t) = Q_3(t) = 20$, $S_1(t) = S_2(t) = S_3(t) = 1$. What should the value of $P_1(t)$ be under the drift-plus-penalty algorithm? (give a numeric value)

c) Suppose $(X(t), Y(t))$ is i.i.d. over slots with $\mathbb{E}\{X(t)\} = \lambda_X$ and $\mathbb{E}\{Y(t)\} = \lambda_Y$. Suppose $(S_1(t), S_2(t), S_3(t))$ is i.i.d. over slots. Suppose there is a stationary and randomized policy that observes $(X(t), Y(t), S_1(t), S_2(t), S_3(t))$ every slot t, and makes randomized decisions $(a_1^*(t), a_2^*(t), P_1^*(t), P_2^*(t), P_3^*(t))$ based only on the observed vector $(X(t), Y(t), S_1(t), S_2(t), S_3(t))$. State desirable properties for the expectations of $\mathbb{E}\left\{a_1^*(t)\right\}$, $\mathbb{E}\left\{a_2^*(t)\right\}$, $\mathbb{E}\left\{\log(1 + P_i^*(t)S_i(t))\right\}$ for $i \in \{1, 2, 3\}$ that would ensure your algorithm of part (a) would make all queues mean rate stable with time average expected power expenditure given by:

$$\overline{P}_1 + \overline{P}_2 + \overline{P}_3 \leq \phi + B/V$$

where ϕ is a desired value for the sum time average power. Your properties should be in the form of desirable inequalities.

4.11 APPENDIX 4.A — PROVING THEOREM 4.5

This appendix characterizes the set of all possible time average expectations for the variables $[(y_l(t)), (e_j(t)), (a_k(t)), (b_k(t))]$ defined in Section 4.2. It concludes with a proof of Theorem 4.5, which shows that optimality for the problem (4.31)-(4.35) can be defined over the class of ω-only policies. The proof involves set theoretic concepts of *convex sets*, *closed sets*, *limit points*, and *convergent subsequences*. In particular, we use the well known fact that if $\{\boldsymbol{x}(t)\}_{t=0}^{\infty}$ is an infinite sequence of vectors that are contained in some bounded set $\mathcal{X} \subseteq \mathbb{R}^k$ (for some finite integer $k > 0$), then there must exist a convergent subsequence $\{\boldsymbol{x}(t_i)\}_{i=1}^{\infty}$ that converges to a point $\boldsymbol{x}$ in the closure of $\mathcal{X}$ (see, for example, A14 of (145)). Specifically, there is a vector $\boldsymbol{x}$ in the closure of $\mathcal{X}$ and an infinite sequence of increasing positive integers $\{t_1, t_2, t_3, \ldots\}$ such that:

$$\lim_{i\to\infty} \boldsymbol{x}(t_i) = \boldsymbol{x}$$

4.11.1 THE REGION Γ

Let Γ represent the region of all $[(\overline{y}_l)_{l=0}^{L}, (\overline{e}_j)_{j=1}^{J}, (\overline{a}_k)_{k=1}^{K}, (\overline{b}_k)_{k=1}^{K}]$ values that can be achieved by ω-only policies. Equivalently, this can be viewed as the region of all one-slot expectations that can be achieved via randomized decisions when the $\omega(t)$ variable takes values according to its stationary distribution. The boundedness assumptions (4.25)-(4.30) ensure that the set Γ is bounded. It is easy to show that Γ is also convex by using an ω-only policy that is a mixture of two other ω-only policies.

Now note that for any slot τ and assuming that $\omega(\tau)$ has its stationary distribution, the one-slot expectation under *any decision* $\alpha(\tau) \in \mathcal{A}_{\omega(\tau)}$ is in the set Γ, even if that decision is from an arbitrary policy that is not an ω-only policy. That is:

$$\mathbb{E}\left\{[(\hat{y}_l(\alpha(\tau), \omega(\tau)), (\hat{e}_j(\alpha(\tau), \omega(\tau)), (\hat{a}_k(\alpha(\tau), \omega(\tau)), (\hat{b}_k(\alpha(\tau), \omega(\tau))]\right\} \in \Gamma$$

where the expectation is with respect to the random $\omega(\tau)$ (which has the stationary distribution) and the possibly random $\alpha(\tau)$ that is made by the policy in reaction to the observed $\omega(\tau)$. This expectation is in Γ because any sample path of events that lead to the policy choosing $\alpha(\tau)$ on slot τ simply affects the conditional distribution of $\alpha(\tau)$ given the observed $\omega(\tau)$, and hence the expectation can be equally achieved by the ω-only policy that uses the same conditional distribution.[9] This observation directly leads to the following simple lemma.

Lemma 4.17 *If $\omega(\tau)$ is in its stationary distribution for all slots τ, then for any policy that chooses $\alpha(\tau) \in \mathcal{A}_{\omega(\tau)}$ over time (including policies that are not ω-only), we have for any slot $t > 0$:*

$$\frac{1}{t}\sum_{\tau=0}^{t-1}\mathbb{E}\left\{[(\hat{y}_l(\alpha(\tau), \omega(\tau)), (\hat{e}_j(\alpha(\tau), \omega(\tau)), (\hat{a}_k(\alpha(\tau), \omega(\tau)), (\hat{b}_k(\alpha(\tau), \omega(\tau))]\right\} \in \Gamma \quad (4.94)$$

[9] We implicitly assume that the decision $\alpha(\tau)$ on slot τ has a well defined conditional distribution.

Thus, if $\boldsymbol{r}^*$ is a limit point of the time average on the left-hand-side of (4.94) over a subsequence of times t_i that increase to infinity, then $\boldsymbol{r}^*$ is in the closure of Γ.

Proof. Each term in the time average is itself in Γ, and so the time average is also in Γ because Γ is convex. □

Thus, the finite horizon time average expectation under any policy cannot escape the set Γ, and any infinite horizon time average that converges to a limit point cannot escape the closure of Γ. If the set Γ is closed, then any limit point $\boldsymbol{r}^*$ is inside Γ and hence (by definition of Γ) can be exactly achieved as the one-slot average under some ω-only policy. If Γ is not closed, then $\boldsymbol{r}^*$ can be achieved arbitrarily closely (i.e., within a distance δ, for any arbitrarily small $\delta > 0$), by an ω-only policy. This naturally leads to the following characterization of optimality in terms of ω-only policies.

4.11.2 CHARACTERIZING OPTIMALITY

Define $\tilde{\Gamma}$ as the set of all points $[(y_l), (e_j), (a_k), (b_k)]$ in the closure of Γ that satisfy:

$$y_l \leq 0 \ \forall l \in \{1, \ldots, L\} \ , \ e_j = 0 \ \forall j \in \{1, \ldots, J\} \ , \ a_k \leq b_k \ \forall k \in \{1, \ldots, K\} \tag{4.95}$$

It can be shown that, if non-empty, $\tilde{\Gamma}$ is closed and bounded. If $\tilde{\Gamma}$ is non-empty, define y_0^* as the minimum value of y_0 for which there is a point $[(y_l), (e_j), (a_k), (b_k)] \in \tilde{\Gamma}$. Intuitively, the set $\tilde{\Gamma}$ is the set of all time averages achievable by ω-only policies that meet the required time average constraints and that have time average expected arrivals less than or equal to time average expected service, and y_0^* is the minimum time average penalty achievable by such ω-only policies. We now show that $y_0^* = y_0^{opt}$.

Theorem 4.18 *Suppose the $\omega(t)$ process is stationary with distribution $\pi(\omega)$, and that the system satisfies the boundedness assumptions (4.25)-(4.30) and the law of large numbers assumption specified in Section 4.2. Suppose the problem (4.31)-(4.35) is feasible. Let $\alpha(t)$ be any control policy that satisfies the constraints (4.32)-(4.35), and let $\boldsymbol{r}(t)$ represent the t-slot expected time average in the left-hand-side of (4.94) under this policy.*

a) Any limit point $[(y_l), (e_j), (a_k), (b_k)]$ of $\{\boldsymbol{r}(t)\}_{t=1}^{\infty}$ is in the set $\tilde{\Gamma}$. In particular, the set $\tilde{\Gamma}$ is non-empty.

b) The time average expected penalty under the algorithm $\alpha(t)$ satisfies:

$$\liminf_{t\to\infty} \frac{1}{t}\sum_{\tau=0}^{t-1} \mathbb{E}\left\{\hat{y}_0(\alpha(t), \omega(t))\right\} \geq y_0^* \tag{4.96}$$

Thus, no algorithm that satisfies the constraints (4.32)-(4.35) can yield a time average expected penalty smaller than y_0^*. Further, $y_0^* = y_0^{opt}$.

Proof. To prove part (a), note from Lemma 4.17 that $\boldsymbol{r}(t)$ is always inside the (bounded) set Γ. Hence, it has a limit point, and any such limit point is in the closure of Γ. Now consider a particular limit point $[(y_l), (e_j), (a_k), (b_k)]$, and let $\{t_i\}_{i=1}^{\infty}$ be the subsequence of non-negative integer time slots that increase to infinity and satisfy:

$$\lim_{i\to\infty} \boldsymbol{r}(t_i) = [(y_l), (e_j), (a_k), (b_k)]$$

Because the constraints (4.32) and (4.33) are satisfied, it must be the case that:

$$y_l \leq 0 \ \forall l \in \{1, \ldots, L\} \ , \ e_j = 0 \ \forall j \in \{1, \ldots, J\} \tag{4.97}$$

Further, by the sample-path inequality (2.5), we have for all $t_i > 0$ and all k:

$$\frac{\mathbb{E}\{Q_k(t_i)\}}{t_i} - \frac{\mathbb{E}\{Q_k(0)\}}{t_i} \geq \frac{1}{t_i}\sum_{\tau=0}^{t_i-1} \mathbb{E}\left\{\hat{a}_k(\alpha(\tau), \omega(\tau)) - \hat{b}_k(\alpha(\tau), \omega(\tau))\right\}$$

Because the control policy makes all queues mean rate stable, taking a limit of the above over the times $t_i \to \infty$ yields $0 \geq a_k - b_k$, and hence we find that:

$$a_k \leq b_k \ \forall k \in \{1, \ldots, K\} \tag{4.98}$$

The results (4.97) and (4.98) imply that the limit point $[(y_l), (e_j), (a_k), (b_k)]$ is in the set $\tilde{\Gamma}$.

To prove part (b), let $\{t_i\}_{i=1}^{\infty}$ be a subsequence of non-negative integer time slots that increase to infinity, that yield the lim inf by:

$$\lim_{i\to\infty} \frac{1}{t_i}\sum_{\tau=0}^{t_i-1} \mathbb{E}\left\{\hat{y}_0(\alpha(\tau), \omega(\tau))\right\} = \liminf_{t\to\infty} \frac{1}{t}\sum_{\tau=0}^{t-1} \mathbb{E}\left\{\hat{y}_0(\alpha(\tau), \omega(\tau))\right\} \tag{4.99}$$

and that yield well defined time averages $[(y_l), (e_j), (a_k), (b_k)]$ for $\boldsymbol{r}(t_i)$ (such a subsequence can be constructed by first taking a subsequence $\{t_i'\}$ that achieves the lim inf, and then taking a convergent subsequence $\{t_i\}$ of $\{t_i'\}$ that ensures the $\boldsymbol{r}(t_i)$ values converge to a limit point). Then by part (a), we know that $[(y_l), (e_j), (a_k), (b_k)] \in \tilde{\Gamma}$, and so its y_0 component (being the lim inf value in (4.99)) is greater than or equal to y_0^* because y_0^*, is the smallest possible y_0 value of all points in $\tilde{\Gamma}$.

It follows that no control algorithm that satisfies the required constraints has a time average expected penalty less than y_0^*. We now show that it is possible to achieve y_0^*, and so $y_0^* = y_0^{opt}$. For simplicity, we consider only the case when Γ is closed. Let $[(y_l^*), (e_j^*), (a_k^*), (b_k^*)]$ be the point in $\tilde{\Gamma}$ that has component y_0^*. Because Γ is closed, $\tilde{\Gamma}$ is a subset of Γ, and so $[(y_l^*), (e_j^*), (a_k^*), (b_k^*)] \in \Gamma$. It follows there is an ω-only algorithm $\alpha^*(t)$ with expectations exactly equal to $[(y_l^*), (e_j^*), (a_k^*), (b_k^*)]$ on every slot t. Thus, the time average penalty is y_0^*, and the constraints (4.32), (4.33) are satisfied because $y_l^* \leq 0$ for all $l \in \{1, \ldots, L\}$, $e_j^* = 0$ for all $j \in \{1, \ldots, J\}$. Further, our "law-of-large-number" assumption on $\omega(t)$ ensures the time averages of $\hat{a}_k(\alpha^*(t), \omega(t))$ and $\hat{b}_k(\alpha^*(t), \omega(t))$, achieved under the ω-only algorithm $\alpha^*(t)$, are equal to a_k^* and b_k^* with probability 1. Because $a_k^* \leq b_k^*$ and the second moments of $a_k(t)$ and $b_k(t)$ are bounded by a finite constant σ^2 for all t, the Rate Stability Theorem (Theorem 2.4) ensures that all queues $Q_k(t)$ are mean rate stable. □

We use this result to prove Theorem 4.5.

Proof. (Theorem 4.5) Let $[(y_l^*), (e_j^*), (a_k^*), (b_k^*)]$ be the point in $\tilde{\Gamma}$ that has component y_0^* (where $y_0^* = y_0^{opt}$ by Theorem 4.18). Note by definition that $\tilde{\Gamma}$ is in the closure of Γ. If Γ is closed, then $[(y_l^*), (e_j^*), (a_k^*), (b_k^*)] \in \Gamma$ and so there exists an ω-only policy $\alpha^*(t)$ that achieves the averages $[(y_l^*), (e_j^*), (a_k^*), (b_k^*)]$ and thus satisfies (4.36)-(4.39) with $\delta = 0$. If Γ is not closed, then $[(y_l^*), (e_j^*), (a_k^*), (b_k^*)]$ is a limit point of Γ and so there is an ω-only policy that gets arbitrarily close to $[(y_l^*), (e_j^*), (a_k^*), (b_k^*)]$, yielding (4.36)-(4.39) for any $\delta > 0$. □

The above proof shows that if the assumptions of Theorem 4.5 hold and if the set Γ is closed, then an ω-only policy exists that satisfies the inequalities (4.36)-(4.39) with $\delta = 0$.

CHAPTER 5

Optimizing Functions of Time Averages

Here we use the drift-plus-penalty technique to develop methods for optimizing convex functions of time averages, and for finding local optimums for non-convex functions of time averages. To begin, consider a discrete time queueing system $\boldsymbol{Q}(t) = (Q_1(t), \ldots, Q_K(t))$ with the standard update equation:

$$Q_k(t+1) = \max[Q_k(t) - b_k(t), 0] + a_k(t) \tag{5.1}$$

Let $\boldsymbol{x}(t) = (x_1(t), \ldots, x_M(t))$, $\boldsymbol{y}(t) = (y_1(t), \ldots, y_L(t))$ be attribute vectors. As before, the arrival, service, and attribute variables are determined by general functions $a_k(t) = \hat{a}_k(\alpha(t), \omega(t))$, $b_k(t) = \hat{b}_k(\alpha(t), \omega(t))$, $x_m(t) = \hat{x}_m(\alpha(t), \omega(t))$ and $y_l(t) = \hat{y}_l(\alpha(t), \omega(t))$. Consider now the following problem:

$$\begin{aligned}
&\text{Maximize:} && \phi(\overline{\boldsymbol{x}}) && (5.2)\\
&\text{Subject to:} && 1)\ \ \overline{y}_l \leq 0 \ \ \forall l \in \{1, \ldots, L\} && (5.3)\\
& && 2)\ \ \text{All queues } Q_k(t) \text{ are mean rate stable} && (5.4)\\
& && 3)\ \ \alpha(t) \in \mathcal{A}_{\omega(t)} \ \ \forall t && (5.5)
\end{aligned}$$

where $\phi(\boldsymbol{x})$ is a concave, continuous, and entrywise non-decreasing *utility function* defined over an appropriate region of $\mathbb{R}^M$ (such as the non-negative orthant when $x_m(t)$ attributes are non-negative, or all $\mathbb{R}^M$ otherwise). A more general problem, without the entrywise non-decreasing assumption, is considered in Section 5.4.

Problems with the structure (5.2)-(5.5) arise, for example, when maximizing network throughput-utility, where $\overline{\boldsymbol{x}}$ represents a vector of achieved throughput and $\phi(\overline{\boldsymbol{x}})$ is a concave function that measures *network fairness*. An example utility function that is useful when attributes $x_m(t)$ are non-negative is:

$$\phi(\boldsymbol{x}) = \sum_{m=1}^{M} \log(1 + \nu_m x_m) \tag{5.6}$$

where ν_m are positive constants. This is useful because each component function $\log(1 + \nu_m x_m)$ has a diminishing returns property as x_m is increased, has maximum derivative ν_m, and is 0 when $x_m = 0$. Another common example is:

$$\phi(\boldsymbol{x}) = \sum_{m=1}^{M} \log(x_m) \tag{5.7}$$

This corresponds to the *proportional fairness objective* (1)(2)(5). The function $\phi(\boldsymbol{x})$ does not need to be differentiable. An example non-differentiable function that is concave, continuous, and entrywise non-decreasing is $\phi(\boldsymbol{x}) = \min[x_1, x_2, \ldots, x_M]$.

The problem (5.2)-(5.5) is different from all of the problems seen in Chapter 4 because it involves a *function* of a time average. It does not conform to the structure required for the drift-plus-penalty framework of Chapter 4 unless the function $\phi(\boldsymbol{x})$ is linear, because a linear function of a time average is equal to the time average of the linear function. In the case when $\phi(\boldsymbol{x})$ is concave but nonlinear, maximizing the time average of $\phi(\boldsymbol{x}(t))$ is typically *not* the same as maximizing $\phi(\overline{\boldsymbol{x}})$ (see Exercise 5.12 for a special case when it *is* the same). Below we transform the problem by adding a *rectangle constraint* and *auxiliary variables* in such a way that the transformed problem involves only time averages (not functions of time averages), so that the drift-plus-penalty framework of Chapter 4 can be applied. The key step in analyzing the transformed problem is *Jensen's inequality*.

5.0.3 THE RECTANGLE CONSTRAINT $\mathcal{R}$

Define ϕ^{opt} as the maximum utility associated with the above problem, augmented with the following rectangle constraint:

$$\overline{\boldsymbol{x}} \in \mathcal{R} \tag{5.8}$$

where $\mathcal{R}$ is defined:

$$\mathcal{R} \triangleq \{(x_1, \ldots, x_M) \in \mathbb{R}^M | \gamma_{m,min} \leq x_m \leq \gamma_{m,max} \ \forall m \in \{1, \ldots, M\}\}$$

where $\gamma_{m,min}$ and $\gamma_{m,max}$ are finite constants (we typically choose $\gamma_{m,min} = 0$ in cases when attributes $x_m(t)$ are non-negative). This rectangle constraint is useful because it limits the $\overline{\boldsymbol{x}}$ vector to a bounded region, and it will ensure that the auxiliary variables that we soon define are also bounded. While this $\overline{\boldsymbol{x}} \in \mathcal{R}$ constraint may limit optimality, it is clear that ϕ^{opt} increases to the maximum utility of the problem without this constraint as the rectangle $\mathcal{R}$ is expanded. Further, ϕ^{opt} is exactly equal to the maximum utility of the original problem (5.2)-(5.5) whenever the rectangle $\mathcal{R}$ is chosen large enough to contain a time average attribute vector $\overline{\boldsymbol{x}}$ that is optimal for the original problem.

5.0.4 JENSEN'S INEQUALITY

Assume the concave utility function $\phi(\boldsymbol{x})$ is defined over the rectangle region $\boldsymbol{x} \in \mathcal{R}$. Let $\boldsymbol{X} = (X_1, \ldots, X_M)$ be a random vector that takes values in $\mathcal{R}$. Jensen's inequality for concave functions states that:

$$\mathbb{E}\{\boldsymbol{X}\} \in \mathcal{R} \ , \quad \text{and} \quad \mathbb{E}\{\phi(\boldsymbol{X})\} \leq \phi(\mathbb{E}\{\boldsymbol{X}\}) \tag{5.9}$$

Indeed, even though we stated Jensen's inequality in Section 1.8 in terms of *convex* functions $f(\boldsymbol{x})$ with a reversed inequality $\mathbb{E}\{f(\boldsymbol{X})\} \geq f(\mathbb{E}\{\boldsymbol{X}\})$, this immediately implies (5.9) by defining $f(\boldsymbol{X}) = -\phi(\boldsymbol{X})$.

Now let $\boldsymbol{\gamma}(\tau) = (\gamma_1(\tau), \ldots, \gamma_M(\tau))$ be an infinite sequence of random vectors that take values in the set $\mathcal{R}$ for $\tau \in \{0, 1, 2, \ldots\}$. It is easy to show that Jensen's inequality for concave functions directly implies the following for all $t > 0$ (see Exercise 5.3):

$$\frac{1}{t}\sum_{\tau=0}^{t-1}\boldsymbol{\gamma}(\tau) \in \mathcal{R} \quad \text{and} \quad \frac{1}{t}\sum_{\tau=0}^{t-1}\phi(\boldsymbol{\gamma}(\tau)) \leq \phi\left(\frac{1}{t}\sum_{\tau=0}^{t-1}\boldsymbol{\gamma}(\tau)\right) \tag{5.10}$$

$$\frac{1}{t}\sum_{\tau=0}^{t-1}\mathbb{E}\{\boldsymbol{\gamma}(\tau)\} \in \mathcal{R} \quad \text{and} \quad \frac{1}{t}\sum_{\tau=0}^{t-1}\mathbb{E}\{\phi(\boldsymbol{\gamma}(\tau))\} \leq \phi\left(\frac{1}{t}\sum_{\tau=0}^{t-1}\mathbb{E}\{\boldsymbol{\gamma}(\tau)\}\right) \tag{5.11}$$

Taking limits of (5.11) as $t \to \infty$ yields:

$$\overline{\boldsymbol{\gamma}} \in \mathcal{R} \quad \text{and} \quad \overline{\phi(\boldsymbol{\gamma})} \leq \phi(\overline{\boldsymbol{\gamma}})$$

where $\overline{\boldsymbol{\gamma}}$ and $\overline{\phi(\boldsymbol{\gamma})}$ are defined as the following limits:

$$\overline{\boldsymbol{\gamma}} \triangleq \lim_{t\to\infty}\frac{1}{t}\sum_{\tau=0}^{t-1}\mathbb{E}\{\boldsymbol{\gamma}(\tau)\} \quad , \quad \overline{\phi(\boldsymbol{\gamma})} \triangleq \lim_{t\to\infty}\frac{1}{t}\sum_{\tau=0}^{t-1}\mathbb{E}\{\phi(\boldsymbol{\gamma}(\tau))\} \tag{5.12}$$

where we temporarily assume the above limits exist. We have used the fact that the rectangle $\mathcal{R}$ is a closed set to conclude that a limit of vectors in $\mathcal{R}$ is also in $\mathcal{R}$.

In summary, whenever the limits of $\overline{\boldsymbol{\gamma}}$ and $\overline{\phi(\boldsymbol{\gamma})}$ exist, we can conclude by Jensen's inequality that $\phi(\overline{\boldsymbol{\gamma}}) \geq \overline{\phi(\boldsymbol{\gamma})}$. *That is, the utility function evaluated at the time average expectation $\overline{\boldsymbol{\gamma}}$ is greater than or equal to the time average expectation of $\phi(\boldsymbol{\gamma}(t))$.*

5.0.5 AUXILIARY VARIABLES

Let $\boldsymbol{\gamma}(t) = (\gamma_1(t), \ldots, \gamma_M(t))$ be a vector of *auxiliary variables* chosen within the set $\mathcal{R}$ every slot. We consider the following modified problem:

$$\begin{aligned}
&\text{Maximize:} && \overline{\phi(\boldsymbol{\gamma})} && (5.13)\\
&\text{Subject to:} && 1)\ \ \overline{y}_l \leq 0 \ \ \forall l \in \{1, \ldots, L\} && (5.14)\\
& && 2)\ \ \overline{\gamma}_m \leq \overline{x}_m \ \ \forall m \in \{1, \ldots, M\} && (5.15)\\
& && 3)\ \ \text{All queues } Q_k(t) \text{ are mean rate stable} && (5.16)\\
& && 4)\ \ \boldsymbol{\gamma}(t) \in \mathcal{R} \ \ \forall t && (5.17)\\
& && 5)\ \ \alpha(t) \in \mathcal{A}_{\omega(t)} \ \ \forall t && (5.18)
\end{aligned}$$

where $\overline{\phi(\boldsymbol{\gamma})}$ and $\overline{\boldsymbol{\gamma}} = (\overline{\gamma}_1, \ldots, \overline{\gamma}_M)$ are defined in (5.12). This transformed problem involves only time averages, rather than functions of time averages, and hence can be solved with the drift-plus-penalty framework of Chapter 4. Indeed, we can define $y_0(t) \triangleq -\phi(\boldsymbol{\gamma}(t))$, and define a new control action $\alpha'(t) = (\alpha(t), \boldsymbol{\gamma}(t))$ subject to $\alpha'(t) \in [\mathcal{A}_{\omega(t)}, \mathcal{R}]$.

This transformed problem (5.13)-(5.18) relates to the original problem as follows: Suppose we have an algorithm that makes decisions $\alpha^*(t)$ and $\boldsymbol{\gamma}^*(t)$ over time $t \in \{0, 1, 2, \ldots\}$ to solve the transformed problem. That is, assume the solution meets all constraints (5.14)-(5.18) and yields a maximum value for the objective (5.13). For simplicity, assume all limiting time average expectations $\overline{\boldsymbol{x}}^*, \overline{y}_l^*, \overline{\boldsymbol{\gamma}}^*, \overline{\phi(\boldsymbol{\gamma}^*)}$ exist, where $\overline{\phi(\boldsymbol{\gamma}^*)}$ is the maximum objective value. Then:

- The decisions $\alpha^*(t)$ produce time averages that satisfy all desired constraints of the original problem (5.2)-(5.5) (so that $\overline{y}_l^* \leq 0$ for all l and all queues $Q_k(t)$ are mean rate stable), and the resulting time average attribute vector $\overline{\boldsymbol{x}}^*$ satisfies $\phi(\overline{\boldsymbol{x}}^*) \geq \overline{\phi(\boldsymbol{\gamma}^*)}$. This is because:

$$\phi(\overline{\boldsymbol{x}}^*) \geq \phi(\overline{\boldsymbol{\gamma}}^*) \geq \overline{\phi(\boldsymbol{\gamma}^*)}$$

  where the first inequality is due to (5.15) and the entrywise non-decreasing property of $\phi(\boldsymbol{x})$, and the second inequality is Jensen's inequality.

- $\overline{\phi(\boldsymbol{\gamma}^*)} \geq \phi^{opt}$. That is, the maximum utility of the transformed problem (5.13)-(5.18) is greater than or equal to ϕ^{opt}. This is shown in Exercise 5.2.

The above two observations imply that $\phi(\overline{\boldsymbol{x}}^*) \geq \phi^{opt}$. *Thus, designing a policy to solve the transformed problem ensures all desired constraints of the original problem (5.2)-(5.5) are satisfied while producing a utility that is at least as good as* ϕ^{opt}.

5.1 SOLVING THE TRANSFORMED PROBLEM

Following the drift-plus-penalty method (using a fixed V), we enforce the constraints $\overline{y}_l \leq 0$ and $\overline{\gamma}_m \leq \overline{x}_m$ in the transformed problem (5.13)-(5.18) with virtual queues $Z_l(t)$ and $G_m(t)$:

$$Z_l(t+1) = \max[Z_l(t) + y_l(t), 0] \quad , \forall l \in \{1, \ldots, L\} \tag{5.19}$$

$$G_m(t+1) = \max[G_m(t) + \gamma_m(t) - x_m(t), 0] \quad , \forall m \in \{1, \ldots, M\} \tag{5.20}$$

Define $\boldsymbol{\Theta}(t) \triangleq [\boldsymbol{Q}(t), \boldsymbol{Z}(t), \boldsymbol{G}(t)]$, and define the Lyapunov function:

$$L(\boldsymbol{\Theta}(t)) \triangleq \frac{1}{2}\left[\sum_{k=1}^K Q_k(t)^2 + \sum_{l=1}^L Z_l(t)^2 + \sum_{m=1}^M G_m(t)^2\right]$$

Assume that $\omega(t)$ is i.i.d., and that $y_l(t), x_m(t), a_k(t), b_k(t)$ satisfy the boundedness assumptions (4.25)-(4.28). It is easy to show the drift-plus-penalty expression satisfies:

$$\Delta(\boldsymbol{\Theta}(t)) - V\mathbb{E}\{\phi(\boldsymbol{\gamma}(t))|\boldsymbol{\Theta}(t)\} \leq D - V\mathbb{E}\{\phi(\boldsymbol{\gamma}(t))|\boldsymbol{\Theta}(t)\} + \sum_{l=1}^L Z_l(t)\mathbb{E}\{y_l(t)|\boldsymbol{\Theta}(t)\}$$

$$+ \sum_{k=1}^K Q_k(t)\mathbb{E}\{a_k(t) - b_k(t)|\boldsymbol{\Theta}(t)\} + \sum_{m=1}^M G_m(t)\mathbb{E}\{\gamma_m(t) - x_m(t)|\boldsymbol{\Theta}(t)\} \tag{5.21}$$

where D is a finite constant related to the worst-case second moments of $y_l(t), x_m(t), a_k(t), b_k(t)$. A C-additive approximation chooses $\boldsymbol{\gamma}(t) \in \mathcal{R}$ and $\alpha(t) \in \mathcal{A}_{\omega(t)}$ such that, given $\boldsymbol{\Theta}(t)$, the right-hand-side of (5.21) is within C of its infimum value. A 0-additive approximation thus performs the following:

- (Auxiliary Variables) For each slot t, observe $\boldsymbol{G}(t)$ and choose $\boldsymbol{\gamma}(t)$ to solve:

$$\text{Maximize:} \quad V\phi(\boldsymbol{\gamma}(t)) - \sum_{m=1}^M G_m(t)\gamma_m(t) \tag{5.22}$$

$$\text{Subject to:} \quad \gamma_{m,min} \leq \gamma_m(t) \leq \gamma_{m,max} \;\; \forall m \in \{1, \ldots, M\} \tag{5.23}$$

- ($\alpha(t)$ Decision) For each slot t, observe $\mathbf{\Theta}(t)$ and $\omega(t)$, and choose $\alpha(t) \in \mathcal{A}_{\omega(t)}$ to minimize:

$$\sum_{l=1}^{L} Z_l(t)\hat{y}_l(\alpha(t), \omega(t)) + \sum_{k=1}^{K} Q_k(t)[\hat{a}_k(\alpha(t), \omega(t)) - \hat{b}_k(\alpha(t), \omega(t))] \\ - \sum_{m=1}^{M} G_m(t)\hat{x}_m(\alpha(t), \omega(t))$$

- (Queue Update) Update the virtual queues $Z_l(t)$ and $G_m(t)$ according to (5.19) and (5.20), and the actual queues $Q_k(t)$ by (5.1).

Define time average expectations $\overline{\boldsymbol{x}}(t), \overline{\boldsymbol{y}}(t), \overline{y}_l(t)$ by:

$$\overline{\boldsymbol{x}}(t) \triangleq \frac{1}{t}\sum_{\tau=0}^{t-1} \mathbb{E}\{\boldsymbol{x}(\tau)\} \ , \ \overline{\boldsymbol{\gamma}}(t) \triangleq \frac{1}{t}\sum_{\tau=0}^{t-1} \mathbb{E}\{\boldsymbol{\gamma}(\tau)\} \ , \ \overline{y}_l(t) \triangleq \frac{1}{t}\sum_{\tau=0}^{t-1} \mathbb{E}\{y_l(\tau)\} \tag{5.24}$$

Define ϕ^{max} as an upper bound on $\phi(\boldsymbol{\gamma}(t))$ for all t, and assume it is finite:

$$\phi^{max} \triangleq \phi(\gamma_{1,max}, \gamma_{2,max}, \ldots, \gamma_{m,max}) < \infty \tag{5.25}$$

Theorem 5.1 *Suppose the boundedness assumptions (4.25)-(4.28), (5.25) hold, the function $\phi(\boldsymbol{x})$ is continuous, concave, and entrywise non-decreasing, the problem (5.2)-(5.5), (5.8) (including the constraint $\overline{\boldsymbol{x}} \in \mathcal{R}$) is feasible, and $\mathbb{E}\{L(\mathbf{\Theta}(0))\} < \infty$. If $\omega(t)$ is i.i.d. over slots and any C-additive approximation is used every slot, then all actual and virtual queues are mean rate stable and:*

$$\liminf_{t\to\infty} \phi(\overline{\boldsymbol{x}}(t)) \geq \phi^{opt} - (D+C)/V \tag{5.26}$$

$$\limsup_{t\to\infty} \overline{y}_l(t) \leq 0 \quad , \forall l \in \{1, \ldots, L\} \tag{5.27}$$

where ϕ^{opt} is the maximum utility of the problem (5.2)-(5.5), (5.8) (including the constraint $\overline{\boldsymbol{x}} \in \mathcal{R}$), and $\overline{\boldsymbol{x}}(t), \overline{y}_l(t)$ are defined in (5.24).

The following extended result provides average queue bounds and utility bounds for all slots t.

Theorem 5.2 *Suppose the assumptions of Theorem 5.1 hold.*

(a) If there is an $\epsilon > 0$, an ω-only policy $\alpha^*(t)$, and a finite constant ϕ_ϵ such that the following Slater-type conditions hold:

$$\mathbb{E}\left\{\hat{y}_l(\alpha^*(t), \omega(t))\right\} \leq 0 \quad \forall l \in \{1, \ldots, L\} \tag{5.28}$$

$$\mathbb{E}\left\{\hat{a}_k(\alpha^*(t), \omega(t)) - \hat{b}_k(\alpha^*(t), \omega(t))\right\} \leq -\epsilon \quad \forall k \in \{1, \ldots, K\} \tag{5.29}$$

$$\gamma_{m,min} \leq \mathbb{E}\left\{\hat{x}_m(\alpha^*(t), \omega(t))\right\} \leq \gamma_{m,max} \quad \forall m \in \{1, \ldots, M\} \tag{5.30}$$

$$\phi(\mathbb{E}\left\{\hat{\boldsymbol{x}}(\alpha^*(t), \omega(t))\right\}) = \phi_\epsilon \tag{5.31}$$

then all queues $Q_k(t)$ are strongly stable and for all $t > 0$, we have:

$$\frac{1}{t}\sum_{\tau=0}^{t-1}\sum_{k=1}^{K}\mathbb{E}\{Q_k(\tau)\} \leq \frac{D + C + V\left(\overline{\phi(\boldsymbol{\gamma}^*)} - \phi_\epsilon\right)}{\epsilon} + \frac{\mathbb{E}\{L(\boldsymbol{\Theta}(0))\}}{\epsilon t}$$

where $\overline{\phi(\boldsymbol{\gamma}^*)}$ is the maximum objective function value for the transformed problem (5.13)-(5.18).

(b) If all virtual and actual queues are initially empty (so that $\boldsymbol{\Theta}(0) = \mathbf{0}$) and if there are finite constants $\nu_m \geq 0$ such that for all $\boldsymbol{\gamma}(t)$ and all $\boldsymbol{x}(t)$, we have:

$$|\phi(\boldsymbol{\gamma}(t)) - \phi(\boldsymbol{x}(t))| \leq \sum_{m=1}^{M}\nu_m|\gamma_m(t) - x_m(t)| \tag{5.32}$$

then for all $t > 0$, we have:

$$\phi(\overline{\boldsymbol{x}}(t)) \geq \phi^{opt} - \frac{D + C}{V} - \sum_{m=1}^{M}\frac{\nu_m\mathbb{E}\{G_m(t)\}}{t} \tag{5.33}$$

where $\mathbb{E}\{G_m(t)\}/t$ is $O(1/\sqrt{t})$ for all $m \in \{1, \ldots, M\}$.

The assumption that all queues are initially empty, made in part (b) of the above theorem, is made only for convenience. The right-hand-side of (5.33) would be modified by subtracting the additional term $\mathbb{E}\{L(\boldsymbol{\Theta}(0))\}/Vt$ otherwise. We note that the ν_m constraint (5.32) needed in part (b) of the above theorem is satisfied for the example utility function in (5.6), but not for the proportionally fair utility function in (5.7). Further, the algorithm developed in this section (or C-additive approximations of the algorithm) often result in deterministically bounded queues, regardless of whether or not the Slater assumptions (5.28)-(5.31) hold (see flow control examples in Sections 5.2-5.3 and Exercises 5.5-5.7). For example, it can be shown that if (5.32) holds, if $\boldsymbol{\gamma}(t)$ is chosen by (5.22)-(5.23), and if $x_m(t) \geq \gamma_{m,min}$ for all t, then $G_m(t) \leq V\nu_m + \gamma_{m,max}$ for all t (provided this holds at $t = 0$). In this case, $\mathbb{E}\{G_m(t)\}/t$ is $O(1/t)$, better than the $O(1/\sqrt{t})$ bound given in the above theorem. As before, the same algorithm can be shown to perform efficiently when the $\omega(t)$ process is non-i.i.d. (38)(39)(136)(42). This is because the auxiliary variables transform the problem to a structure that is the same as that covered by the ergodic theory and universal scheduling theory of Section 4.9.

Proof. (Theorem 5.1) Because the C-additive approximation comes within C of minimizing the right-hand-side of (5.21), we have:

$$\Delta(\boldsymbol{\Theta}(t)) - V\mathbb{E}\{\phi(\boldsymbol{\gamma}(t))|\boldsymbol{\Theta}(t)\} \leq D + C - V\phi(\boldsymbol{\gamma}^*) + \sum_{l=1}^{L}Z_l(t)\mathbb{E}\left\{y_l^*(t)|\boldsymbol{\Theta}(t)\right\}$$
$$+\sum_{k=1}^{K}Q_k(t)\mathbb{E}\left\{a_k^*(t) - b_k^*(t)|\boldsymbol{\Theta}(t)\right\} + \sum_{m=1}^{M}G_m(t)\mathbb{E}\left\{\gamma_m^* - x_m^*(t)|\boldsymbol{\Theta}(t)\right\} \tag{5.34}$$

where $\boldsymbol{\gamma}^* = (\gamma_1^*, \ldots, \gamma_M^*)$ is any vector in $\mathcal{R}$, and $y_l^*(t), a_k^*(t), b_k^*(t), x_m^*(t)$ are from any alternative (possibly randomized) policy $\alpha^*(t) \in \mathcal{A}_{\omega(t)}$. Now note that feasibility of the problem (5.2)-(5.5), (5.8) implies feasibility of the transformed problem (5.13)-(5.18).[1] This together with Theorem 4.5 implies that for any $\delta > 0$, there is an ω-only policy $\alpha^*(t) \in \mathcal{A}_{\omega(t)}$ and a vector $\boldsymbol{\gamma}^* \in \mathcal{R}$ such that:

$$\begin{aligned} -\phi(\boldsymbol{\gamma}^*) &\leq -\phi^{opt} + \delta \\ \mathbb{E}\left\{\hat{y}_l(\alpha^*(t), \omega(t))\right\} &\leq \delta \quad \forall l \in \{1, \ldots, L\} \\ \mathbb{E}\left\{\hat{a}_k(\alpha^*(t), \omega(t)) - \hat{b}_k(\alpha^*(t), \omega(t))\right\} &\leq \delta \quad \forall k \in \{1, \ldots, K\} \\ \mathbb{E}\left\{\gamma_m^* - \hat{x}_m(\alpha^*(t), \omega(t))\right\} &\leq \delta \quad \forall m \in \{1, \ldots, M\} \end{aligned}$$

Assuming that $\delta = 0$ for convenience and plugging the above into (5.34) gives:[2]

$$\Delta(\boldsymbol{\Theta}(t)) - V\mathbb{E}\left\{\phi(\boldsymbol{\gamma}(t))|\boldsymbol{\Theta}(t)\right\} \leq D + C - V\phi^{opt} \tag{5.35}$$

This is in the exact form for application of the Lyapunov Optimization Theorem (Theorem 4.2) and hence by that theorem (or, equivalently, by using iterated expectations and telescoping sums in the above inequality), for all $t > 0$, we have:

$$\frac{1}{t}\sum_{\tau=0}^{t-1}\mathbb{E}\left\{\phi(\boldsymbol{\gamma}(\tau))\right\} \geq \phi^{opt} - (D + C)/V - \mathbb{E}\left\{L(\boldsymbol{\Theta}(0))\right\}/(Vt)$$

By Jensen's inequality for the concave function $\phi(\boldsymbol{\gamma})$, we have for all $t > 0$:

$$\phi(\overline{\boldsymbol{\gamma}}(t)) \geq \phi^{opt} - (D + C)/V - \mathbb{E}\left\{L(\boldsymbol{\Theta}(0))\right\}/(Vt) \tag{5.36}$$

Taking a lim inf of both sides yields:

$$\liminf_{t\to\infty} \phi(\overline{\boldsymbol{\gamma}}(t)) \geq \phi^{opt} - (D + C)/V \tag{5.37}$$

On the other hand, rearranging (5.35) yields:

$$\Delta(\boldsymbol{\Theta}(t)) \leq D + C + V(\phi^{max} - \phi^{opt})$$

Thus, by the Lyapunov Drift Theorem (Theorem 4.1), we know that all queues $Q_k(t), Z_l(t), G_m(t)$ are mean rate stable (in fact, we know that $\mathbb{E}\left\{Q_k(t)\right\}/t, \mathbb{E}\left\{G_m(t)\right\}/t$, and $\mathbb{E}\left\{Z_l(t)\right\}/t$ are $O(1/\sqrt{t})$). Mean rate stability of $Z_l(t)$ and $G_m(t)$ together with Theorem 2.5 implies that (5.27) holds, and that for all $m \in \{1, \ldots, M\}$:

$$\limsup_{t\to\infty}[\overline{\gamma}_m(t) - \overline{x}_m(t)] \leq 0$$

Using this with the continuity and entrywise non-decreasing properties of $\phi(\boldsymbol{x})$, it can be shown that:

$$\liminf_{t\to\infty} \phi(\overline{\boldsymbol{\gamma}}(t)) \leq \liminf_{t\to\infty} \phi(\overline{\boldsymbol{x}}(t))$$

Using this in (5.37) proves (5.26). □

[1] To see this, the transformed problem can just use the same $\alpha(t)$ decisions, and it can choose $\boldsymbol{\gamma}(t) = \overline{\boldsymbol{x}}$ for all t.

[2] The same can be derived using $\delta > 0$ and then taking a limit as $\delta \to 0$.

Proof. (Theorem 5.2) We first prove part (b). We have:

$$\begin{aligned}\phi(\overline{\boldsymbol{y}}(t)) &= \phi(\overline{\boldsymbol{x}}(t) + [\overline{\boldsymbol{y}}(t) - \overline{\boldsymbol{x}}(t)]) \\ &\leq \phi(\overline{\boldsymbol{x}}(t) + \max[\overline{\boldsymbol{y}}(t) - \overline{\boldsymbol{x}}(t), \mathbf{0}]) && (5.38) \\ &\leq \phi(\overline{\boldsymbol{x}}(t)) + \sum_{m=1}^{N} \nu_m \max[\overline{y}_m(t) - \overline{x}_m(t), 0] && (5.39)\end{aligned}$$

where (5.38) follows by the entrywise non-decreasing property of $\phi(\boldsymbol{x})$ (where the $\max[\cdot]$ represents an entrywise max), and (5.39) follows by (5.32). Substituting this into (5.36) and using $\mathbb{E}\{L(\boldsymbol{\Theta}(0))\} = 0$ yields:

$$\phi(\overline{\boldsymbol{x}}(t)) \geq \phi^{opt} - (D + C)/V - \sum_{m=1}^{M} \nu_m \max[\overline{y}_m(t) - \overline{x}_m(t), 0] \quad (5.40)$$

By definition of $G_m(t)$ in (5.20) and the sample path queue property (2.5) together with the fact that $G_m(0) = 0$, we have for all $m \in \{1, \ldots, M\}$ and any $t > 0$:

$$\frac{G_m(t)}{t} \geq \frac{1}{t}\sum_{\tau=0}^{t-1} \gamma_m(\tau) - \frac{1}{t}\sum_{\tau=0}^{t-1} x_m(\tau)$$

Taking expectations above yields for all $t > 0$:

$$\frac{\mathbb{E}\{G_m(t)\}}{t} \geq \overline{\gamma}_m(t) - \overline{x}_m(t) \implies \frac{\mathbb{E}\{G_m(t)\}}{t} \geq \max[\overline{\gamma}_m(t) - \overline{x}_m(t), 0]$$

Using this in (5.40) proves part (b) of the theorem.

To prove part (a), we plug the ω-only policy $\alpha^*(t)$ from (5.28)-(5.31) (using $\boldsymbol{\gamma}^*(t) = \mathbb{E}\left\{\hat{\boldsymbol{x}}(\alpha^*(t), \omega(t))\right\}$) into (5.34). This directly leads to a version of part (a) of the theorem with $\overline{\phi(\boldsymbol{\gamma}^*)}$ replaced with ϕ^{max}. A more detailed analysis shows this can be replaced with $\overline{\phi(\boldsymbol{\gamma}^*)}$ because all constraints of the transformed problem are satisfied and so the lim sup time average objective can be no bigger than $\overline{\phi(\boldsymbol{\gamma}^*)}$ (recall (4.96) of Theorem 4.18). □

5.2 A FLOW-BASED NETWORK MODEL

Here we apply the stochastic utility maximization framework to a simple flow based network model, where we neglect the actual network queueing and develop a flow control policy that simply ensures the flow rate over each link is no more than the link capacity (similar to the flow based models for internet and wireless systems in (2)(23)(29)(149)(150)). Section 5.3 treats a more extensive network model that explicitly accounts for all queues.

Suppose there are N nodes and L links, where each link $l \in \{1, \ldots, L\}$ has a possibly time-varying link capacity $b_l(t)$, for slotted time $t \in \{0, 1, 2, \ldots\}$. Suppose there are M sessions, and let

$A_m(t)$ represent the new arrivals to session m on slot t. Each session $m \in \{1, \ldots, M\}$ has a particular source node and a particular destination node. The random network event $\omega(t)$ is thus:

$$\omega(t) \triangleq [(b_1(t), \ldots, b_L(t)); (A_1(t), \ldots, A_M(t))] \tag{5.41}$$

The control action taken every slot is to first choose $x_m(t)$, the amount of type m traffic admitted into the network on slot t, according to:

$$0 \leq x_m(t) \leq A_m(t) \;\; \forall m \in \{1, \ldots, M\}, \forall t \tag{5.42}$$

The constraint (5.42) is just one example of a flow control constraint. We can easily modify this to the constraint $x_m(t) \in \{0, A_m(t)\}$, which either admits all newly arriving data, or drops all of it. Alternatively, the flow controller could place all non-admitted data into a *transport layer storage reservoir* (rather than dropping it), as in (18)(22)(19)(17) (see also Section 5.6). One can model a network where all sources always have data to send by $A_m(t) = \gamma_{m,max}$ for all t, for some finite value $\gamma_{m,max}$ used to limit the amount of data admitted to the network on any slot.

Next, we must specify a *path* for the newly arriving data from a collection of paths $\mathcal{P}_m$ associated with path options of session m on slot t (possibly being the set of all possible paths in the network from the source of session m to its destination). Here, a path is defined in the usual sense, being a sequence of links starting at the source, ending at the destination, and being such that the end node of each link is the start node of the next link. Let $1_{l,m}(t)$ be an indicator variable that is 1 if the data $x_m(t)$ is selected to use a path that contains link l, and is 0 else. The $(1_{l,m}(t))$ values completely specify the chosen paths for slot t, and hence the decision variable for slot t is given by:

$$\alpha(t) \triangleq [(x_1(t), \ldots, x_M(t)); (1_{l,m}(t))|_{l\in\{1,\ldots,L\}, m\in\{1,\ldots,M\}}]$$

Let $\overline{\boldsymbol{x}} = (\overline{x}_1, \ldots, \overline{x}_M)$ be a vector of the infinite horizon time average admitted flow rates. Let $\phi(\boldsymbol{x}) = \sum_{m=1}^{M} \phi_m(x_m)$ be a *separable utility function*, where each $\phi_m(x)$ is a continuous, concave, non-decreasing function in x. Our goal is to maximize the throughput-utility $\phi(\overline{\boldsymbol{x}})$ subject to the constraint that the time average flow over each link l is less than or equal to the time average capacity of that link. The infinite horizon utility optimization problem of interest is thus:

$$\text{Maximize:} \quad \sum_{m=1}^{M} \phi_m(\overline{x}_m) \tag{5.43}$$

$$\text{Subject to:} \quad \sum_{m=1}^{M} \overline{1_{l,m} x_m} \leq \overline{b}_l \;\; \forall l \in \{1, \ldots, L\} \tag{5.44}$$

$$0 \leq x_m(t) \leq A_m(t) \;\; , \;\; (1_{l,m}(t)) \in \mathcal{P}_m \;\; \forall m \in \{1, \ldots, M\}, \forall t \tag{5.45}$$

where the time averages are defined:

$$\begin{aligned}
\overline{x}_m &\triangleq \lim_{t\to\infty} \frac{1}{t}\sum_{\tau=0}^{t-1}\mathbb{E}\{x_m(\tau)\} \\
\overline{1_{l,m}x_m} &\triangleq \lim_{t\to\infty} \frac{1}{t}\sum_{\tau=0}^{t-1}\mathbb{E}\left\{1_{l,m}(\tau)x_m(\tau)\right\} \\
\overline{b}_l &\triangleq \lim_{t\to\infty} \frac{1}{t}\sum_{\tau=0}^{t-1}\mathbb{E}\{b_l(\tau)\}
\end{aligned}$$

We emphasize that while the actual network can queue data at each link l, we are not explicitly accounting for such queueing dynamics. Rather, we are only ensuring the time average flow rate on each link l satisfies (5.44).

Define ϕ^{opt} as the maximum utility associated with the above problem and subject to the additional constraint that:

$$0 \leq \overline{x}_m \leq \gamma_{m,max} \;\; \forall m \in \{1, \ldots, M\} \tag{5.46}$$

for some finite values $\gamma_{m,max}$. This fits the framework of the utility maximization problem (5.2)-(5.5) with $y_l(t) \triangleq \sum_{m=1}^{M} 1_{l,m}(t)x_m(t) - b_l(t)$, $K = 0$, and with $\mathcal{R}$ being all $\boldsymbol{\gamma}$ vectors that satisfy $0 \leq \gamma_m \leq \gamma_{m,max}$ for all $m \in \{1, \ldots, M\}$ (we choose $\gamma_{m,min} = 0$ because attributes $x_m(t)$ are non-negative). As there are no actual queues $Q_k(t)$ in this model, we use only virtual queues $Z_l(t)$ and $G_m(t)$, defined by update equations:

$$Z_l(t+1) = \max\left[Z_l(t) + \sum_{m=1}^{M} 1_{l,m}(t)x_m(t) - b_l(t), 0\right] \tag{5.47}$$

$$G_m(t+1) = \max[G_m(t) + \gamma_m(t) - x_m(t), 0] \tag{5.48}$$

where $\gamma_m(t)$ are auxiliary variables for $m \in \{1, \ldots, M\}$. The algorithm given in Section 5.0.5 thus reduces to:

- (Auxiliary Variables) Every slot t, each session $m \in \{1, \ldots, M\}$ observes $G_m(t)$ and chooses $\gamma_m(t)$ as the solution to:

$$\text{Maximize:} \quad V\phi_m(\gamma_m(t)) - G_m(t)\gamma_m(t) \tag{5.49}$$

$$\text{Subject to:} \quad 0 \leq \gamma_m(t) \leq \gamma_{m,max} \tag{5.50}$$

- (Routing and Flow Control) For each slot t and each session $m \in \{1, \ldots, M\}$, observe the new arrivals $A_m(t)$, the virtual queue backlogs $G_m(t)$, and the link queues $Z_l(t)$, and choose $x_m(t)$ and a path to maximize:

$$\begin{aligned}
\text{Maximize:} \quad & x_m(t)G_m(t) - x_m(t)\textstyle\sum_{l=1}^{L} 1_{l,m}(t)Z_l(t) \\
\text{Subject to:} \quad & 0 \leq x_m(t) \leq A_m(t) \\
& \text{The path specified by } (1_{l,m}(t)) \text{ is in } \mathcal{P}_m
\end{aligned}$$

This reduces to the following: First find a shortest path from the source of session m to the destination of session m, using link weights $Z_l(t)$ as link costs. If the total weight of the shortest path is less than or equal to $G_m(t)$, choose $x_m(t) = A_m(t)$ and route this data over this single shortest path. Else, there is too much congestion in the network, and so we choose $x_m(t) = 0$ (thereby dropping all data $A_m(t)$).

- (Virtual Queue Updates) Update the virtual queues according to (5.47) and (5.48).

The shortest path routing in this algorithm is similar to that given in (149), which treats a flow-based network stability problem under the assumption that arriving traffic is admissible (so that flow control is not used). This problem with flow control was introduced in (39) using the universal scheduling framework of Section 4.9.2, where there are no probabilistic assumptions on the arrivals or time varying link capacities.

5.2.1 PERFORMANCE OF THE FLOW-BASED ALGORITHM

To apply Theorems 5.1 and 5.2, assume $\omega(t) = [(b_1(t), \ldots, b_L(t)); (A_1(t), \ldots, A_M(t))]$ is i.i.d. over slots, and that the $b_l(t)$ and $A_m(t)$ processes have bounded second moments. Note that the problem (5.43)-(5.46) is trivially feasible because it is always possible to satisfy the constraints by admitting no new arrivals on any slot. Suppose we use any C-additive approximation (where a 0-additive approximation is an exact implementation of the above algorithm). It follows from Theorem 5.1 that all virtual queues are mean rate stable, and so the time average constraints (5.44) are satisfied, and the achieved utility satisfies:

$$\liminf_{t\to\infty} \phi(\overline{x}(t)) \geq \phi^{opt} - (D + C)/V \tag{5.51}$$

where D is a finite constant related to the maximum second moments of $A_m(t)$ and $b_l(t)$. Thus, utility can be pushed arbitrarily close to optimal by increasing V.

We now show that, under some mild additional assumptions, the flow control structure of this algorithm yields tight *deterministic bounds* of size $O(V)$ on the virtual queues. Suppose that $A_m(t) \leq A_{m,max}$ for all t, for some finite constant $A_{m,max}$. Further, to satisfy the constraints (5.32) needed for Theorem 5.2, assume the utility functions $\phi_m(x)$ have finite right derivatives at $x = 0$, given by constants $\nu_m \geq 0$, so that for any non-negative x and y we have:

$$|\phi_m(x) - \phi_m(y)| \leq \nu_m|x - y| \tag{5.52}$$

It can be shown that if $G_m(t) > V\nu_m$, then the solution to (5.49)-(5.50) is $\gamma_m(t) = 0$ (see Exercise 5.5). Because $\gamma_m(t)$ acts as the arrival to virtual queue $G_m(t)$ defined in (5.48), it follows that $G_m(t)$ cannot increase on the next slot. Therefore, for all $m \in \{1, \ldots, M\}$:

$$0 \leq G_m(t) \leq V\nu_m + \gamma_{m,max} \quad \forall t \in \{0, 1, 2, \ldots\} \tag{5.53}$$

provided that this is true for $G_m(0)$ (which is indeed the case if $G_m(0) = 0$). This allows one to deterministically bound the queue sizes $Z_l(t)$ for all $l \in \{1, \ldots, L\}$:

$$0 \leq Z_l(t) \leq V\nu^{max} + \gamma^{max} + MA^{max} \quad \forall t \tag{5.54}$$

provided this holds at time 0, and where ν^{max}, γ^{max}, A^{max} are defined as the maximum of all ν_m, $\gamma_{m,max}$, $A_{m,max}$ values:

$$\nu^{max} \triangleq \max_{m\in\{1,\ldots,M\}} \nu_m \quad , \quad \gamma^{max} \triangleq \max_{m\in\{1,\ldots,M\}} \gamma_{m,max} \quad , \quad A^{max} \triangleq \max_{m\in\{1,\ldots,M\}} A_{m,max}$$

To prove this fact, note that if a link l satisfies $Z_l(t) \leq V\nu^{max} + \gamma^{max}$, then on the next slot, we have $Z_l(t+1) \leq V\nu^{max} + \gamma^{max} + MA^{max}$ because the queue can increase by at most MA^{max} on any slot (see update equation (5.47)). Else, if $Z_l(t) > V\nu^{max} + \gamma^{max}$, then any path that uses this link incurs a cost larger than $V\nu^{max} + \gamma^{max}$, and thus would incur a cost larger than $G_m(t)$ for any session m. Thus, by the routing and flow control algorithm, no session will choose a path that uses this link on the current slot, and so $Z_l(t)$ cannot increase on the next slot.

Using the sample path inequality (2.3) with the deterministic bound on $Z_l(t)$ in (5.54), it follows that over any interval of T slots (for any positive integer T and any initial slot t_0), the data injected for use over link l is no more than $V\nu^{max} + \gamma^{max} + MA^{max}$ beyond the total capacity offered by the link over that interval:

$$\sum_{\tau=t_0}^{t_0+T-1} \sum_{m=1}^{M} 1_{l,m}(\tau)x_m(\tau) \leq \sum_{\tau=t_0}^{t_0+T-1} b_l(\tau) + V\nu^{max} + \gamma^{max} + MA^{max} \tag{5.55}$$

5.2.2 DELAYED FEEDBACK

We note that it may be difficult to use the exact queue values $Z_l(t)$ when solving for the shortest path, as these values change every slot. Hence, a practical implementation may use out-of-date values $Z_l(t - \tau_{l,t})$ for some time delay $\tau_{l,t}$ that may depend on l and t. Further, the virtual queue updates for $Z_l(t)$ in (5.47) are most easily done at each link l, in which case, the actual admitted data $x_m(t)$ for that link may not be known until some time delay, arriving as a process $x_m(t - \tau_{l,m,t})$. However, as the virtual queue size cannot change by more than a fixed amount every slot, the queue value used differs from the ideal queue value by no more than an additive constant that is proportional to the maximum time delay. In this case, provided that the maximum time delay is bounded, we are simply using a C-additive approximation and the utility and queue bounds are adjusted accordingly (see Exercise 4.10 and also Section 6.1.1). A more extensive treatment of delayed feedback for the case of networks without dynamic arrivals or channels is found in (150), which uses a differential equation method.

5.2.3 LIMITATIONS OF THIS MODEL

While (5.55) is a very strong deterministic bound that says no link is given more data than it can handle, it does not directly imply anything about the *actual* network queues (other than the links are not overloaded). The (unproven) understanding is that, because the links are not overloaded, the actual network queues will be stable and all data can arrive to its destination with (hopefully small) delay.

One might approximate average congestion or delay on a link as a convex function of the time average flow rate over the link, as in (151)(129)(150).[3] However, we emphasize that this is only an approximation and does not represent the actual network delay, or even a bound on delay. Indeed, while it is known that average queue congestion and delay is convex if a general stream of traffic is probabilistically split (152), this is not necessarily true (or relevant) for dynamically controlled networks, particularly when the control depends on the queue backlogs and delays themselves. Most problems involving optimization of actual network delay are difficult and unsolved. Such problems involve not only optimization of rate based utility functions, but engineering of the Lagrange multipliers (which are related to queue backlogs) associated with those utility functions.

Finally, observe that the update equation for $Z_l(t)$ in (5.47) can be interpreted as a queueing model where all admitted data on slot t is placed immediately on all links l of its path. Similar models are used in (23)(29)(150)(31). However, this is clearly an approximation because data in an actual network will traverse its path one link at a time. It is assumed that the actual network stamps all data with its intended path, so that there is no dynamic re-routing mid-path. Section 5.3 treats an actual multi-hop queueing network and allows such dynamic routing.

5.3 MULTI-HOP QUEUEING NETWORKS

Here we consider a general multi-hop network, treating the actual queueing rather than using the flow-based model of the previous section. Suppose the network has N nodes and operates in slotted time. There are M sessions, and we let $\boldsymbol{A}(t) = (A_1(t), \ldots, A_M(t))$ represent the vector of data that exogenously arrives to the transport layer for each session on slot t (measured either in integer units of *packets* or real units of *bits*).

Each session $m \in \{1, \ldots, M\}$ has a particular source node and destination node. Data delivery takes place by transmissions over possibly multi-hop paths. We assume that a *transport layer flow controller* observes $A_m(t)$ every slot and decides how much of this data to add to the network layer at its source node and how much to drop (flow control decisions are made to limit queue buffers and ensure the network is stable). Let $(x_m(t))|_{m=1}^{M}$ be the collection of *flow control decision variables* on slot t. These decisions are made subject to the constraints $0 \leq x_m(t) \leq A_m(t)$ (see also discussion after (5.42) on modifications of this constraint).

All data that is intended for destination node $c \in \{1, \ldots, N\}$ is called *commodity c data*, regardless of its particular session. For each $n \in \{1, \ldots, N\}$ and $c \in \{1, \ldots, N\}$, let $\mathcal{M}_n^{(c)}$ denote the set of all sessions $m \in \{1, \ldots, M\}$ that have source node n and commodity c. All data is queued according to its commodity, and we define $Q_n^{(c)}(t)$ as the amount of commodity c data in node n on slot t. We assume that $Q_n^{(n)}(t) = 0$ for all t, as data that reaches its destination is removed from the network. Let $\boldsymbol{Q}(t)$ denote the matrix of current queue backlogs for all nodes and commodities.

[3]Convex constraints can be incorporated using the generalized structure of Section 5.4.

The queue backlogs change from slot to slot as follows:

$$Q_n^{(c)}(t+1) = Q_n^{(c)}(t) - \sum_{j=1}^{N} \tilde{\mu}_{nj}^{(c)}(t) + \sum_{i=1}^{N} \tilde{\mu}_{in}^{(c)}(t) + \sum_{m\in\mathcal{M}_n^{(c)}} x_m(t)$$

where $\tilde{\mu}_{ij}^{(c)}(t)$ denotes the actual amount of commodity c data transmitted from node i to node j (i.e., over link (i, j)) on slot t. It is useful to define *transmission decision variables* $\mu_{ij}^{(c)}(t)$ as the bit rate *offered* by link (i, j) to commodity c data, where this full amount is used if there is that much commodity c data available at node i, so that:

$$\tilde{\mu}_{ij}^{(c)}(t) \leq \mu_{ij}^{(c)}(t) \;\; \forall i, j, c \in \{1, \ldots, N\}, \forall t$$

For simplicity, we assume that if there is not enough data to send at the offered rate, then *null data* is sent, so that:[4]

$$Q_n^{(c)}(t+1) \;\; = \;\; \max\left[Q_n^{(c)}(t) - \sum_{j=1}^{N} \mu_{nj}^{(c)}(t), 0\right] + \sum_{i=1}^{N} \mu_{in}^{(c)}(t) + \sum_{m\in\mathcal{M}_n^{(c)}} x_m(t) \qquad (5.56)$$

This satisfies (5.1) if we relate index k (for $Q_k(t)$ in (5.1)) to index (n, c) (for $Q_n^{(c)}(t)$ in (5.56)), and if we define:

$$b_n^{(c)}(t) \triangleq \sum_{j=1}^{N} \mu_{nj}^{(c)}(t) \quad , \quad a_n^{(c)}(t) \triangleq \sum_{i=1}^{N} \mu_{in}^{(c)}(t) + \sum_{m\in\mathcal{M}_n^{(c)}} x_m(t)$$

5.3.1 TRANSMISSION VARIABLES

Let $S(t)$ represent the *topology state* of the network on slot t, observed on each slot t as in (22). The value of $S(t)$ is an abstract and possibly multi-dimensional quantity that describes the current link conditions between all nodes under the current slot. The collection of all transmission rates that can be offered over each link (i, j) of the network is given by a general transmission rate function $\boldsymbol{b}(I(t), S(t))$:[5]

$$\boldsymbol{b}(I(t), S(t)) = (b_{ij}(I(t), S(t)))_{i,j\in\{1,\ldots,N\}, i\neq j}$$

where $I(t)$ is a general network-wide resource allocation decision (such as link scheduling, bandwidth selection, modulation, etc.) and takes values in some abstract set $\mathcal{I}_{S(t)}$ that possibly depends on the current $S(t)$.

[4] All results hold exactly as stated if this null data is not sent, so that "=" in (5.56) is modified to "$\leq$" (22).

[5] It is worth noting now that for networks with orthogonal channels, our "max-weight" transmission algorithm (to be defined in the next subsection) decouples to allow nodes to make transmission decisions based only on those components of the current topology state $S(t)$ that relate to their own local channels. Of course, for wireless interference networks, all channels are coupled, although distributed approximations of max-weight transmission exist in this case (see Chapter 6).

Every slot the network controller observes the current $S(t)$ and makes a resource allocation decision $I(t) \in \mathcal{I}_{S(t)}$. The controller then chooses $\mu_{ij}^{(c)}(t)$ variables subject to the following constraints:

$$\mu_{ij}^{(c)}(t) \geq 0 \quad \forall i, j, c \in \{1, \ldots, N\} \tag{5.57}$$

$$\mu_{ii}^{(c)}(t) = \mu_{ij}^{(i)}(t) = 0 \quad \forall i, j, c \in \{1, \ldots, N\} \tag{5.58}$$

$$\sum_{c=1}^{N} \mu_{ij}^{(c)} \leq b_{ij}(I(t), S(t)) \qquad \forall i, j \in \{1, \ldots, N\} \tag{5.59}$$

Constraints (5.58) are due to the common-sense observation that it makes no sense to transmit data from a node to itself, or to keep transmitting data that has already arrived to its destination. One can easily incorporate additional constraints that restrict the set of allowable links that certain commodities are allowed to use, as in (22).

5.3.2 THE UTILITY OPTIMIZATION PROBLEM

This problem fits our general framework by defining the random event $\omega(t) \triangleq [\boldsymbol{A}(t); S(t)]$. The control action $\alpha(t)$ is defined by:

$$\alpha(t) \triangleq [I(t); (\mu_{ij}^{(c)}(t))|_{i,j,c\in\{1,\ldots,N\}}; (x_m(t))|_{m=1}^{M}]$$

representing the resource allocation, transmission, and flow control decisions. The action space $\mathcal{A}_{\omega(t)}$ is defined by the set of all $I(t) \in \mathcal{I}_{S(t)}$, all $(\mu_{ij}^{(c)}(t))$ that satisfy (5.57)-(5.59), and all $(x_m(t))$ that satisfy $0 \leq x_m(t) \leq A_m(t)$ for all $m \in \{1, \ldots, M\}$.

Define $\overline{\boldsymbol{x}}$ as the time average expectation of the vector $\boldsymbol{x}(t)$. Our objective is to solve the following problem:

$$\text{Maximize:} \quad \phi(\overline{\boldsymbol{x}}) \tag{5.60}$$

$$\text{Subject to:} \quad \alpha(t) \in \mathcal{A}_{\omega(t)} \ \forall t \tag{5.61}$$

$$\text{All queues } Q_n^{(c)}(t) \text{ are mean rate stable} \tag{5.62}$$

where $\phi(\overline{\boldsymbol{x}}) = \sum_{m=1}^{M} \phi_m(x_m)$ is a continuous, concave, and entrywise non-decreasing utility function.

5.3.3 MULTI-HOP NETWORK UTILITY MAXIMIZATION

The rectangle $\mathcal{R}$ is defined by all $(\gamma_1, \ldots, \gamma_M)$ vectors such that $0 \leq \gamma_m \leq \gamma_{m,max}$. Define ϕ^{opt} as the maximum utility for the problem (5.60)-(5.62) augmented with the additional constraint $\overline{\boldsymbol{x}} \in \mathcal{R}$. Because we have not specified any additional constraints, there are no $Z_l(t)$ queues. However, we have auxiliary variables $\gamma_m(t)$ and virtual queues $G_m(t)$ for $m \in \{1, \ldots, M\}$, with update:

$$G_m(t+1) = \max[G_m(t) + \gamma_m(t) - x_m(t), 0] \tag{5.63}$$

The algorithm of Section 5.0.5 is thus:

- (Auxiliary Variables) For each slot t, each session $m \in \{1, \ldots, M\}$ observes the current virtual queue $G_m(t)$ and chooses auxiliary variable $\gamma_m(t)$ to solve:

$$\begin{aligned} \text{Maximize:} \quad & V\phi_m(\gamma_m(t)) - G_m(t)\gamma_m(t) \\ \text{Subject to:} \quad & 0 \leq \gamma_m(t) \leq \gamma_{m,max} \end{aligned} \tag{5.64}$$

- (Flow Control) For each slot t, each session m observes $A_m(t)$ and the queue values $G_m(t)$, $Q_{n_m}^{(c_m)}(t)$ (where n_m denotes the source node of session m, and c_m represents its destination). Note that these queues are all local to the source node of the session, and hence they can be observed easily. It then chooses $x_m(t)$ to solve:

$$\begin{aligned} \text{Maximize:} \quad & G_m(t)x_m(t) - Q_{n_m}^{(c_m)}(t)x_m(t) \\ \text{Subject to:} \quad & 0 \leq x_m(t) \leq A_m(t) \end{aligned} \tag{5.65}$$

  This reduces to the "bang-bang" flow control decision of choosing $x_m(t) = A_m(t)$ if $Q_{n_m}^{(c_m)}(t) \leq G_m(t)$, and $x_m(t) = 0$ otherwise.

- (Resource Allocation and Transmission) For each slot t, the network controller observes queue backlogs $\{Q_n^{(c)}(t)\}$ and the topology state $S(t)$ and chooses $I(t) \in \mathcal{I}_{S(t)}$ and $\{\mu_{ij}^{(c)}(t)\}$ to solve:

$$\begin{aligned} \text{Maximize:} \quad & \textstyle\sum_{n,c} Q_n^{(c)}(t)[\sum_{j=1}^N \mu_{nj}^{(c)}(t) - \sum_{i=1}^N \mu_{in}^{(c)}(t)] \\ \text{Subject to:} \quad & I(t) \in \mathcal{I}_{S(t)} \text{ and (5.57)-(5.59)} \end{aligned} \tag{5.66}$$

- (Queue Updates) Update the virtual queues $G_m(t)$ according to (5.63) and the actual queues $Q_n^{(c)}(t)$ according to (5.56).

The resource allocation and transmission decisions that solve (5.66) are described in Subsection 5.3.4 below. Before covering this, we state the performance of the algorithm under a general C-additive approximation. Assuming that second moments of arrivals and service variables are finite, and that $\omega(t)$ is i.i.d. over slots, by Theorem 5.1, we have that all virtual and actual queues are mean rate stable, and:

$$\liminf_{t\to\infty} \phi(\overline{x}(t)) \geq \phi^{opt} - (D + C)/V \tag{5.67}$$

where D is a constant related to the maximum second moments of arrivals and transmission rates. The queues $Q_n^{(c)}(t)$ can be shown to be strongly stable with average size $O(V)$ under an additional Slater-type condition. If the $\phi_m(x)$ functions are bounded with bounded right derivatives, it can be shown that the queues $G_m(t)$ are deterministically bounded. A slight modification of the algorithm that results in a C-additive approximation can deterministically bound all actual queues by a constant of size $O(V)$ (38)(42)(153), even without the Slater condition. The theory of Section 4.9 can be used to show that the same algorithm operates efficiently for non-i.i.d. traffic and channel processes, including processes that arise from arbitrary node mobility (38).

5.3.4 BACKPRESSURE-BASED ROUTING AND RESOURCE ALLOCATION

By switching the sums in (5.66), it is easy to show that the resource allocation and transmission maximization reduces to the following generalized "max-weight" and "backpressure" algorithms (see (7)(22)): Every slot t, choose $I(t) \in \mathcal{I}_{S(t)}$ to maximize:

$$\sum_{i=1}^{N}\sum_{j=1}^{N} b_{ij}(I(t), S(t))W_{ij}(t) \tag{5.68}$$

where $W_{ij}(t)$ are weights defined by:

$$W_{ij}(t) \triangleq \max_{c\in\{1,\ldots,N\}} \max[W_{ij}^{(c)}(t), 0] \tag{5.69}$$

where $W_{ij}^{(c)}(t)$ are differential backlogs:

$$W_{ij}^{(c)}(t) \triangleq Q_i^{(c)}(t) - Q_j^{(c)}(t)$$

The transmission decision variables are then given by:

$$\mu_{ij}^{(c)}(t) = \begin{cases} b_{ij}(I(t), S(t)) & \text{if } c = c_{ij}^*(t) \text{ and } W_{ij}^{(c)}(t) \geq 0 \\ 0 & \text{otherwise} \end{cases} \tag{5.70}$$

where $c_{ij}^*(t)$ is defined as the commodity $c \in \{1, \ldots, N\}$ that maximizes the differential backlog $W_{ij}^{(c)}(t)$ (breaking ties arbitrarily).

This backpressure approach achieves throughput optimality, but, because it explores all possible routes, may incur large delay. A useful C-additive approximation that experimentally *improves* delay is to combine the queue differential with a shortest path estimate for each link. This is proposed in (15) as an enhancement to backpressure routing, and it is shown to perform quite well in simulations given in (154)(22) ((154) extends to networks with unreliable channels). Related work that combines shortest paths and backpressure using the drift-plus-penalty method is developed in (155) to treat maximum hop count constraints. A theory of more aggressive place-holder packets for delay improvement in backpressure is developed in (37), although the algorithm ideally requires knowledge of Lagrange multiplier information in advance. A related and very simple Last-In-First-Out (LIFO) implementation of backpressure that does not need Lagrange multiplier information is developed in (54), where experiments on wireless sensor networks show delay improvements by more than an order of magnitude over FIFO implementations (for all but 2% of the packets) while preserving efficient throughput (note that LIFO does not change the dynamics of (5.1) or (5.56)). Analysis of the LIFO rule and its connection to place-holders and Lagrange multipliers is in (55).

5.4 GENERAL OPTIMIZATION OF CONVEX FUNCTIONS OF TIME AVERAGES

Here we provide a recipe for the following more general problem of optimizing convex functions of time averages:

$$\begin{aligned}
&\text{Minimize:} && \overline{y}_0 + f(\overline{\boldsymbol{x}}) && (5.71)\\
&\text{Subject to:} & 1)\ & \overline{y}_l + g_l(\overline{\boldsymbol{x}}) \leq 0 \ \forall l \in \{1, \ldots, L\} && (5.72)\\
& & 2)\ & \overline{\boldsymbol{x}} \in \mathcal{X} \cap \mathcal{R} && (5.73)\\
& & 3)\ & \text{All queues } Q_k(t) \text{ are mean rate stable} && (5.74)\\
& & 4)\ & \alpha(t) \in \mathcal{A}_{\omega(t)} \ \forall t && (5.75)
\end{aligned}$$

where $f(\boldsymbol{x})$ and $g_l(\boldsymbol{x})$ are continuous and convex functions of $\boldsymbol{x} \in \mathbb{R}^M$, $\mathcal{X}$ is a closed and convex subset of $\mathbb{R}^M$, and $\mathcal{R}$ is an M-dimensional hyper-rectangle defined as:

$$\mathcal{R} = \{(x_1, \ldots, x_M) \in \mathbb{R}^M | \gamma_{m,min} \leq x_m \leq \gamma_{m,max} \ \ \forall m \in \{1, \ldots, M\}\}$$

where $\gamma_{m,min}$ and $\gamma_{m,max}$ are finite constants (this rectangle set $\mathcal{R}$ is only added to bound the auxiliary variables that we use, as in the previous sections).

Let $\boldsymbol{\gamma}(t) = (\gamma_1(t), \ldots, \gamma_M(t))$ be a vector of *auxiliary variables* that can be chosen within the set $\mathcal{X} \cap \mathcal{R}$ every slot t. We transform the problem (5.71)-(5.75) to:

$$\begin{aligned}
&\text{Minimize:} && \overline{y}_0 + \overline{f(\boldsymbol{\gamma})} && (5.76)\\
&\text{Subject to:} & 1)\ & \overline{y}_l + \overline{g_l(\boldsymbol{\gamma})} \leq 0 \ \forall l \in \{1, \ldots, L\} && (5.77)\\
& & 2)\ & \overline{\gamma}_m = \overline{x}_m \ \forall m \in \{1, \ldots, M\} && (5.78)\\
& & 3)\ & \text{All queues } Q_k(t) \text{ are mean rate stable} && (5.79)\\
& & 4)\ & \boldsymbol{\gamma}(t) \in \mathcal{X} \cap \mathcal{R} \ \forall t && (5.80)\\
& & 5)\ & \alpha(t) \in \mathcal{A}_{\omega(t)} \ \forall t && (5.81)
\end{aligned}$$

where we define:

$$\overline{f(\boldsymbol{\gamma})} \triangleq \lim_{t\to\infty} \frac{1}{t} \sum_{\tau=0}^{t-1} \mathbb{E}\{f(\boldsymbol{\gamma}(\tau))\} \quad , \quad \overline{g_l(\boldsymbol{\gamma})} \triangleq \lim_{t\to\infty} \frac{1}{t} \sum_{\tau=0}^{t-1} \mathbb{E}\{g_l(\boldsymbol{\gamma}(\tau))\}$$

It is not difficult to show that this transformed problem is *equivalent* to the problem (5.71)-(5.75), in that the maximum utility values are the same, and any solution to one can be used to construct a solution to the other (see Exercise 5.9).

We solve the transformed problem (5.76)-(5.81) simply by re-stating the drift-plus-penalty algorithm for this context. While a variable-V implementation can be developed, we focus here on the fixed V algorithm as specified in (4.48)-(4.49). For each inequality constraint (5.77), define a virtual queue $Z_l(t)$ with update equation:

$$Z_l(t+1) = \max[Z_l(t) + \hat{y}_l(\alpha(t), \omega(t)) + g_l(\boldsymbol{\gamma}(t)), 0] \ \ \forall l \in \{1, \ldots, L\} \qquad (5.82)$$

For each equality constraint (5.78), define a virtual queue $H_m(t)$ with update equation:

$$H_m(t+1) = H_m(t) + \gamma_m(t) - \hat{x}_m(\alpha(t), \omega(t)) \;\; \forall m \in \{1, \ldots, M\} \tag{5.83}$$

Define $\boldsymbol{\Theta}(t) = [\boldsymbol{Q}(t), \boldsymbol{Z}(t), \boldsymbol{H}(t)]$. Assume the boundedness assumptions (4.25)-(4.30) hold, and that $\omega(t)$ is i.i.d. over slots. For the Lyapunov function (4.43), we have the following drift bound:

$$\begin{aligned}\Delta(\boldsymbol{\Theta}(t)) + V\mathbb{E}\{y_0(t) + f(\boldsymbol{\gamma}(t))|\boldsymbol{\Theta}(t)\} \leq\; & D + V\mathbb{E}\{y_0(t) + f(\boldsymbol{\gamma}(t))|\boldsymbol{\Theta}(t)\} \\ & + \sum_{l=1}^{L} Z_l(t)\mathbb{E}\{y_l(t) + g_l(\boldsymbol{\gamma}(t))|\boldsymbol{\Theta}(t)\} \\ & + \sum_{k=1}^{K} Q_k(t)\mathbb{E}\{a_k(t) - b_k(t)|\boldsymbol{\Theta}(t)\} \\ & + \sum_{m=1}^{M} H_m(t)\mathbb{E}\{\gamma_m(t) - x_m(t)|\boldsymbol{\Theta}(t)\} \end{aligned} \tag{5.84}$$

where D is a finite constant related to the worst case second moments of the arrival, service, and attribute vectors. Now define a C-additive approximation as any algorithm for choosing $\boldsymbol{\gamma}(t) \in \mathcal{X} \cap \mathcal{R}$ and $\alpha(t) \in \mathcal{A}_{\omega(t)}$ every slot t that, subject to a given $\boldsymbol{\Theta}(t)$, yields a right-hand-side in (5.84) that is within a distance C from its infimum value.

Theorem 5.3 *(Algorithm Performance) Suppose the boundedness assumptions (4.25)-(4.30) hold, the problem (5.71)-(5.75) is feasible, and $\mathbb{E}\{L(\boldsymbol{\Theta}(0))\} < \infty$. Suppose the functions $f(\boldsymbol{\gamma})$ and $g_l(\boldsymbol{\gamma})$ are upper and lower bounded by finite constants over $\boldsymbol{\gamma} \in \mathcal{X} \cap \mathcal{R}$. If $\omega(t)$ is i.i.d. over slots and any C-additive approximation is used every slot, then:*

$$\limsup_{t\to\infty} \left[\overline{y}_0(t) + f(\overline{\boldsymbol{x}}(t))\right] \leq y_0^{opt} + f^{opt} + \frac{D+C}{V} \tag{5.85}$$

where $y_0^{opt} + f^{opt}$ represents the infimum cost metric of the problem (5.71)-(5.75) over all feasible policies. Further, all actual and virtual queues are mean rate stable, and:

$$\limsup_{t\to\infty} \left[\overline{y}_l(t) + g_l(\overline{\boldsymbol{x}}(t))\right] \leq 0 \quad \forall l \in \{1, \ldots, L\} \tag{5.86}$$

$$\lim_{t\to\infty} \text{dist}\left(\overline{\boldsymbol{x}}(t), \mathcal{X} \cap \mathcal{R}\right) = 0 \tag{5.87}$$

where $\text{dist}(\overline{\boldsymbol{x}}(t), \mathcal{X} \cap \mathcal{R})$ *represents the distance between the vector $\overline{\boldsymbol{x}}(t)$ and the set $\mathcal{X} \cap \mathcal{R}$, being zero if and only if $\overline{\boldsymbol{x}}(t)$ is in the (closed) set $\mathcal{X} \cap \mathcal{R}$.*

Proof. See Exercise 5.10. □

As before, an $O(V)$ backlog bound can also be derived under a Slater assumption.

5.5 NON-CONVEX STOCHASTIC OPTIMIZATION

Consider now the problem:

$$\text{Minimize:} \quad f(\overline{\boldsymbol{x}}) \tag{5.88}$$
$$\text{Subject to:} \quad \overline{y}_l \leq 0 \;\; \forall l \in \{1, \ldots, L\} \tag{5.89}$$
$$\alpha(t) \in \mathcal{A}_{\omega(t)} \tag{5.90}$$
$$\text{All queues } Q_k(t) \text{ are mean rate stable} \tag{5.91}$$

where $f(\boldsymbol{x})$ is a possibly *non-convex* function that is assumed to be continuously differentiable with upper and lower bounds f_{min} and f_{max}, and with partial derivatives $\partial f(\boldsymbol{x})/\partial x_m$ having bounded magnitudes $\nu_m \geq 0$. Applications of such problems include throughput-utility maximization with $f(\boldsymbol{x})$ given by -1 times a sum of non-concave "sigmoidal" functions that give low utility until throughput exceeds a certain threshold (see Fig. 5.1). Such problems are treated in a non-stochastic (static) network optimization setting in (156)(157). A related utility-proportional fairness objective is studied for static networks in (158), which treats a convex optimization problem that has a fairness interpretation with respect to a non-concave utility function. The stochastic problem we present here is developed in (43). An application to risk management in network economics is given in Exercise 5.11.

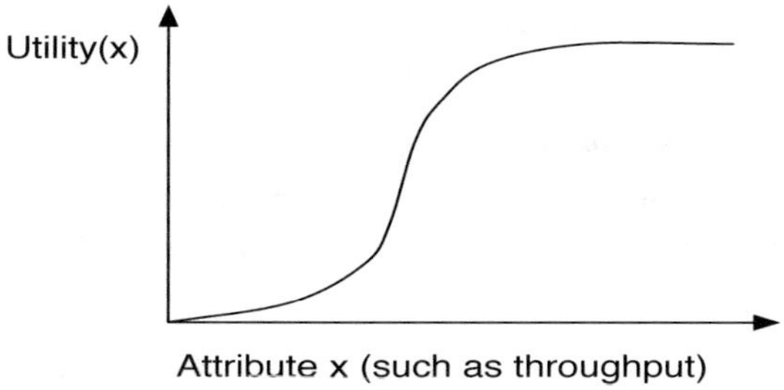

Figure 5.1: An example non-concave utility function of a time average attribute.

Performing such a general non-convex optimization is, in some cases, as hard as combinatorial bin-packing, and so we do not expect to find a global optimum. Rather, we seek an algorithm that satisfies the constraints (5.89)-(5.91) and that yields a *local optimum* of $f(\overline{\boldsymbol{x}})$.

We use the drift-plus-penalty framework with the same virtual queues as before:

$$Z_l(t+1) = \max[Z_l(t) + \hat{y}_l(\alpha(t), \omega(t)), 0] \tag{5.92}$$

The actual queues $Q_k(t)$ are assumed to satisfy (5.1). Define $\boldsymbol{\Theta}(t) \triangleq [\boldsymbol{Q}(t), \boldsymbol{Z}(t), \boldsymbol{x}_{av}(t)]$, where $\boldsymbol{x}_{av}(t)$ is defined as an empirical running time average of the attribute vector:

$$\boldsymbol{x}_{av}(t) \triangleq \begin{cases} \frac{1}{t}\sum_{\tau=0}^{t-1} x_m(\tau) & \text{if } t > 0 \\ \hat{x}_m(\alpha(-1), \omega(-1)) & \text{if } t = 0 \end{cases}$$

where $\hat{x}_m(\alpha(-1), \omega(-1))$ can be viewed as an initial sample taken at time "$t = -1$" before the network implementation begins. Define $L(\boldsymbol{\Theta}(t)) \triangleq \frac{1}{2}[\sum_{k=1}^{K} Q_k(t)^2 + \sum_{l=1}^{L} Z_l(t)^2]$. Assume $\omega(t)$ is i.i.d. over slots. We thus have:

$$\begin{aligned}\Delta(\boldsymbol{\Theta}(t)) + V\mathbb{E}\{Penalty(t)|\boldsymbol{\Theta}(t)\} \leq D + V\mathbb{E}\{Penalty(t)|\boldsymbol{\Theta}(t)\} \\ + \sum_{k=1}^{K} Q_k(t)\mathbb{E}\left\{\hat{a}_k(\alpha(t), \omega(t)) - \hat{b}_k(\alpha(t), \omega(t))|\boldsymbol{\Theta}(t)\right\} \\ + \sum_{l=1}^{L} Z_l(t)\mathbb{E}\left\{\hat{y}_l(\alpha(t), \omega(t))|\boldsymbol{\Theta}(t)\right\}\end{aligned} \tag{5.93}$$

The penalty we use is:

$$Penalty(t) \triangleq \sum_{m=1}^{M} \hat{x}_m(\alpha(t), \omega(t)) \frac{\partial f(\boldsymbol{x}_{av}(t))}{\partial x_m}$$

Below we state the performance of the algorithm that observes queue backlogs every slot t and takes an action $\alpha(t) \in \mathcal{A}_{\omega(t)}$ that comes within C of minimizing the right-hand-side of the drift expression (5.93).

Theorem 5.4 *(Non-Convex Stochastic Network Optimization (43)) Suppose $\omega(t)$ is i.i.d. over slots, the boundedness assumptions (4.25)-(4.28) hold, the function $f(\boldsymbol{x})$ is bounded and continuously differentiable with partial derivatives bounded in magnitude by finite constants $\nu_m \geq 0$, and the problem (5.88)-(5.91) is feasible. For simplicity, assume that $\boldsymbol{\Theta}(0) = \boldsymbol{0}$. For any $V \geq 0$, and for any C-additive approximation of the above algorithm that is implemented every slot, we have:*

(a) All queues $Q_k(t)$ and $Z_l(t)$ are mean rate stable and:

$$\limsup_{t\to\infty} \overline{y}_l(t) \leq 0 \ \forall l \in \{1, \ldots, L\}$$

(b) For all $t > 0$ and for any alternative vector $\boldsymbol{x}^*$ that can be achieved as the time average of a policy that makes all queues mean rate stable and satisfies all required constraints, we have:

$$\frac{1}{t}\sum_{\tau=0}^{t-1}\sum_{m=1}^{M}\mathbb{E}\left\{\frac{x_m(\tau)\partial f(\boldsymbol{x}_{av}(\tau))}{\partial x_m}\right\} \leq \frac{1}{t}\sum_{\tau=0}^{t-1}\sum_{m=1}^{M} x_m^*\mathbb{E}\left\{\frac{\partial f(\boldsymbol{x}_{av}(\tau))}{\partial x_m}\right\} + \frac{D+C}{V}$$

where D is a finite constant related to second moments of the $a_k(t)$, $b_k(t)$, $y_l(t)$ processes.

c) If all time averages converge, so that there is a constant vector $\overline{\boldsymbol{x}}$ such that $\boldsymbol{x}_{av}(t) \to \overline{\boldsymbol{x}}$ with probability 1 and $\overline{\boldsymbol{x}}(t) \to \overline{\boldsymbol{x}}$, then the achieved limit is a near local optimum/critical point, *in the sense that for any alternative vector $\boldsymbol{x}^*$ that can be achieved as the time average of a policy that makes all queues mean rate stable and satisfies all required constraints, we have:*

$$\sum_{m=1}^{M}(x_m^* - \overline{x}_m)\frac{\partial f(\overline{\boldsymbol{x}})}{\partial x_m} \geq -\frac{D+C}{V}$$

d) Suppose there is an $\epsilon > 0$ and an ω-only policy $\alpha^*(t)$ such that:

$$\mathbb{E}\left\{\hat{y}_l(\alpha^*(t), \omega(t))\right\} \leq 0 \quad \forall l \in \{1, \ldots, L\} \tag{5.94}$$

$$\mathbb{E}\left\{\hat{a}_k(\alpha^*(t), \omega(t)) - \hat{b}_k(\alpha^*(t), \omega(t))\right\} \leq -\epsilon \quad \forall k \in \{1, \ldots, K\} \tag{5.95}$$

Then all queues $Q_k(t)$ are strongly stable with average size $O(V)$.

e) Suppose we use a variable $V(t)$ algorithm with $V(t) \triangleq V_0 \cdot (1+t)^d$ for $V_0 > 0$ and $0 < d < 1$, and use any C-additive approximation (where C is constant for all t). Then all virtual and actual queues are mean rate stable (and so all constraints $\overline{y}_l \leq 0$ are satisfied), and under the convergence assumptions of part (c), the limiting $\overline{\boldsymbol{x}}$ is a local optimum/critical point, *in that:*

$$\sum_{m=1}^{M}(x_m^* - \overline{x}_m)\frac{\partial f(\overline{\boldsymbol{x}})}{\partial x_m} \geq 0$$

where $\boldsymbol{x}^*$ is any alternative vector as specified in part (c).

The inequality guarantee in part (e) can be understood as follows: Suppose we start at our achieved time average attribute vector $\overline{\boldsymbol{x}}$, and we want to shift this in any feasible direction by moving towards another feasible vector $\boldsymbol{x}^*$ by an amount ϵ (for some $\epsilon > 0$). Then:

$$f\left(\overline{\boldsymbol{x}} + \epsilon(\boldsymbol{x}^* - \overline{\boldsymbol{x}})\right) \approx f(\overline{\boldsymbol{x}}) + \sum_{m=1}^{M}\epsilon(x_m^* - \overline{x}_m)\frac{\partial f(\overline{\boldsymbol{x}})}{\partial x_m} \geq f(\overline{\boldsymbol{x}})$$

Hence, the new cost achieved by taking a small step in any feasible direction is no less than the cost $f(\overline{\boldsymbol{x}})$ that we are already achieving. More precisely, the change in cost $\Delta_{cost}(\epsilon)$ satisfies:

$$\lim_{\epsilon \to 0}\frac{\Delta_{cost}(\epsilon)}{\epsilon} \geq 0$$

Proof. (Theorem 5.4) Our proof uses the same drift-plus-penalty technique as described in previous sections. Analogous to Theorem 4.5, it can be shown that for any $\boldsymbol{x}^* = (x_1^*, \ldots, x_M^*)$ that is a limit point of $\overline{\boldsymbol{x}}(t)$ under any policy that makes all queues mean rate stable and satisfies all constraints, and for any $\delta > 0$, there exists an ω-only policy $\alpha^*(t)$ such that (43):

$$\mathbb{E}\left\{\hat{y}_l(\alpha^*(t), \omega(t))\right\} \leq \delta \ \forall l \in \{1, \ldots, L\}$$

$$\mathbb{E}\left\{\hat{a}_k(\alpha^*(t), \omega(t)) - \hat{b}_k(\alpha^*(t), \omega(t))\right\} \leq \delta \ \forall k \in \{1, \ldots, K\}$$

$$\text{dist}(\mathbb{E}\left\{\hat{\boldsymbol{x}}(\alpha^*(t), \omega(t))\right\}, \boldsymbol{x}^*) \leq \delta$$

For simplicity of the proof, assume the above holds with $\delta = 0$. Plugging the above into the right-hand-side of (5.93) with $\delta = 0$ yields:[6]

$$\Delta(\boldsymbol{\Theta}(t)) + V\mathbb{E}\left\{\sum_{m=1}^{M}\hat{x}_m(\alpha(t), \omega(t))\frac{\partial f(\boldsymbol{x}_{av}(t))}{\partial x_m}|\boldsymbol{\Theta}(t)\right\} \leq D + C + V\sum_{m=1}^{M}x_m^*\frac{\partial f(\boldsymbol{x}_{av}(t))}{\partial x_m}$$

[6]The same result can be derived by plugging in with $\delta > 0$ and then taking a limit as $\delta \to 0$.

Taking expectations of the above drift bound (using the law of iterated expectations), summing the telescoping series over $\tau \in \{0, 1, \ldots, t-1\}$, and dividing by Vt immediately yields the result of part (b).

On the other hand, this drift expression can also be rearranged as:

$$\Delta(\boldsymbol{\Theta}(t)) \leq D + C + V \sum_{m=1}^{M} \nu_m (x_m^* - x_{m,min})$$

where $x_{m,min}$ is a bound on the expectation of $x_m(t)$ under any policy, known to exist by the boundedness assumptions. Hence, the drift is less than or equal to a finite constant, and so by Theorem 4.2, we know all queues are mean rate stable, proving part (a). The proof of part (d) follows similarly by plugging in the policy $\alpha^*(t)$ of (5.94)-(5.95).

The proof of part (c) follows by taking a limit of the result in part (b), where the limits can be pushed through by the boundedness assumptions and the continuity assumption on the derivatives of $f(\boldsymbol{x})$. The proof of part (e) is similar to that of Theorem 4.9 and is omitted for brevity. □

Using a penalty given by partial derivatives of the function evaluated at the empirical average attribute vector can be viewed as a "primal-dual" operation that differs from our "pure-dual" approach for convex problems. Such a primal-dual approach was first used in context of *convex* network utility maximization problems in (32)(33)(34). Specifically, the work (32)(33) used a partial derivative evaluated at the time average $\boldsymbol{x}_{av}(t)$ to maximize a concave function of throughput in a multi-user wireless downlink with time varying channels. However, the system in (32)(33) assumed *infinite backlog* in all queues (similar to Exercise 5.6), so that there were no queue stability constraints. This was extended in (34) to consider the primal-dual technique for joint stability and performance optimization, again for convex problems, but using an exponential weighted average, rather than a running time average $\boldsymbol{x}_{av}(t)$. There, it was shown that a related "fluid limit" of the system has an optimal utility, and that this limit is "weakly" approached under appropriately scaled systems. It was also conjectured in (34) that the actual network will have utility that is close to this fluid limit as a parameter β related to the exponential weighting is scaled (see Section 4.9 in (34)). However, the analysis does not specify the size of β needed to achieve a near-optimal utility. Recent work in (36) considers related primal-dual updates for convex problems, and it shows the long term utility of the actual network is close to optimal as a parameter is scaled.

For the special case of convex problems, Theorem 5.4 above shows that, if the algorithm is assumed to converge to well defined time averages, and if we use a running time average $\boldsymbol{x}_{av}(t)$ rather than an exponential average, the primal-dual algorithm achieves a similar $[O(1/V), O(V)]$ performance-congestion tradeoff as the dual algorithm. Unfortunately, it is not clear how long the system must run to approach convergence. The pure dual algorithm seems to provide stronger analytical guarantees for convex problems because: (i) It does not need a running time average $\boldsymbol{x}_{av}(t)$ and hence can be shown to be robust to changes in system parameters (as in Section 4.9 and (42)(38)(17)), (ii) It does not require additional assumptions about convergence, (iii) It provides results for all $t > 0$ that show how long we must run the system to be close to the infinite horizon limit guarantees. However, if one applies the pure dual technique with a non-convex cost function $f(\boldsymbol{x})$, one would get a global optimum of the time average $\overline{f(\boldsymbol{x})}$, which may not even be a local optimum of $f(\overline{\boldsymbol{x}})$. This is where the primal-dual technique shows its real potential, as it can achieve a local optimum for non-convex problems.

5.6 WORST CASE DELAY

Here we extend the utility optimization framework to enable $O(V)$ tradeoffs in *worst case delay*. Related problems are treated in (76)(159). Consider a 1-hop network with K queues $\boldsymbol{Q}(t) = (Q_1(t), \ldots, Q_K(t))$. In addition to these queues, we keep *transport layer queues* $\boldsymbol{L}(t) = (L_1(t), \ldots, L_K(t))$, where $L_k(t)$ stores incoming data before it is admitted to the network layer queue $Q_k(t)$ (as in (17)). Let $\omega(t) = [\boldsymbol{A}(t), \boldsymbol{S}(t)]$, where $\boldsymbol{A}(t) = (A_1(t), \ldots, A_K(t))$ is a vector of new arrivals to the transport layer, and $\boldsymbol{S}(t) = (S_1(t), \ldots, S_K(t))$ is a vector of channel conditions that affect transmission. Assume that $\omega(t)$ is i.i.d. over slots.

Every slot t, choose *admission variables* $\boldsymbol{a}(t) = (a_1(t), \ldots, a_K(t))$ subject to the constraints:

$$0 \leq a_k(t) \leq \min[L_k(t) + A_k(t), A_{max}] \tag{5.96}$$

where A_{max} is a finite constant. This means that $a_k(t)$ is chosen from the $L_k(t) + A_k(t)$ amount of data available on slot t, and is no more than A_{max} per slot (which limits the amount we can send into the network layer). It is assumed that $A_k(t) \leq A_{max}$ for all k and all t. Newly arriving data $A_k(t)$ that is not immediately admitted into the network layer is stored in the transport layer queue $L_k(t)$. The controller also chooses a *channel-aware transmission decision* $I(t) \in \mathcal{I}_{\boldsymbol{S}(t)}$, where $\mathcal{I}_{\boldsymbol{S}(t)}$ is an abstract set that defines transmission options under channel state $\boldsymbol{S}(t)$. The transmission rates are given by deterministic functions of $I(t)$ and $\boldsymbol{S}(t)$:

$$b_k(t) = \hat{b}_k(I(t), \boldsymbol{S}(t))$$

Second moments of $b_k(t)$ are assumed to be uniformly bounded.

In addition, define *packet drop decisions* $\boldsymbol{d}(t) = (d_1(t), \ldots, d_K(t))$. These allow packets already admitted to the network layer queues $Q_k(t)$ to be dropped if their delay is too large. Drop decisions $d_k(t)$ are chosen subject to the constraints:

$$0 \leq d_k(t) \leq A_{max}$$

The resulting queue update equation is thus:

$$Q_k(t+1) = \max[Q_k(t) - b_k(t) - d_k(t), 0] + a_k(t) \;\; \forall k \in \{1, \ldots, K\} \tag{5.97}$$

For each $k \in \{1, \ldots, K\}$, let $\phi_k(a)$ be a continuous, concave, and non-decreasing utility function defined over the interval $0 \leq a \leq A_{max}$. Let ν_k be the maximum right-derivative of $\phi_k(a)$ (which occurs at $a = 0$), and assume $\nu_k < \infty$. Example utility functions that have this form are:

$$\phi_k(a) = \log(1 + \nu_k a)$$

where $\log(\cdot)$ denotes the natural logarithm. We desire a solution to the following problem, defined in terms of a parameter $\epsilon > 0$:

$$\begin{aligned}
\text{Maximize:} \quad & \sum_{k=1}^{K} \phi_k(\overline{a}_k) - \sum_{k=1}^{K} \beta \nu_k \overline{d}_k && (5.98)\\
\text{Subject to:} \quad & \text{All queues } Q_k(t) \text{ are mean rate stable} && (5.99)\\
& \overline{b}_k \geq \epsilon \;\; \forall k \in \{1, \ldots, K\} && (5.100)\\
& 0 \leq a_k(t) \leq A_k(t) \;\; \forall k \in \{1, \ldots, K\}, \forall t && (5.101)\\
& I(t) \in \mathcal{I}_{S(t)} \;\; \forall k \in \{1, \ldots, K\}, \forall t && (5.102)
\end{aligned}$$

where β is a constant that satisfies $1 \leq \beta < \infty$. This problem does not specify anything about worst-case delay, but we soon develop an algorithm with worst case delay of $O(V)$ that comes within $O(1/V)$ of optimizing the utility associated with the above problem (5.98)-(5.102). Note the following:

- The constraint (5.101) is different from the constraint (5.96). Thus, the less stringent constraint (5.96) is used for the actual algorithm, but performance is measured with respect to the optimum utility achievable in the problem (5.98)-(5.102). It turns out that optimal utility is the same with either constraint (5.101) or (5.96), and in particular, it is the same if there are no transport layer queues, so that $L_k(t) = 0$ for all t and all data is either admitted or dropped upon arrival. We include the $L_k(t)$ queues as they are useful in situations where it is preferable to store data for later transmission than to drop it.

- An optimal solution to (5.98)-(5.102) has $\overline{d}_k = 0$ for all k. That is, the objective (5.98) can equivalently be replaced by the objective of maximizing $\sum_{k=1}^{K} \phi_k(\overline{a}_k)$ and by adding the constraint $\overline{d}_k = 0$ for all k. This is because the penalty for dropping is $\beta\nu_k$, which is greater than or equal to the largest derivative of the utility function $\phi_k(a)$. Thus, it can be shown that it is always better to restrict data at the transport layer rather than admitting it and later dropping it. We recommend choosing β such that $1 \leq \beta \leq 2$. A larger value of β will trade packet drops at the network layer for packet non-admissions at the flow controller.

- The constraint (5.100) requires each queue to transmit with a time-average rate of at least ϵ. This constraint ensures all queues are getting at least a minimum rate ϵ of service. If the input rate $\mathbb{E}\{A_k(t)\}$ is less than ϵ, then this constraint is wasteful. However, we shall not enforce this constraint. Rather, we simply measure utility of our system with respect to the optimal utility of the problem (5.98)-(5.102), which includes this constraint. It is assumed throughout that this constraint is feasible, and so the problem (5.98)-(5.102) is feasible. If one prefers to enforce constraint (5.100), this is easily done with an appropriate virtual queue.

5.6.1 THE ϵ-PERSISTENT SERVICE QUEUE

To ensure worst-case delay is bounded, we define an *ϵ-persistent service queue*, being a virtual queue $Z_k(t)$ for each $k \in \{1, \ldots, K\}$ with $Z_k(0) = 0$ and with dynamics:

$$Z_k(t+1) = \begin{cases} \max[Z_k(t) - b_k(t) - d_k(t) + \epsilon, 0] & \text{if } Q_k(t) > b_k(t) + d_k(t) \\ 0 & \text{if } Q_k(t) \leq b_k(t) + d_k(t) \end{cases} \quad (5.103)$$

where $\epsilon > 0$. We assume throughout that $\epsilon \leq A_{max}$. The condition $Q_k(t) \leq b_k(t) + d_k(t)$ is satisfied whenever the backlog $Q_k(t)$ is cleared (by service and/or drops) on slot t. If this constraint is not active, then $Z_k(t)$ has a departure process that is the same as $Q_k(t)$, but it has an arrival of size ϵ every slot. The size of the queue $Z_k(t)$ can provide a bound on the delay of the head-of-line data in queue $Q_k(t)$ in a first-in-first-out (FIFO) system. This is similar to (76) (where explicit delays are kept for each packet) and (159) (which uses a slightly different update). If a scheduling algorithm is used that ensures $Z_k(t) \leq Z_{k,max}$ and $Q_k(t) \leq Q_{k,max}$ for all t, for some finite constants $Z_{k,max}$ and $Q_{k,max}$, then worst-case delay is also bounded, as shown in the following lemma:

Lemma 5.5 *Suppose $Q_k(t)$ and $Z_k(t)$ evolve according to (5.97) and (5.103), and that an algorithm is used that ensures $Q_k(t) \leq Q_{k,max}$ and $Z_k(t) \leq Z_{k,max}$ for all slots $t \in \{0, 1, 2, \ldots\}$. Assume service and drops are done in FIFO order. Then the worst-case delay of all non-dropped data in queue k is $W_{k,max}$, defined:*

$$W_{k,max} \triangleq \lceil (Q_{k,max} + Z_{k,max})/\epsilon \rceil \quad (5.104)$$

Proof. Fix a slot t. We show that all arrivals $a(t)$ are either served or dropped on or before slot $t + W_{k,max}$. Suppose this is *not* true. We reach a contradiction. Note by (5.97) that arrivals $a(t)$ are added to the queue backlog $Q_k(t+1)$ and are first available for service on slot $t+1$. It must be that $Q_k(\tau) > b_k(\tau) + d_k(\tau)$ for all $\tau \in \{t+1, \ldots, t+W_{k,max}\}$ (else, the backlog on slot τ would be cleared). Therefore, by (5.103), we have for all slots $\tau \in \{t+1, \ldots, t+W_{k,max}\}$:

$$Z_k(\tau+1) = \max[Z_k(\tau) - b_k(\tau) - d_k(\tau) + \epsilon, 0]$$

In particular, for all slots $\tau \in \{t+1, \ldots, t+W_{k,max}\}$:

$$Z_k(\tau+1) \geq Z_k(\tau) - b_k(\tau) - d_k(\tau) + \epsilon$$

Summing the above over $\tau \in \{t+1, \ldots, t+W_{k,max}\}$ yields:

$$Z_k(t+W_{k,max}+1) - Z_k(t+1) \geq -\sum_{\tau=t+1}^{t+W_{k,max}} [b_k(\tau) + d_k(\tau)] + W_{k,max}\epsilon$$

Rearranging terms in the above inequality and using the fact that $Z_k(t+1) \geq 0$ and $Z_k(t+W_{k,max}+1) \leq Z_{k,max}$ yields:

$$W_{k,max}\epsilon \leq \sum_{\tau=t+1}^{t+W_{k,max}} [b_k(\tau) + d_k(\tau)] + Z_{k,max} \tag{5.105}$$

On the other hand, the sum of $b_k(\tau) + d_k(\tau)$ over the interval $\tau \in \{t+1, \ldots, t+W_{k,max}\}$ must be strictly less than $Q_k(t+1)$ (else, by the FIFO service, all data $a(t)$, which is included at the end of the backlog $Q_k(t+1)$, would have been cleared during this interval). Thus:

$$\sum_{\tau=t+1}^{t+W_{k,max}} [b_k(\tau) + d_k(\tau)] < Q_k(t+1) \leq Q_{k,max} \tag{5.106}$$

Combining (5.106) and (5.105) yields:

$$W_{k,max}\epsilon < Q_{k,max} + Z_{k,max}$$

which implies:

$$W_{k,max} < (Q_{k,max} + Z_{k,max})/\epsilon$$

This contradicts (5.104), proving the result. □

5.6.2 THE DRIFT-PLUS-PENALTY FOR WORST-CASE DELAY

As usual, we transform the problem (5.98)-(5.102) using auxiliary variables $\boldsymbol{\gamma}(t) = (\gamma_1(t), \ldots, \gamma_K(t))$ by:

$$\begin{aligned}
\text{Maximize:} \quad & \sum_{k=1}^{K} \overline{\phi_k(\gamma_k)} - \sum_{k=1}^{K} \beta \nu_k \overline{d}_k & (5.107)\\
\text{Subject to:} \quad & \overline{a}_k \geq \overline{\gamma}_k \ \ \forall k \in \{1, \ldots, K\} & (5.108)\\
& \text{All queues } Q_k(t) \text{ are mean rate stable} & (5.109)\\
& \overline{b}_k \geq \epsilon \ \ \forall k \in \{1, \ldots, K\} & (5.110)\\
& 0 \leq \gamma_k(t) \leq A_{max} \ \ \forall k \in \{1, \ldots, K\} & (5.111)\\
& 0 \leq a_k(t) \leq A_k(t) \ \ \forall k \in \{1, \ldots, K\} & (5.112)\\
& I(t) \in \mathcal{I}_{\boldsymbol{S}(t)} \ \ \forall k \in \{1, \ldots, K\} & (5.113)
\end{aligned}$$

To enforce the constraints (5.108), define virtual queues $G_k(t)$ by:

$$G_k(t+1) = \max[G_k(t) - a_k(t) + \gamma_k(t), 0] \tag{5.114}$$

Now define $\boldsymbol{\Theta}(t) \triangleq [\boldsymbol{Q}(t), \boldsymbol{Z}(t), \boldsymbol{G}(t)]$ as the combined queue vector, and define the Lyapunov function $L(\boldsymbol{\Theta}(t))$ by:

$$L(\boldsymbol{\Theta}(t)) \triangleq \frac{1}{2}\sum_{k=1}^{K}[Q_k(t)^2 + Z_k(t)^2 + G_k(t)^2]$$

Using the fact that $Z_k(t+1) \leq \max[Z_k(t) - b_k(t) - d_k(t) + \epsilon, 0]$, it can be shown (as usual) that the Lyapunov drift satisfies:

$$\begin{aligned}
\Delta(\boldsymbol{\Theta}(t)) - V\mathbb{E}\left\{\sum_{k=1}^{K}[\phi_k(\gamma_k(t)) - \beta\nu_k d_k(t)]|\boldsymbol{\Theta}(t)\right\} &\leq B \\
&-V\mathbb{E}\left\{\sum_{k=1}^{K}[\phi_k(\gamma_k(t)) - \beta\nu_k d_k(t)]|\boldsymbol{\Theta}(t)\right\} \\
&+\sum_{k=1}^{K} Z_k(t)\mathbb{E}\left\{\epsilon - \hat{b}_k(I(t), \boldsymbol{S}(t)) - d_k(t)|\boldsymbol{\Theta}(t)\right\} \\
&+\sum_{k=1}^{K} Q_k(t)\mathbb{E}\left\{a_k(t) - \hat{b}_k(I(t), \boldsymbol{S}(t)) - d_k(t)|\boldsymbol{\Theta}(t)\right\} \\
&+\sum_{k=1}^{K} G_k(t)\mathbb{E}\left\{\gamma_k(t) - a_k(t)|\boldsymbol{\Theta}(t)\right\}
\end{aligned} \tag{5.115}$$

where B is a constant that satisfies:

$$\begin{aligned}
B &\geq \frac{1}{2}\sum_{k=1}^{K}[\mathbb{E}\left\{(\epsilon - b_k(t) - d_k(t))^2|\boldsymbol{\Theta}(t)\right\} \\
&+\frac{1}{2}\sum_{k=1}^{K}\mathbb{E}\left\{a_k(t)^2 + (b_k(t) - d_k(t))^2 + (\gamma_k(t) - a_k(t))^2|\boldsymbol{\Theta}(t)\right\}
\end{aligned} \tag{5.116}$$

Such a constant B exists by the boundedness assumptions on the processes.

The algorithm that minimizes the right-hand-side of (5.115) thus observes $\boldsymbol{Z}(t), \boldsymbol{Q}(t), \boldsymbol{G}(t), \boldsymbol{S}(t)$ every slot t, and does the following:

- (Auxiliary Variables) For each $k \in \{1, \ldots, K\}$, choose $\gamma_k(t)$ to solve:

$$\text{Maximize:} \quad V\phi_k(\gamma_k(t)) - G_k(t)\gamma_k(t) \tag{5.117}$$
$$\text{Subject to:} \quad 0 \leq \gamma_k(t) \leq A_{max} \tag{5.118}$$

- (Flow Control) For each $k \in \{1, \ldots, K\}$, choose $a_k(t)$ by:

$$a_k(t) = \begin{cases} \min[L_k(t) + A_k(t), A_{max}] & \text{if } Q_k(t) \leq G_k(t) \\ 0 & \text{if } Q_k(t) > G_k(t) \end{cases} \tag{5.119}$$

- (Transmission) Choose $I(t) \in \mathcal{I}_{\boldsymbol{S}(t)}$ to maximize:

$$\sum_{k=1}^{K}[Q_k(t) + Z_k(t)]\hat{b}_k(I(t), \boldsymbol{S}(t)) \tag{5.120}$$

- (Packet Drops) For each $k \in \{1, \ldots, K\}$, choose $d_k(t)$ by:

$$d_k(t) = \begin{cases} A_{max} & \text{if } Q_k(t) + Z_k(t) > \beta V \nu_k \\ 0 & \text{if } Q_k(t) + Z_k(t) \leq \beta V \nu_k \end{cases} \tag{5.121}$$

- (Queue Update) Update $Q_k(t)$, $Z_k(t)$, $G_k(t)$ by (5.97), (5.103), (5.114).

In some cases, the above algorithm may choose a drop variable $d_k(t)$ such that $Q_k(t) < b_k(t) + d_k(t)$. In this case, all queue updates are kept the same (so the algorithm is unchanged), but it is useful to *first* transmit data with offered rate $b_k(t)$ on slot t, and then drop only what remains.

5.6.3 ALGORITHM PERFORMANCE

Define $Z_{k,max}$ and $Q_{k,max}$ as follows:

$$Z_{k,max} \triangleq \beta V \nu_k + \epsilon \tag{5.122}$$
$$Q_{k,max} \triangleq \min[\beta V \nu_k + A_{max}, V \nu_k + 2A_{max}] \tag{5.123}$$
$$G_{k,max} \triangleq V \nu_k + A_{max} \tag{5.124}$$

Theorem 5.6 *If $\epsilon \leq A_{max}$, then for arbitrary sample paths the above algorithm ensures:*

$$Z_k(t) \leq Z_{k,max} \ , \ Q_k(t) \leq Q_{k,max} \ , \ G_k(t) \leq G_{k,max} \forall t$$

where $Z_{k,max}$, $Q_{k,max}$, $G_{k,max}$ are defined in (5.122)-(5.124), provided that these inequalities hold for $t = 0$. Thus, worst-case delay $W_{k,max}$ is given by:

$$W_{k,max} \triangleq \lceil (Z_{k,max} + Q_{k,max})/\epsilon \rceil = O(V)$$

Proof. That $G_k(t) \leq G_{k,max}$ for all t follows by an argument similar to that given in Section 5.2.1, showing that the auxiliary variable update (5.117)-(5.118) chooses $\gamma_k(t) = 0$ whenever $G_k(t) > V \nu_k$.

To show the $Q_{k,max}$ bound, it is clear that the packet drop decision (5.121) yields $d_k(t) = A_{max}$ whenever $Q_k(t) > \beta V \nu_k$. Because $a_k(t) \leq A_{max}$, the arrivals are less than or equal to the offered drops whenever $Q_k(t) > \beta V \nu_k$, and so $Q_k(t) \leq \beta V \nu_k + A_{max}$ for all t. However, we also see that if $Q_k(t) > G_{k,max}$, then the flow control decision will choose $a_k(t) = 0$, and so $Q_k(t)$ also cannot increase. It follows that $Q_k(t) \leq G_{k,max} + A_{max}$ for all t. This proves the $Q_{k,max}$ bound. The $Z_{k,max}$ bound is proven similarly. The worst-case-delay result then follows immediately from Lemma 5.5.
□

The above theorem only uses the fact that packet drops $d_k(t)$ take place according to the rule (5.121), flow control decisions $a_k(t)$ take place according to the rule (5.119), and auxiliary variable decisions satisfy $\gamma_k(t) = 0$ whenever $G_k(t) > V\nu_k$ (a property of the solution to (5.117)-(5.118)). The fact that $\gamma_k(t) = 0$ whenever $G_k(t) > V\nu_k$ can be hard-wired into the auxiliary variable decisions, even when they are chosen to approximately solve (5.117)-(5.118) otherwise. Further, the $I(t)$ decisions can be arbitrary and are not necessarily those that maximize (5.120). The next theorem holds for any C-additive approximation for minimizing the right-hand-side of (5.115) that preserves the above basic properties. A 0-additive approximation performs the exact algorithm given above.

Theorem 5.7 *Suppose $\omega(t)$ is i.i.d. over slots and any C-additive approximation for minimizing the right-hand-side of (5.115) is used such that (5.121), (5.119) hold exactly, and $\gamma_k(t) = 0$ whenever $G_k(t) > V\nu_k$. Suppose $Q_k(0) \leq Q_{k,max}$, $Z_k(0) \leq Z_{k,max}$, $G_k(0) \leq G_{k,max}$ for all k, and $\epsilon \leq A_{max}$. Then the worst-case queue backlog and delay bounds given in Theorem 5.6 hold, and achieved utility satisfies:*

$$\liminf_{t\to\infty} \left[\sum_{k=1}^{K} \phi_k(\overline{a}_k(t)) - \sum_{k=1}^{K} \beta \overline{d}_k(t)\right] \geq \phi^* - B/V$$

where B is defined in (5.116), $\overline{a}_k(t)$ and $\overline{d}_k(t)$ are defined:

$$\overline{a}_k(t) \triangleq \frac{1}{t}\sum_{\tau=0}^{t-1} \mathbb{E}\left\{a_k(\tau)\right\} \quad , \quad \overline{d}_k(t) \triangleq \frac{1}{t}\sum_{\tau=0}^{t-1} \mathbb{E}\left\{d_k(\tau)\right\}$$

and where ϕ^* is the optimal utility associated with the problem (5.98)-(5.102).

The theorem relies on the following fact, which can be proven using Theorem 4.5: For all $\delta > 0$, there exists a vector $\boldsymbol{\gamma}^* = (\gamma_1^*, \ldots, \gamma_K^*)$ and an ω-only policy $[\boldsymbol{a}^*(t), I^*(t), \boldsymbol{d}^*(t)]$ that chooses $\boldsymbol{a}^*(t)$ as a random function of $\boldsymbol{A}(t)$, $I^*(t)$ as a random function of $\boldsymbol{S}(t)$, and $\boldsymbol{d}^*(t) = \boldsymbol{0}$ (so that it does not drop any data) such that:

$$\sum_{k=1}^{K} \phi_k(\gamma_k^*) = \phi^* \tag{5.125}$$

$$\mathbb{E}\left\{a_k^*(t)\right\} = \gamma_k^* \quad \forall k \in \{1, \ldots, K\} \tag{5.126}$$

$$\mathbb{E}\left\{\hat{b}_k(I^*(t), \boldsymbol{S}(t))\right\} \geq \epsilon - \delta \quad \forall k \in \{1, \ldots, K\} \tag{5.127}$$

$$\mathbb{E}\left\{\hat{b}_k(I^*(t), \boldsymbol{S}(t))\right\} \geq \mathbb{E}\left\{a_k^*(t)\right\} - \delta \quad \forall k \in \{1, \ldots, K\} \tag{5.128}$$

$$I^*(t) \in \mathcal{I}_{\boldsymbol{S}(t)} \ , \ 0 \leq \gamma_k^* \leq A_{max} \ , \ 0 \leq a_k^*(t) \leq A_k(t) \quad \forall k \in \{1, \ldots, K\}, \forall t \tag{5.129}$$

where ϕ^* is the optimal utility associated with the problem (5.98)-(5.102).

Proof. (Theorem 5.7) The C-additive approximation ensures by (5.115):

$$\begin{aligned}\Delta(\boldsymbol{\Theta}(t)) - V\mathbb{E}\left\{\sum_{k=1}^{K}[\phi_k(\gamma_k(t)) - \beta\nu_k d_k(t)]|\boldsymbol{\Theta}(t)\right\} &\leq B + C \\ &\quad -V\mathbb{E}\left\{\sum_{k=1}^{K}[\phi_k(\gamma_k^*) - \beta\nu_k d_k^*(t)]|\boldsymbol{\Theta}(t)\right\} \\ &\quad + \sum_{k=1}^{K} Z_k(t)\mathbb{E}\left\{\epsilon - \hat{b}_k(I^*(t), \boldsymbol{S}(t)) - d_k^*(t)|\boldsymbol{\Theta}(t)\right\} \\ &\quad + \sum_{k=1}^{K} Q_k(t)\mathbb{E}\left\{a_k^*(t) - \hat{b}_k(I^*(t), \boldsymbol{S}(t)) - d_k^*(t)|\boldsymbol{\Theta}(t)\right\} \\ &\quad + \sum_{k=1}^{K} G_k(t)\mathbb{E}\left\{\gamma_k^*(t) - a_k^*(t)|\boldsymbol{\Theta}(t)\right\}\end{aligned}$$

where $\boldsymbol{d}^*(t)$, $I^*(t)$, $\boldsymbol{a}^*(t)$ are any alternative decisions that satisfy $I^*(t) \in \mathcal{I}_{\boldsymbol{S}(t)}, 0 \leq d_k^*(t) \leq A_{max}$, and $0 \leq a_k^*(t) \leq \min[L_k(t) + A_k(t), A_{max}]$ for all $k \in \{1, \ldots, K\}$ and all t. Substituting the ω-only policy from (5.125)-(5.129) in the right-hand-side of the above inequality and taking $\delta \to 0$ yields:

$$\Delta(\boldsymbol{\Theta}(t)) - V\mathbb{E}\left\{\sum_{k=1}^{K}[\phi_k(\gamma_k(t)) - \beta\nu_k d_k(t)]|\boldsymbol{\Theta}(t)\right\} \leq B + C - V\phi^*$$

Using iterated expectations and telescoping sums as usual yields for all $t > 0$:

$$\frac{1}{t}\sum_{\tau=0}^{t-1}\mathbb{E}\left\{\sum_{k=1}^{K}[\phi_k(\gamma_k(\tau)) - \beta\nu_k d_k(\tau)]\right\} \geq \phi^* - (B + C)/V - \mathbb{E}\{L(\boldsymbol{\Theta}(0))\}/(Vt)$$

Using Jensen's inequality for the concave functions $\phi_k(\gamma)$ yields for all $t > 0$:

$$\sum_{k=1}^{K}[\phi_k(\overline{\gamma}_k(t)) - \overline{d}_k(t)] \geq \phi^* - (B + C)/V - \mathbb{E}\{L(\boldsymbol{\Theta}(0))\}/(Vt) \tag{5.130}$$

However, because $G_k(t) \leq G_{k,max}$ for all t, it is easy to show (via (5.114) and (2.5)) that for all k and all slots $t > 0$:

$$\overline{a}_k(t) \geq \max[\overline{\gamma}_k(t) - G_{k,max}/t, 0]$$

Therefore, since $\phi_k(\gamma)$ is continuous and non-decreasing, it can be shown:

$$\liminf_{t\to\infty}\sum_{k=1}^{K}[\phi_k(\overline{a}_k(t)) - \overline{d}_k(t)] \geq \liminf_{t\to\infty}\sum_{k=1}^{K}[\phi_k(\overline{\gamma}_k(t)) - \overline{d}_k(t)]$$

Using this in (5.130) proves the result. □

Because the network layer packet drops $d_k(t)$ are inefficient, it can be shown that:

$$\limsup_{t\to\infty} \sum_{k=1}^{K} \nu_k \overline{d}_k(t) \leq \frac{B+C}{V(\beta-1)} + \frac{(\phi^*_{\epsilon=0} - \phi^*)}{\beta-1}$$

where ϕ^* is the optimal solution to (5.98)-(5.102) for the given $\epsilon > 0$, and $\phi^*_{\epsilon=0}$ is the solution to (5.98)-(5.102) with $\epsilon = 0$ (which removes constraint (5.100)). Thus, if $\phi^*_\epsilon = \phi^*_{\epsilon=0}$, network layer drops can be made arbitrarily small by either increasing β or V.[7]

The above analysis allows for an arbitrary operation of the transport layer queues $L_k(t)$. Indeed, the above theorems only assume that $L_k(t) \geq 0$ for all t. Thus, as in (17), these can have either infinite buffer space, finite buffer space, or 0 buffer space. With 0 buffer space, all data that is not immediately admitted to the network layer is dropped.

5.7 ALTERNATIVE FAIRNESS METRICS

One type of fairness used in the literature is the so-called *max-min fairness* (see, for example, (129)(3)(5)(6)). Let $(\overline{x}_1, \ldots, \overline{x}_M)$ represent average throughputs achieved by users $\{1, \ldots, M\}$ under some stabilizing control algorithm, and let Λ denote the set of all possible $(\overline{x}_1, \ldots, \overline{x}_M)$ vectors. A vector $(x_1, \ldots, x_M) \in \Lambda$ is *max-min fair* if:

- It maximizes the lowest entry of $(x_1, \ldots, x_M)$ over all possible vectors in Λ.
- It maximizes the second lowest entry over all vectors in Λ that satisfy the above condition.
- It maximizes the third lowest entry over all vectors in Λ that satisfy the above two conditions, and so on.

This can be viewed as a sequence of nested optimizations, much different from the utility optimization framework treated in this chapter. For flow-based networks with capacitated links, one can reach a max-min fair allocation by starting from 0 and gradually increasing all flows equally until a bottleneck link is found, then increasing all non-bottlenecked flows equally, and so on (see Chapter 6.5.2 in (129)). A token-based scheduling scheme is developed in (160) for achieving max-min fairness in one-hop wireless networks on graphs with link selections defined by matchings.

One can approximate max-min fairness using a concave utility function in a network with capacitated links. Indeed, it is shown in (3) that optimizing a sum of concave functions of the form $g_\alpha(x) = \frac{-1}{x^\alpha}$ approaches a max-min fair point as $\alpha \to \infty$. It is likely that such an approach also holds for more general wireless networks with transmission rate allocation and scheduling. However, such functions are non-singular at $x = 0$ (preventing worst-case backlog bounds as in Exercises 5.6-5.7),

[7] If $\overline{b}_k \geq \epsilon$ for all k then the final term $(\phi^*_{\epsilon=0} - \phi^*)/(\beta-1)$ can be removed. Alternatively, if virtual queues $H_k(t+1) = \max[H_k(t) - \mu_k(t) + \epsilon, 0]$ are added to enforce these constraints, then $\limsup_{t\to\infty}[\nu_1\overline{d}_1(t) + \ldots + \nu_K\overline{d}_K(t)] \leq (\tilde{B} + C)/(V(\beta-1))$, where $\tilde{B}$ adds second moment terms $(\mu_k(t) - \epsilon)^2$ to (5.116).

and for large α they have very large values of $|g'_\alpha(x)/g_\alpha(x)|$ for $x > 0$, which typically results in large queue backlog if used in conjunction with the drift-plus-penalty method.

A simpler *hard fairness* approach seeks only to maximize the minimum throughput (161). This easily fits into the concave utility based drift-plus-penalty framework using the concave function $g(\boldsymbol{x}) = \min[x_1, \ldots, x_M]$:

$$\begin{aligned}
&\text{Maximize:} && \min[\overline{x}_1, \overline{x}_2, \ldots, \overline{x}_M] && (5.131)\\
&\text{Subject to:} \quad 1) && \text{All queues are mean rate stable} && (5.132)\\
&\quad 2) && \alpha(t) \in \mathcal{A}_{\omega(t)} \ \forall t \in \{0, 1, 2, \ldots\} && (5.133)
\end{aligned}$$

See also Exercise 5.4. A "mixed" approach can also be considered, which seeks to maximize $\beta \min[\overline{x}_1, \ldots, \overline{x}_M] + \sum_{m=1}^{M} \log(1 + \overline{x}_m)$. The constant β is a large weight that ensures maximizing the minimum throughput has a higher priority than maximizing the logarithmic terms.

5.8 EXERCISES

Exercise 5.1. (Using Logarithmic Utilities) Give a closed form solution to the auxiliary variable update of (5.49)-(5.50) when:

a) $\phi(\boldsymbol{\gamma}) = \sum_{m=1}^{M} \log(\gamma_m)$, where $\log(\cdot)$ denotes the natural logarithm.
b) $\phi(\boldsymbol{\gamma}) = \sum_{m=1}^{M} \log(1 + \nu_m \gamma_m)$, where $\log(\cdot)$ denotes the natural logarithm.

Exercise 5.2. (Transformed Problem with Auxiliary Variables) Let $\alpha^\star(t)$ be a policy that yields well defined averages $\overline{\boldsymbol{x}}^\star, \overline{y}_l^\star$, and that satisfies all constraints of problem (5.2)-(5.5),(5.8) (including the constraint $\overline{\boldsymbol{x}} \in \mathcal{R}$), with utility $\phi(\overline{\boldsymbol{x}}^\star) = \phi^{opt}$. Construct a policy that satisfies all constraints of problem (5.13)-(5.18) and that yields the same utility value $\phi(\overline{\boldsymbol{x}}^\star)$. Hint: Use $\boldsymbol{\gamma}(t) = \overline{\boldsymbol{x}}^\star$ for all t.

Exercise 5.3. (Jensen's Inequality) Let $\phi(\boldsymbol{\gamma})$ be a concave function defined over a convex set $\mathcal{R} \subseteq \mathbb{R}^M$. Let $\boldsymbol{\gamma}(\tau)$ be a sequence of random vectors in $\mathcal{R}$ for $\tau \in \{0, 1, 2, \ldots\}$. Fix an integer $t > 0$, and define T as an independent and random time that is uniformly distributed over the integers $\{0, 1, \ldots, t-1\}$. Define the random vector $\boldsymbol{X} = \boldsymbol{\gamma}(T)$. Use (5.9) to prove (5.10)-(5.11).

Exercise 5.4. (Hard Fairness (161)) Consider a system with M attributes $\boldsymbol{x}(t) = (x_1(t), \ldots, x_M(t))$, where $x_m(t) = \hat{x}_m(\alpha(t), \omega(t))$ for $m \in \{1, \ldots, M\}$. Assume there is a positive constant θ_{max} such that:

$$0 \leq \hat{x}_m(\alpha, \omega) \leq \theta_{max} \ \ \forall m \in \{1, \ldots, M\}, \forall \omega, \forall \alpha \in \mathcal{A}_\omega$$

a) State the drift-plus-penalty algorithm for solving the following problem, with $\theta(t)$ as a new variable:

$$\begin{array}{lll} \text{Maximize:} & & \overline{\theta} \\ \text{Subject to:} & 1) & \overline{x}_m \geq \overline{\theta} \;\; \forall m \in \{1, \ldots, M\} \\ & 2) & 0 \leq \theta(t) \leq \theta_{max} \;\; \forall t \in \{0, 1, 2, \ldots\} \\ & 3) & \alpha(t) \in \mathcal{A}_{\omega(t)} \;\; \forall t \in \{0, 1, 2, \ldots\} \end{array}$$

b) State the utility-based drift-plus-penalty algorithm for solving the problem:

$$\begin{array}{ll} \text{Maximize:} & \min[\overline{x}_1, \overline{x}_2, \ldots, \overline{x}_M] \\ \text{Subject to:} & \alpha(t) \in \mathcal{A}_{\omega(t)} \;\; \forall t \in \{0, 1, 2, \ldots\} \end{array}$$

which is solved with auxiliary variables $\gamma_m(t)$ with $0 \leq \gamma_m(t) \leq \theta_{max}$.

c) The problems in (a) and (b) both seek to maximize the minimum throughput. Show that if both algorithms "break ties" when choosing auxiliary variables by choosing the lowest possible values, then they are exactly the same algorithm. Show they are slightly different if ties are broken to choose the *largest* possible auxiliary variables, particularly in cases when some virtual queues are zero.

Exercise 5.5. (Bounded Virtual Queues) Consider the auxiliary variable optimization for $\gamma_m(t)$ in (5.49)-(5.50), where $\phi_m(x)$ has the property that:

$$\phi_m(x) \leq \phi_m(0) + \nu_m x \quad \text{whenever } 0 \leq x \leq \gamma_{m,max}$$

for a constant $\nu_m > 0$. Show that if $0 \leq \gamma_m(t) \leq \gamma_{m,max}$, we have:

$$V\phi_m(\gamma_m(t)) - G_m(t)\gamma_m(t) \leq V\phi_m(0) + (V\nu_m - G_m(t))\gamma_m(t)$$

Use this to prove that $\gamma_m(t) = 0$ is the unique optimal solution to (5.49)-(5.50) whenever $G_m(t) > V\nu_m$. Conclude from (5.48) that $G_m(t) \leq V\nu_m + \gamma_{m,max}$ for all t, provided this is true at $t = 0$.

Exercise 5.6. (1-Hop Wireless System with Infinite Backlog) Consider a wireless system with M channels. Transmission rates on slot t are given by $\boldsymbol{b}(t) = (b_1(t), \ldots, b_M(t))$ with $b_m(t) = \hat{b}_m(\alpha(t), \omega(t))$, where $\omega(t) = (S_1(t), \ldots, S_M(t))$ is an observed *channel state vector* for slot t (assumed to be i.i.d. over slots), and $\alpha(t)$ is a control action chosen within a set $\mathcal{A}_{\omega(t)}$. Assume that each channel has an infinite backlog of data, so that there is always data to send. The goal is to choose $\alpha(t)$ every slot to maximize $\phi(\overline{\boldsymbol{b}})$, where $\phi(\boldsymbol{b})$ is a concave and entrywise non-decreasing utility function.

a) Verify that the algorithm of Section 5.0.5 in this case is:

- (Auxiliary Variables) Choose $\boldsymbol{\gamma}(t) = (\gamma_1(t), \ldots, \gamma_M(t))$ to solve:

$$\begin{array}{ll} \text{Maximize:} & V\phi(\boldsymbol{\gamma}(t)) - \sum_{m=1}^{M} G_m(t)\gamma_m(t) \\ \text{Subject to:} & 0 \leq \gamma_m(t) \leq \gamma_{m,max} \ \forall m \in \{1, \ldots, M\} \end{array}$$

- (Transmission) Observe $\omega(t)$ and choose $\alpha(t) \in \mathcal{A}_{\omega(t)}$ to maximize $\sum_{m=1}^{M} G_m(t)\hat{b}_m(\alpha(t), \omega(t))$.

- (Virtual Queue Update) Update $G_m(t)$ for all $m \in \{1, \ldots, M\}$ according to:

$$G_m(t+1) = \max[G_m(t) + \gamma_m(t) - \hat{b}_m(\alpha(t), \omega(t)), 0]$$

b) Suppose that $\phi(\boldsymbol{b}) = \sum_{m=1}^{M} \phi_m(b_m)$, where the functions $\phi_m(b_m)$ are continuous, concave, non-decreasing, with maximum right-derivative $\nu_m < \infty$, so that $\phi_m(\gamma) \leq \phi_m(0) + \nu_m\gamma$ for all $\gamma \geq 0$. Prove that the auxiliary variable decisions above yield $\gamma_m(t) = 0$ if $G_m(t) > V\nu_m$ (see also Exercise 5.5). Conclude that $0 \leq G_m(t) \leq V\nu_m + \gamma_{m,max}$ for all t, provided that this holds at $t = 0$.

c) Use (5.33) to conclude that if the conditions of part (b) hold, if all virtual queues are initially empty, and if any C-additive approximation is used, then:

$$\phi(\overline{\boldsymbol{b}}(t)) \geq \phi^{opt} - \frac{D+C}{V} - \sum_{m=1}^{M} \frac{\nu_m(V\nu_m + \gamma_{m,max})}{t} \quad , \ \forall t > 0$$

Exercise 5.7. (1-Hop Wireless System with Random Arrivals) Consider the same system as Exercise 5.6, with the exception that we have random arrivals $A_m(t)$ and:

$$Q_m(t+1) = \max[Q_m(t) - \hat{b}_m(\alpha(t), \omega(t)), 0] + x_m(t)$$

where $x_m(t)$ is a flow control decision, made subject to $0 \leq x_m(t) \leq A_m(t)$. We want to maximize $\phi(\overline{\boldsymbol{x}})$.

a) State the new algorithm for this case.

b) Suppose $0 \leq A_m(t) \leq A_{m,max}$ for some finite constant $A_{m,max}$. Suppose $\phi(\boldsymbol{b})$ has the structure of Exercise 5.6(b). Using a similar argument, show that all queues $G_m(t)$ and $Q_k(t)$ are deterministically bounded.

Exercise 5.8. (Imperfect Channel Knowledge) Consider the general problem of Theorem 5.3, but under the assumption that $\omega(t)$ provides only a *partial* understanding of the channel for each queue $Q_k(t)$, so that $\hat{b}_k(\alpha(t), \omega(t))$ is a *random* function of $\alpha(t)$ and $\omega(t)$, assumed to be i.i.d. over all slots with the same $\alpha(t)$ and $\omega(t)$, and assumed to have finite second moments regardless of the choice of $\alpha(t)$. Define:

$$\beta_k(\alpha, \omega) \triangleq \mathbb{E}\left\{\hat{b}_k(\alpha(t), \omega(t))|\alpha(t) = \alpha, \omega(t) = \omega\right\}$$

Assume that the function $\beta_k(\alpha, \omega)$ is known. Assume the other functions $\hat{x}_m(\cdot)$, $\hat{y}_l(\cdot)$, $\hat{a}_k(\cdot)$ are deterministic as before. State the modified algorithm that minimizes the right-hand-side of (5.84) in this case. Hint:

$$\mathbb{E}\left\{b_k(t)|\boldsymbol{\Theta}(t)\right\} = \mathbb{E}\left\{\mathbb{E}\left\{b_k(t)|\boldsymbol{\Theta}(t), \alpha(t), \omega(t)\right\}|\boldsymbol{\Theta}(t)\right\} = \mathbb{E}\left\{\beta_k(\alpha(t), \omega(t))|\boldsymbol{\Theta}(t)\right\}$$

Note: Related problems with randomized service outcomes and Lyapunov drift are considered in (162)(163)(164)(154)(165)(161), where knowledge of the channel statistics is needed for computing the $\beta_k(\alpha, \omega)$ functions and their generalizations, and a max-weight learning framework is developed in (166) for the case of unknown statistics.

Exercise 5.9. (Equivalence of the Transformed Problem Using Auxiliary Variables)

a) Suppose that $\alpha^*(t)$ is a policy that satisfies all constraints of the problem (5.71)-(5.75), yielding time averages $\overline{\boldsymbol{x}}^*$ and $\overline{y}_l^*$ and a cost value of $\overline{y}_0^* + f(\overline{\boldsymbol{x}}^*)$. Show that this policy also satisfies all constraints of the problem (5.76)-(5.81), and yields the same cost value, if we define the auxiliary variable decisions to be $\boldsymbol{\gamma}(t) = \overline{\boldsymbol{x}}^*$ for all t.

b) Suppose that $\alpha^\star(t)$, $\boldsymbol{\gamma}^\star(t)$ is a policy that satisfies all constraints of problem (5.76)-(5.81), yielding time averages $\overline{\boldsymbol{x}}^\star$, $\overline{y}_l^\star$ and a cost value in (5.76) given by some value v. Show that this same policy also satisfies all constraints of problem (5.71)-(5.75), with a cost $\overline{y}_0^\star + f(\overline{\boldsymbol{x}}^\star) \leq v$.

Exercise 5.10. (Proof of Theorem 5.3) We make use of the following fact, analogous to Theorem 4.5: If problem (5.71)-(5.75) is feasible, then for all $\delta > 0$ there exists an ω-only policy $\alpha^*(t) \in \mathcal{A}_{\omega(t)}$ such that $\mathbb{E}\left\{\hat{\boldsymbol{x}}(\alpha^*(t), \omega(t))\right\} = \boldsymbol{\gamma}^*$ for some vector $\boldsymbol{\gamma}^*$, and:

$$\begin{aligned}
\mathbb{E}\left\{\hat{y}_0(\alpha^*(t), \omega(t))\right\} + f(\boldsymbol{\gamma}^*) &\leq y_0^{opt} + f^{opt} + \delta \\
\mathbb{E}\left\{\hat{y}_l(\alpha^*(t), \omega(t))\right\} + g_l(\boldsymbol{\gamma}^*) &\leq \delta \ , \forall l \in \{1, \ldots, L\} \\
\mathbb{E}\left\{\hat{a}_k(\alpha^*(t), \omega(t)) - \hat{b}_k(\alpha^*(t), \omega(t))\right\} &\leq \delta \ , \forall k \in \{1, \ldots, K\} \\
\text{dist}\left(\boldsymbol{\gamma}^*, \mathcal{X} \cap \mathcal{R}\right) &\leq \delta
\end{aligned}$$

For simplicity, in this proof, we assume the above holds for $\delta = 0$, and that all actual and virtual queues are initially empty. Further assume that the functions $f(\boldsymbol{\gamma})$ and $g_l(\boldsymbol{\gamma})$ are *Lipschitz continuous*, so that there are positive constants ν_m, $\beta_{l,m}$ such that for all $\boldsymbol{x}(t)$ and $\boldsymbol{\gamma}(t)$, we have:

$$\begin{aligned}
|f(\boldsymbol{\gamma}(t)) - f(\boldsymbol{x}(t))| &\leq \textstyle\sum_{m=1}^{M} \nu_m |\gamma_m(t) - x_m(t)| \\
|g_l(\boldsymbol{\gamma}(t)) - g_l(\boldsymbol{\gamma}(t))| &\leq \textstyle\sum_{m=1}^{M} \beta_{l,m} |\gamma_m(t) - x_m(t)| \ , \forall l \in \{1, \ldots, L\}
\end{aligned}$$

a) Plug the above policy $\alpha^*(t)$, together with the constant auxiliary vector $\boldsymbol{\gamma}(t) = \boldsymbol{\gamma}^*$, into the right-hand-side of the drift bound (5.84) and add C (because of the C-additive approximation) to derive a simpler bound on the drift expression. The resulting right-hand-side should be: $D + C + V(y_0^{opt} + f^{opt})$.

b) Use the Lyapunov optimization theorem to prove that for all $t > 0$:

$$\frac{1}{t}\sum_{\tau=0}^{t-1} \mathbb{E}\left\{y_0(\tau) + f(\boldsymbol{\gamma}(\tau))\right\} \leq y_0^{opt} + f^{opt} + (D + C)/V$$

and hence, by Jensen's inequality (with $\overline{y}_0(t)$ and $\overline{\boldsymbol{\gamma}}(t)$ defined by (5.24)):

$$\overline{y}_0(t) + f(\overline{\boldsymbol{\gamma}}(t)) \leq y_0^{opt} + f^{opt} + (D + C)/V$$

c) Manipulate the drift bound of part (a) to prove that $\Delta(\boldsymbol{\Theta}(t)) \leq W$ for some finite constant W. Conclude that all virtual and actual queues are mean rate stable, and that (4.7) holds for all $t > 0$ and so $\mathbb{E}\left\{|H_m(t)|\right\}/t \leq \sqrt{2W/t}$.

d) Use (5.83) and (4.42) to prove that for all $m \in \{1, \ldots, M\}$:

$$0 \leq \lim_{t\to\infty} |\overline{x}_m(t) - \overline{\gamma}_m(t)| = \lim_{t\to\infty} \frac{|\mathbb{E}\left\{H_m(t)\right\}|}{t} \leq \lim_{t\to\infty} \frac{\mathbb{E}\left\{|H_m(t)|\right\}}{t} = 0$$

Argue that $\overline{\boldsymbol{\gamma}}(t) \in \mathcal{X} \cap \mathcal{R}$ for all t, and hence (5.87) holds.

e) Use part (b) and the Lipschitz assumptions to prove (5.85).

f) Use (5.82), Theorem 2.5, and the Lipschitz conditions to prove (5.86).

Exercise 5.11. (Profit Risk and Non-Convexity) Consider a K-queue system described by (5.1), with arrival and service functions $\hat{a}_k(\alpha(t), \omega(t))$ and $\hat{b}_k(\alpha(t), \omega(t))$. Let $p(t) = \hat{p}(\alpha(t), \omega(t))$ be a *random profit variable* that is i.i.d. over all slots for which we have $\alpha(t)$ and $\omega(t)$, and that has finite second moment regardless of the policy. Define:

$$\begin{aligned}
\phi(\alpha, \omega) &\triangleq \mathbb{E}\left\{\hat{p}(\alpha(t), \omega(t))|\alpha(t) = \alpha, \omega(t) = \omega\right\} \\
\psi(\alpha, \omega) &\triangleq \mathbb{E}\left\{\hat{p}(\alpha(t), \omega(t))^2|\alpha(t) = \alpha, \omega(t) = \omega\right\}
\end{aligned}$$

and assume the functions $\phi(\alpha, \omega)$, $\psi(\alpha, \omega)$ are known. The goal is to stabilize all queues while maximizing a linear combination of the profit minus the variance of the profit (where variance is a proxy for "risk"). Specifically, define the variance as $Var(p) \triangleq \overline{p^2} - \overline{p}^2$, where the notation $\overline{h}$ represents a time average expectation of a given process $h(t)$, as usual. We want to maximize $\theta_1\overline{p} - \theta_2 Var(p)$, where θ_1 and θ_2 are positive constants.

a) Define attributes $p_1(t) = p(t)$, $p_2(t) = p(t)^2$. Write the problem using $\overline{p}_1$ and $\overline{p}_2$ in the form of (5.88)-(5.91), and show this is a non-convex stochastic network optimization problem.

b) State the "primal-dual" algorithm that minimizes the right-hand-side of (5.93) in this context. Hint: Note that:

$$\mathbb{E}\{p_1(t)|\boldsymbol{\Theta}(t)\} = \mathbb{E}\{\mathbb{E}\{p_1(t)|\boldsymbol{\Theta}(t), \alpha(t), \omega(t)\}|\boldsymbol{\Theta}(t)\} = \mathbb{E}\{\phi(\alpha(t), \omega(t))|\boldsymbol{\Theta}(t)\}$$

Exercise 5.12. (Optimization without Auxiliary Variables (17)(18)) Consider the problem (5.2)-(5.5). Assume there is a vector $\boldsymbol{\gamma}^\star = (\gamma_1^\star, \ldots, \gamma_M^\star)$, called the *optimal operating point*, such that $\phi(\boldsymbol{\gamma}^\star) = \phi^\star$, where $\phi^\star$ is the maximum utility for the problem. Assume that there is an ω-only policy $\alpha^\star(t)$ such that for all possible values of $\omega(t)$, we have:

$$\hat{x}_m(\alpha^\star(t), \omega(t)) = \gamma_m^\star \quad \forall m \in \{1, \ldots, M\} \tag{5.134}$$

$$\mathbb{E}\left\{\hat{a}_k(\alpha^\star(t), \omega(t))\right\} \leq \mathbb{E}\left\{\hat{b}_k(\alpha^\star(t), \hat{\omega}(t))\right\} \quad \forall k \in \{1, \ldots, K\} \tag{5.135}$$

$$\mathbb{E}\left\{\hat{y}_l(\alpha^\star(t), \omega(t))\right\} \leq 0 \quad \forall l \in \{1, \ldots, L\} \tag{5.136}$$

The assumptions (5.134)-(5.136) are restrictive, particularly because (5.134) must hold *deterministically for all $\omega(t)$ realizations*. However, these assumptions can be shown to hold for the special case when $x_m(t)$ represents the amount of data admitted to a network from a source m when: (i) All sources are "infinitely backlogged" and hence always have data to send, and (ii) Data can be admitted as a real number.

The Lyapunov drift can be shown to satisfy the following for some constant $B > 0$:

$$\Delta(\boldsymbol{\Theta}(t)) - V\mathbb{E}\left\{\phi(\hat{\boldsymbol{x}}(\alpha(t), \omega(t))) \mid \boldsymbol{\Theta}(t)\right\} \leq B + \sum_{l=1}^{L} Z_l(t)\mathbb{E}\left\{\hat{y}_l(\alpha(t), \omega(t))|\boldsymbol{\Theta}(t)\right\}$$

$$+\sum_{k=1}^{K} Q_k(t)\mathbb{E}\left\{\hat{a}_k(\alpha(t), \omega(t)) - \hat{b}_k(\alpha(t), \omega(t)) \mid \boldsymbol{\Theta}(t)\right\} - V\mathbb{E}\left\{\phi(\hat{\boldsymbol{x}}(\alpha(t), \omega(t))) \mid \boldsymbol{\Theta}(t)\right\}$$

Suppose every slot we observe $\boldsymbol{\Theta}(t)$ and $\omega(t)$ and choose an action $\alpha(t)$ that minimizes the right-hand-side of the above drift inequality.

a) Assume $\omega(t)$ is i.i.d. over slots. Plug the alternative policy $\alpha^\star(t)$ into the right-hand-side above to get a greatly simplified drift expression.

b) Conclude from part (a) that $\Delta(\boldsymbol{\Theta}(t)) \leq D + V(\phi^{max} - \phi^\star)$ for all t, for some finite constant D and where ϕ^{max} is an upper bound on the instantaneous value of $\phi(\hat{\boldsymbol{x}}(\cdot))$ (assumed to be finite). Conclude that all actual and virtual queues are mean rate stable, and hence all desired inequality constraints are satisfied.

c) Use Jensen's inequality and part (a) (with iterated expectations and telescoping sums) to conclude that for all $t > 0$, we have:

$$\phi(\overline{\boldsymbol{x}}(t)) \geq \frac{1}{t}\sum_{\tau=0}^{t-1} \mathbb{E}\{\phi(\boldsymbol{x}(\tau))\} \geq \phi^\star - B/V - \mathbb{E}\{L(\boldsymbol{\Theta}(0))\}/(Vt)$$

where $\overline{\boldsymbol{x}}(t)\triangleq\frac{1}{t}\sum_{\tau=0}^{t-1}\mathbb{E}\{\boldsymbol{x}(\tau)\}$ and $\boldsymbol{x}(\tau)\triangleq\hat{\boldsymbol{x}}(\alpha(\tau),\omega(\tau))$.

Exercise 5.13. (Delay-Limited Transmission (71)) Consider a K-user wireless system with arrival vector $\boldsymbol{A}(t) = (A_1(t), \ldots, A_K(t))$ and channel state vector $\boldsymbol{S}(t) = (S_1(t), \ldots, S_K(t))$ for each slot $t \in \{0, 1, 2, \ldots\}$. There is no queueing, and all data must either be transmitted in 1 slot or dropped (similar to the *delay-limited capacity* formulation of (70)). Thus, there are no actual queues in the system. Define $\omega(t)\triangleq[\boldsymbol{A}(t), \boldsymbol{S}(t)]$ as the random network event observed every slot. Define $\alpha(t) \in \mathcal{A}_{\omega(t)}$ as a general control action, which affects how much of the data to transmit and the amount of power used according to general functions $\hat{\mu}_k(\alpha, \omega)$ and $\hat{p}(\alpha, \omega)$:

$$\boldsymbol{\mu}(t) = (\hat{\mu}_1(\alpha(t), \omega(t)), \ldots, \hat{\mu}_K(\alpha(t), \omega(t))) \ , \ \ p(t) = \hat{p}(\alpha(t), \omega(t))$$

where $\boldsymbol{\mu}(t) = (\mu_1(t), \ldots, \mu_K(t))$ is the transmission vector and $p(t)$ is the power used on slot t. Assume these are constrained as follows for all slots t:

$$0 \leq \mu_k(t) \leq A_k(t) \ \forall k \in \{1, \ldots, K\} \ , \ \ 0 \leq p(t) \leq p^{max}$$

for some finite constant p^{max}. Assume that $A_k(t) \leq A_k^{max}$ for all t, for some finite constants A_k^{max} for $k \in \{1, \ldots, K\}$. Let $\overline{\boldsymbol{\mu}}$ be the time average expectation of the transmission vector $\boldsymbol{\mu}(t)$, and let $\phi(\overline{\boldsymbol{\mu}})$ be a continuous, concave, and entrywise non-decreasing utility function of $\overline{\boldsymbol{\mu}}$. The goal is to solve the following problem:

$$\begin{aligned} \text{Maximize:} \quad & \phi(\overline{\boldsymbol{\mu}}) \\ \text{Subject to:} \quad & \overline{p} \leq P_{av} \end{aligned}$$

where $\overline{p}$ is the time average expected power expenditure, and P_{av} is a pre-specified average power constraint. This is a special case of the general problem (5.2)-(5.5).

a) Use auxiliary variables $\boldsymbol{\gamma}(t) = (\gamma_1(t), \ldots, \gamma_K(t))$ subject to $0 \leq \gamma_k(t) \leq A_k^{max}$ for all t, k to write the corresponding transformed problem (5.13)-(5.18) for this case.

b) State the drift-plus-penalty algorithm that solves this transformed problem. Hint: Use a virtual queue $Z(t)$ to enforce the constraint $\overline{p} \leq P_{av}$, and use virtual queues $G_k(t)$ to enforce the constraints $\overline{\mu}_k \geq \overline{\gamma}_k$ for all $k \in \{1, \ldots, K\}$.

Exercise 5.14. (Delay-Limited Transmission with Errors (71)) Consider the same system as Exercise 5.13, but now assume that transmissions can have errors, so that $\mu_k(t) = \hat{\mu}_k(\alpha(t), \omega(t))$ is a *random* transmission outcome (as in Exercise 5.8), assumed to be i.i.d. over all slots with the same $\alpha(t)$ and $\omega(t)$, with known expectations $\beta_k(\alpha(t), \omega(t))\triangleq\mathbb{E}\{\mu_k(t)|\alpha(t), \omega(t)\}$ for all $k \in \{1, \ldots, K\}$. Use iterated expectations (as in Exercise 5.8) to redesign the drift-plus-penalty algorithm for this case. Multi-slot versions of this problem are treated in Section 7.6.1.

CHAPTER 6

Approximate Scheduling

This chapter focuses on the max-weight problem that arises when scheduling for stability or maximum throughput-utility in a wireless network with interference. Previous chapters showed the key step is maximizing the expectation of a weighted sum of link transmission rates, or coming within an additive constant C of the maximum. Specifically, consider a (possibly multi-hop) network with L links, and let $\boldsymbol{b}(t) = (b_1(t), \ldots, b_L(t))$ be the transmission rate offered over link $l \in \{1, \ldots, L\}$ on slot t. The goal is to make (possibly randomized) decisions for $\boldsymbol{b}(t)$ to come within an additive constant C of maximizing the following expectation:

$$\sum_{l=1}^{L} W_l(t)\mathbb{E}\{b_l(t)|\boldsymbol{W}(t)\} \tag{6.1}$$

where the expectation is with respect to the possibly random decision, and where $\boldsymbol{W}(t) = (W_1(t), \ldots, W_L(t))$ is a vector of weights for slot t. The weights are related to queue backlogs for single-hop problems and differential backlogs for multi-hop problems. Algorithms that accomplish this for a given constant $C \geq 0$ every slot are called *C-additive approximations*. For problems of network stability, previous chapters showed that C-additive approximations can be used to stabilize the network whenever arrival rates are inside the network capacity region, with average backlog and delay bounds that grow linearly with C. For problems of maximum throughput-utility, Chapter 5 showed that C-additive approximations can be used with a simple flow control rule to give utility that is within $(B + C)/V$ of optimality (where B is a fixed constant and V is any non-negative parameter chosen as desired), with average backlog that grows linearly in both V and C. Thus, C-additive approximations can be used to push network utility arbitrarily close to optimal, as determined by the parameter V.

Such max-weight problems can be very complex for wireless networks with interference. This is because a transmission on one link can affect transmissions on many other links. Thus, transmission decisions are coupled throughout the network. In this chapter, we first consider a class of interference networks without time varying channels and develop two C-additive approximation algorithms for this context. The first is a simple algorithm based on trading off computation complexity and delay. The second is a more elegant *randomized* transmission technique that admits a simple distributed implementation. We then present a *multiplicative* approximation theorem that holds for general networks with possibly time-varying channels. It guarantees constant factor throughput results for algorithms that schedule transmissions within a multiplicative constant of the max-weight solution every slot.

6.1 TIME-INVARIANT INTERFERENCE NETWORKS

Suppose the network is *time invariant*, in that the channel conditions do not change and the transmission rate options are the same for all slots $t \in \{0, 1, 2, \ldots\}$. Assume that all transmissions are in units of packets, and each link can transmit at most one packet per slot. The transmission rate vector $\boldsymbol{b}(t) = (b_1(t), \ldots, b_L(t))$ is a binary vector with $b_l(t) = 1$ if link l transmits a packet on slot t, and $b_l(t) = 0$ otherwise. We say that a binary vector $\boldsymbol{b}(t)$ is *feasible* if the set of links that correspond to "1" entries can be simultaneously activated for successful transmission. Define $\mathcal{B}$ as the collection of all feasible binary vectors, called the *link activation set* (7). The set $\mathcal{B}$ depends on the interference properties of the network. Every slot t, the network controller observes the current link weights $\boldsymbol{W}(t) = (W_1(t), \ldots, W_L(t))$ and chooses a (possibly random) $\boldsymbol{b}(t) \in \mathcal{B}$, with the goal of maximizing the max-weight value (6.1). It is easy to show that the maximum is achieved by a deterministic choice $\boldsymbol{b}^{opt}(t)$, where:

$$\boldsymbol{b}^{opt}(t) \triangleq \arg\max_{\boldsymbol{b} \in \mathcal{B}} \left[\sum_{l=1}^{L} W_l(t) b_l \right]$$

The amount of computation required to find an optimal vector $\boldsymbol{b}^{opt}(t)$ depends on the structure of the set $\mathcal{B}$. If this set is defined by all links that satisfy *matching constraints*, so that no two active links share a node, then $\boldsymbol{b}^{opt}(t)$ can be found in polynomial time (via a centralized algorithm). However, the problem may be NP-hard for general sets $\mathcal{B}$, so that no polynomial time solution is available.

Let C be a given non-negative constant. A C-additive approximation to the max-weight problem finds a vector $\boldsymbol{b}(t)$ every slot t that satisfies:

$$\sum_{l=1}^{L} W_l(t) \mathbb{E}\left\{ b_l(t) | \boldsymbol{W}(t) \right\} \geq \max_{\boldsymbol{b} \in \mathcal{B}} \left[\sum_{l=1}^{L} W_l(t) b_l \right] - C$$

6.1.1 COMPUTING OVER MULTIPLE SLOTS

We first consider the following simple technique for obtaining a C-additive approximation with arbitrarily low per-time slot computation complexity. Fix a positive integer $T > 0$, and divide the timeline into successive intervals of T-slot frames. Define $t_r \triangleq rT$ as the start of frame r, for $r \in \{0, 1, 2, \ldots\}$. At the beginning of each frame $r \in \{0, 1, 2, \ldots\}$, the network controller observes the weights $\boldsymbol{W}(t_r)$ and begins a computation to find $\boldsymbol{b}^{opt}(t_r)$. We assume the computation is completed within the T slot frame, possibly by exhaustively searching through all options in the set $\mathcal{B}$. The network controller then allocates the constant rate vector $\boldsymbol{b}(t_r)$ for all slots of frame $r + 1$, while also computing $\boldsymbol{b}^{opt}(t_{r+1})$ during that frame. Thus, every frame $r \in \{1, 2, 3, \ldots\}$ the algorithm allocates the constant rate vector that was computed on the previous frame. Meanwhile, it also computes the optimal solution to the max-weight problem for the current frame (see Fig. 6.1). Thus, for any frame $r \in \{1, 2, 3, \ldots\}$, we have:

$$\boldsymbol{b}(t) = \boldsymbol{b}^{opt}(t_{r-1}) \quad \forall t \in \{t_r, \ldots, t_r + T - 1\}$$

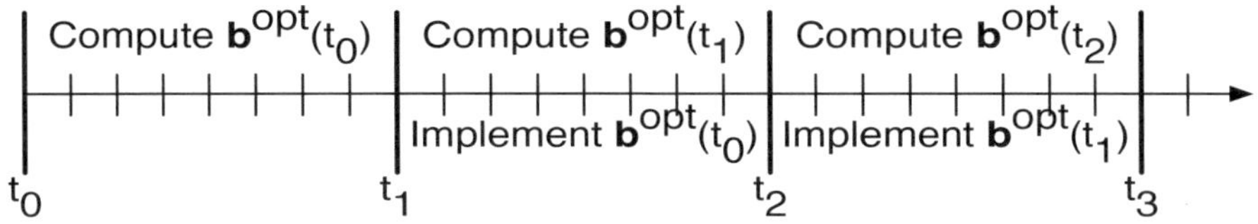

Figure 6.1: An illustration of the frame structure for the algorithm of Section 6.1.1.

Now assume the maximum change in queue backlog over one slot is deterministically bounded, as is the maximum change in each link weight. Specifically, assume that no link weight can change by an amount more than θ, where θ is some positive constant. It follows that for any two slots $t_1 < t_2$:

$$|W_l(t_1) - W_l(t_2)| \leq \theta(t_2 - t_1)$$

Under this assumption, we now compute a value C such that the above algorithm is a C-additive approximation for all slots $t \geq T$. Fix any slot $t \geq T$. Let r represent the frame containing this slot. Note that $|t - t_{r-1}| \leq 2T - 1$. We have:

$$\begin{aligned}
\sum_{l=1}^{L} W_l(t)b_l(t) &= \sum_{l=1}^{L} W_l(t)b_l^{opt}(t_{r-1}) \\
&= \sum_{l=1}^{L} W_l(t_{r-1})b_l^{opt}(t_{r-1}) + \sum_{l=1}^{L}(W_l(t) - W_l(t_{r-1}))b_l^{opt}(t_{r-1}) \\
&\geq \sum_{l=1}^{L} W_l(t_{r-1})b_l^{opt}(t_{r-1}) - \sum_{l=1}^{L}\theta|t - t_{r-1}|b_l^{opt}(t_{r-1}) \\
&\geq \sum_{l=1}^{L} W_l(t_{r-1})b_l^{opt}(t_{r-1}) - L\theta(2T-1) \qquad (6.2)
\end{aligned}$$

Further, because $\boldsymbol{b}^{opt}(t_{r-1})$ solves the max-weight problem for links $\boldsymbol{W}(t_{r-1})$, we have:

$$\begin{aligned}
\sum_{l=1}^{L} W_l(t_{r-1})b_l^{opt}(t_{r-1}) &= \max_{b \in \mathcal{B}}\left[\sum_{l=1}^{L} W_l(t_{r-1})b_l\right] \\
&\geq \sum_{l=1}^{L} W_l(t_{r-1})b_l^{opt}(t)
\end{aligned}$$

$$
\begin{aligned}
&= \sum_{l=1}^{L} W_l(t) b_l^{opt}(t) - \sum_{l=1}^{L} [W_l(t) - W_l(t_{r-1})] b_l^{opt}(t) \\
&\geq \sum_{l=1}^{L} W_l(t) b_l^{opt}(t) - L\theta(2T-1) \\
&= \max_{\boldsymbol{b} \in \mathcal{B}} \left[\sum_{l=1}^{L} W_l(t) b_l \right] - L\theta(2T-1) \qquad (6.3)
\end{aligned}
$$

Combining (6.2) and (6.3) yields:

$$
\sum_{l=1}^{L} W_l(t) b_l(t) \geq \max_{\boldsymbol{b} \in \mathcal{B}} \left[\sum_{l=1}^{L} W_l(t) b_l \right] - 2L\theta(2T-1)
$$

Taking conditional expectations gives:

$$
\sum_{l=1}^{L} W_l(t) \mathbb{E}\left\{ b_l(t) | \boldsymbol{W}(t) \right\} \geq \max_{\boldsymbol{b} \in \mathcal{B}} \left[\sum_{l=1}^{L} W_l(t) b_l \right] - 2L\theta(2T-1)
$$

It follows that this algorithm yields a C-additive approximation for $C \triangleq 2L\theta(2T-1)$. The constant C is linear in the number of links L and in the frame size T.

Now let *complexity* represent the number of operations required to compute the max-weight solution (assuming for simplicity that this number is independent of the size of the weights $W_l(t)$). Because this complexity is amortized over T slots, the algorithm yields a per-slot computation complexity of $complexity/T$. This can be made as small as desired by increasing T, with a tradeoff of increasing the value of C linearly in T. This shows that maximum throughput can be achieved with arbitrarily low per-time slot complexity, with a tradeoff in average queue backlog and average delay.

This technique was used in (167)(168) to reduce the per-slot complexity of scheduling in $N \times N$ packet switches. The max-weight problem for $N \times N$ packet switches is a *max-weight matching problem* that can be computed in time that is polynomial in N. The work (168) uses this to provide a smooth complexity-delay tradeoff for switches, showing average delay of $O(N^{4-\alpha})$ is possible with per-slot complexity $O(N^{\alpha})$, for any α such that $0 \leq \alpha \leq 3$.

Unfortunately, the max-weight problem for networks with general activation sets $\mathcal{B}$ may be NP-hard, so that the only available computation algorithms have complexity that is exponential in the network size L. This means the frame size T must be chosen to be at least exponential in L to achieve polynomial per-slot complexity, which in turn incurs delay that is exponential in L.

6.1.2 RANDOMIZED SEARCHING FOR THE MAX-WEIGHT SOLUTION

The first low-complexity algorithm for full-throughput scheduling in time-invariant interference networks was perhaps (169), where new link activations are tried randomly and compared in the max-weight metric against the previously tried activation. This is analyzed with a different Markov chain argument in (169). However, intuitively this works for the same reason as the frame-based scheme presented in the previous subsection: The randomized selection can be viewed as a (randomized) computation algorithm that solves the max-weight problem over a (variable length) frame. The optimal solution is computed in some random number of T slots, where T is geometric with success probability equal to the number of optimal vectors in $\mathcal{B}$ divided by the size of the set $\mathcal{B}$. While the implementation of the algorithm is more elegant than the deterministic computation method described in the previous subsection, its resulting delay bounds can be worse. For example, in a $N \times N$ packet switch, the randomized method yields complexity that is $O(N)$ and an average delay bound of $O(N!)$. However, the deterministic method of (168) can achieve complexity that is $O(N)$ with an average delay bound of $O(N^3)$. This is achieved by using $\alpha = 1$ in the smooth complexity-delay tradeoff curve described in the previous subsection. A variation on the randomized algorithm of (169) for more complex networks is developed in (170).

All known methods for achieving throughput-utility within ϵ of optimality for networks with general interference constraints (and for arbitrary $\epsilon > 0$) have either non-polynomial per-slot complexity, or non-polynomial delays and/or convergence times. This is not surprising: Suppose the problem of maximizing the number of activated links is NP-hard. If we can design an algorithm that, after a polynomial time T, has produced a throughput that is within $1/2$ from the maximum sum throughput with high probability, then this algorithm (with high probability) must have selected a vector $\boldsymbol{b}(t)$ that is a max-size vector during some slot $t \in \{0, \ldots, T\}$ (else, the throughput would be at least 1 away from optimal). Thus, this could be used as a randomized algorithm for finding a max-size vector in polynomial time. Related NP-hardness results are developed in (171) for pure stability problems with low delay, even when arrival rates are very low.

6.1.3 THE JIANG-WALRAND THEOREM

Here we present a randomized algorithm that produces a C-additive approximation by allocating a link vector $\boldsymbol{b}(t)$ according to the steady state solution of a particular *reversible Markov chain*. The Markov chain can easily be simulated, and it has a simple relation to distributed scheduling in a carrier sense multiple access (CSMA) system. Further, if the vector is chosen according to the desired distribution every slot t, the value of C that this algorithm produces is *linear* in the network size, and hence this yields maximum throughput with polynomial delay. We first present the result, and then discuss the complexity associated with generating a vector with the desired distribution, related to the convergence time required for the Markov chain to approach steady state.

The following randomized algorithm for choosing $\boldsymbol{b}(t) \in \mathcal{B}$ was developed in (172) for wireless systems with general interference constraints, and in (173) for scheduling in optical networks:

Max Link Weight Plus Entropy Algorithm: Every slot t, observe the current link weights $\boldsymbol{W}(t) = (W_1(t), \ldots, W_L(t))$ and choose $\boldsymbol{b}(t)$ by randomly selecting a binary vector $\boldsymbol{b} =$

$(b_1, \ldots, b_L) \in \mathcal{B}$ with probability distribution:

$$p^*(\boldsymbol{b}) \triangleq Pr[\boldsymbol{b}(t) = \boldsymbol{b}] = \frac{\Pi_{l=1}^{L} \exp(W_l(t) b_l)}{A} \tag{6.4}$$

where A is a normalizing constant that makes the distribution sum to 1.

The work (172) motivates this algorithm by the modified problem that computes a probability distribution $p(\boldsymbol{b})$ over the set $\mathcal{B}$ to solve the following:

$$\text{Maximize:} \quad -\textstyle\sum_{\boldsymbol{b}\in\mathcal{B}} p(\boldsymbol{b}) \log(p(\boldsymbol{b})) + \sum_{\boldsymbol{b}\in\mathcal{B}} p(\boldsymbol{b}) \sum_{l=1}^{L} W_l(t) b_l \tag{6.5}$$

$$\text{Subject to:} \quad 0 \leq p(\boldsymbol{b}) \ \forall \boldsymbol{b} \in \mathcal{B} \ , \ \textstyle\sum_{\boldsymbol{b}\in\mathcal{B}} p(\boldsymbol{b}) = 1 \tag{6.6}$$

where $\log(\cdot)$ denotes the natural logarithm. This problem is equivalent to maximizing $H(p(\cdot)) + \sum_{l=1}^{L} W_l(t) \mathbb{E}\{b_l(t)|\boldsymbol{W}(t)\}$, where $H(p(\cdot))$ is the entropy (in nats) associated with the probability distribution $p(\boldsymbol{b})$, and $\mathbb{E}\{b_l(t)|\boldsymbol{W}(t)\}$ is the expected transmission rate over link l given that $\boldsymbol{b}(t)$ is selected according to the probability distribution $p(\boldsymbol{b})$. However, note that because the set $\mathcal{B}$ contains at most 2^L link activation sets, and the entropy of any probability distribution that contains at most k probabilities is at most $\log(k)$, we have for any probability distribution $p(\boldsymbol{b})$:

$$0 \leq -\sum_{\boldsymbol{b}\in\mathcal{B}} p(\boldsymbol{b}) \log(p(\boldsymbol{b})) \leq L \log(2)$$

It follows that if we can find a probability distribution $p(\boldsymbol{b})$ to solve the problem (6.5)-(6.6), then *this produces a C-additive approximation to the max-weight problem (6.1), with* $C = L\log(2)$. It follows that such an algorithm can yield full throughput optimality, and can come arbitrarily close to utility optimality, with an average backlog and delay expression that is polynomial in the network size. Remarkably, the next theorem, developed in (172), shows that the probability distribution (6.4) is the desired distribution, in that it exactly solves the problem (6.5)-(6.6). Thus, the max link-weight-plus-entropy algorithm is a C-additive approximation for the max-weight problem.

Theorem 6.1 *(Jiang-Walrand Theorem (172)) The probability distribution $p^*(\boldsymbol{b})$ that solves (6.5) and (6.6) is given by (6.4).*

Proof. The proof follows directly from the analysis techniques used in (172), although we organize the proof differently below. We first compute the value of the maximization objective under the particular distribution $p^*(\boldsymbol{b})$ given in (6.4). We have:

$$\begin{aligned}
-\sum_{\boldsymbol{b}\in\mathcal{B}} p^*(\boldsymbol{b}) \log(p^*(\boldsymbol{b})) + \sum_{\boldsymbol{b}\in\mathcal{B}} p^*(\boldsymbol{b}) \sum_{l=1}^{L} W_l(t) b_l &= \sum_{\boldsymbol{b}\in\mathcal{B}} p^*(\boldsymbol{b}) \log(A) - \sum_{\boldsymbol{b}\in\mathcal{B}} p^*(\boldsymbol{b}) \sum_{l=1}^{L} W_l(t) b_l \\
&\quad + \sum_{\boldsymbol{b}\in\mathcal{B}} p^*(\boldsymbol{b}) \sum_{l=1}^{L} W_l(t) b_l \\
&= \log(A)
\end{aligned}$$

where we have used the fact that $p^*(\boldsymbol{b})$ is a probability distribution and hence sums to 1. We now show that the expression in the objective of (6.5) for any other distribution $p(\boldsymbol{b})$ is no larger than $\log(A)$, so that $p^*(\boldsymbol{b})$ is optimal for this objective. To this end, consider any other distribution $p(\boldsymbol{b})$. We have:

$$\begin{aligned}
&-\sum_{\boldsymbol{b}\in\mathcal{B}} p(\boldsymbol{b})\log(p(\boldsymbol{b})) + \sum_{\boldsymbol{b}\in\mathcal{B}} p(\boldsymbol{b})\sum_{l=1}^{L} W_l(t)b_l \\
&\quad= -\sum_{\boldsymbol{b}\in\mathcal{B}} p(\boldsymbol{b})\log\left(p^*(\boldsymbol{b})\frac{p(\boldsymbol{b})}{p^*(\boldsymbol{b})}\right) + \sum_{\boldsymbol{b}\in\mathcal{B}} p(\boldsymbol{b})\sum_{l=1}^{L} W_l(t)b_l \\
&\quad= -\sum_{\boldsymbol{b}\in\mathcal{B}} p(\boldsymbol{b})\log\left(\frac{p(\boldsymbol{b})}{p^*(\boldsymbol{b})}\right) - \sum_{\boldsymbol{b}\in\mathcal{B}} p(\boldsymbol{b})\log(p^*(\boldsymbol{b})) \\
&\qquad+\sum_{\boldsymbol{b}\in\mathcal{B}} p(\boldsymbol{b})\sum_{l=1}^{L} W_l(t)b_l \\
&\quad\leq -\sum_{\boldsymbol{b}\in\mathcal{B}} p(\boldsymbol{b})\log(p^*(\boldsymbol{b})) + \sum_{\boldsymbol{b}\in\mathcal{B}} p(\boldsymbol{b})\sum_{l=1}^{L} W_l(t)b_l \qquad (6.7)\\
&\quad= -\sum_{\boldsymbol{b}\in\mathcal{B}} p(\boldsymbol{b})\log(1/A) - \sum_{\boldsymbol{b}\in\mathcal{B}} p(\boldsymbol{b})\sum_{l=1}^{L} W_l(t)b_l \\
&\qquad+\sum_{\boldsymbol{b}\in\mathcal{B}} p(\boldsymbol{b})\sum_{l=1}^{L} W_l(t)b_l \\
&\quad= \log(A)
\end{aligned}$$

where in (6.7), we have used the well known Kullback-Leibler divergence result, which states that the divergence between any two distributions $p^*(\boldsymbol{b})$ and $p(\boldsymbol{b})$ is non-negative (174):

$$d_{KL}(p||p^*) \triangleq \sum_{\boldsymbol{b}\in\mathcal{B}} p(\boldsymbol{b})\log\left(\frac{p(\boldsymbol{b})}{p^*(\boldsymbol{b})}\right) \geq 0$$

Thus, the maximum value of the objective function (6.5) is $\log(A)$, which is achieved by the distribution $p^*(\boldsymbol{b})$, proving the result. □

Assume now the set $\mathcal{B}$ of all valid link activation vectors has a *connectedness property*, so that it is possible to get from any $\boldsymbol{b}_1 \in \mathcal{B}$ to any other $\boldsymbol{b}_2 \in \mathcal{B}$ by a sequence of adding or removing single links, where each step of the sequence produces another valid activation vector in $\mathcal{B}$ (this holds in the reasonable case when removing any activated link from an activation vector in $\mathcal{B}$ yields another activation vector in $\mathcal{B}$). In this case, the distribution (6.4) is particularly interesting because it is the exact stationary distribution associated with a continuous time ergodic Markov chain with state $\boldsymbol{b}(v)$ (where v is a continuous time variable that is not related to the discrete time index t for the current slot). Transitions for this Markov chain take place by having each link l such that $b_l(v) = 1$ deactivate at times according to an independent exponential distribution with rate $\mu = 1$, and having each link l such that $b_l(v) = 0$ independently activate according to an exponential distribution with rate $\lambda_l = \exp(W_l(t))$, provided that turning this link ON does not violate the link constraints $\mathcal{B}$. That the resulting steady state is given by (6.4) can be shown by state space truncation arguments as in (129)(131). This has the form of a simple distributed algorithm where links independently turn ON or OFF, with Carrier Sense Multiple Access (CSMA) telling us if it is possible to turn a new link ON (see also (175)(172)(173)(176)(177) for details on this).

Unfortunately, we need to run such an algorithm in continuous time for a long enough time to reach a near steady state, and this all needs to be done within one slot to implement the result. Of course, we can use a T-slot argument as in Section 6.1.1 to allow more time to reach the steady state, with the understanding that the queue backlog changes by an amount $O(T)$ that yields an additional additive term in our C-additive approximation (see (176) for an argument in this direction using stochastic approximation theory). However, for general networks, the convergence of the Markov chain to near-steady-state takes a non-polynomial amount of time (else, we could solve NP-hard problems with efficient randomized algorithms). This is because the Markov chain can get "trapped" for long durations of time in certain sub-optimal link activations (this is compensated for in the steady state distribution by getting "trapped" in a max-weight link activation for an even longer duration of time). Even computing the normalizing A constant for the distribution in (6.4) is known to be a "#P-complete" problem (178) (see also factor graph approximations in (179)). However, it is known that for link activation sets with certain degree-2 properties, such as those formed by networks on rings, similar Markov chains require only a small (polynomial) time to reach near steady state (180)(181). This may explain why the simulations in (172) for networks with small degree provide good performance.

6.2 MULTIPLICATIVE FACTOR APPROXIMATIONS

While C-additive approximations can push throughput and throughput-utility arbitrarily close to optimal, they may have large convergence times and delays as discussed in the previous section. It is often possible to provide low complexity decisions for $\boldsymbol{b}(t)$ that come within a multiplicative factor of the max-weight solution. This section shows that such algorithms immediately lead to constant-factor stability and throughput-utility guarantees. The result holds for general networks, possibly with time-varying channels, and possibly with non-binary rate vectors.

Let $S(t)$ describe the channel randomness on slot t (i.e., the *topology state*), and let $I(t)$ be the transmission action on slot t, chosen within an abstract set $\mathcal{I}_{\boldsymbol{S}(t)}$. The rate vector $\boldsymbol{b}(t) = (b_1(t), \ldots, b_L(t))$ is determined by a general function of $I(t)$ and $S(t)$:

$$b_l(t) = \hat{b}_l(I(t), S(t)) \;\; \forall l \in \{1, \ldots, L\} \tag{6.8}$$

Definition 6.2 Let β, C be constants such that $0 < \beta \leq 1$ and $C \geq 0$. A (β, C)-*approximation* is an algorithm that makes (possibly randomized) decisions $I(t) \in \mathcal{I}_{S(t)}$ every slot t to satisfy:

$$\sum_{l=1}^{L} W_l(t)\mathbb{E}\left\{\hat{b}_l(I(t), S(t))|\boldsymbol{W}(t)\right\} \geq \beta \sup_{I\in\mathcal{I}_{S(t)}} \left[\sum_{l=1}^{L} W_l(t)\hat{b}_l(I, S(t))\right] - C$$

Under this definition, a $(1, C)$ approximation is the same as a C-additive approximation. It is known that (β, C)-approximations can provide stability in single or multi-hop networks whenever the arrival rates are interior to $\beta\Lambda$, being a β-scaled version of the capacity region (17)(22)(19)(182). For example, if $\beta = 1/2$, then stability is only guaranteed when arrival rates are at most half the distance to the capacity region boundary (so that the region where we can provide stability guarantees shrinks by 50%). Related constant-factor guarantees are available for joint scheduling and flow control to maximize throughput-utility, where the β-scaling goes inside the utility function (see (22)(19) for a precise scaled-utility statement, (137) for applications to cognitive radio, and (154) for applications to channels with errors). Here, we prove this result only for the special case of achieving stability in a 1-hop network. This provides all of the necessary insight with the least amount of notation, and the reader is referred to the above references for proofs of the more general versions.

Consider a 1-hop network with L queues with dynamics:

$$Q_l(t+1) = \max[Q_l(t) - b_l(t), 0] + a_l(t) \ \forall l \in \{1, \ldots, L\}$$

where the service variables $b_l(t)$ are determined by $I(t)$ and $S(t)$ by (6.8), and $\boldsymbol{a}(t) = (a_1(t), \ldots, a_L(t))$ is the random vector of new data arrivals on slot t. Define $\omega(t)\triangleq[S(t), \boldsymbol{a}(t)]$, and assume that $\omega(t)$ is i.i.d. over slots with some probability distribution. Define $\lambda_l = \mathbb{E}\{a_l(t)\}$ as the arrival rate to queue l.

Define an *S-only policy* as a policy that independently chooses $I(t) \in \mathcal{I}_{S(t)}$ based only on a (possibly randomized) function of the observed $S(t)$. Define Γ as the set of all vectors $(\overline{b}_1, \ldots, \overline{b}_L)$ that can be achieved as 1-slot expectations under S-only policies. That is, $(\overline{b}_1, \ldots, \overline{b}_L) \in \Gamma$ if and only if there is a S-only policy $I^*(t)$ that satisfies $I^*(t) \in \mathcal{I}_{S(t)}$ and:

$$\mathbb{E}\left\{\hat{b}_l(I^*(t), S(t))\right\} = \overline{b}_l \ \forall l \in \{1, \ldots, L\}$$

where the expectation in the left-hand-side is with respect to the distribution of $S(t)$ and the possibly randomized decision for $I^*(t)$ that is made in reaction to the observed $S(t)$. For simplicity, assume the set Γ is closed. Recall that for any rate vector $(\lambda_1, \ldots, \lambda_N)$ in the capacity region Λ, there exists a S-only policy $I^*(t)$ that satisfies:

$$\mathbb{E}\left\{\hat{b}_l(I^*(t), S(t))\right\} \geq \lambda_l \ \forall l \in \{1, \ldots, L\}$$

We say that a vector $(\lambda_1, \ldots, \lambda_L)$ is interior to the scaled capacity region $\beta\Lambda$ if there is an $\epsilon > 0$ such that:

$$(\lambda_1 + \epsilon, \ldots, \lambda_L + \epsilon) \in \beta\Lambda$$

Assume second moments of the arrival and service rate processes are bounded. Define $L(\boldsymbol{Q}(t)) = \frac{1}{2}\sum_{l=1}^{L} Q_l(t)^2$, and recall that Lyapunov drift satisfies (see (3.16)):

$$\Delta(\boldsymbol{Q}(t)) \leq B + \sum_{l=1}^{L} Q_l(t)\lambda_l - \sum_{l=1}^{L} Q_l(t)\mathbb{E}\left\{\hat{b}_l(I(t), S(t))|\boldsymbol{Q}(t)\right\} \tag{6.9}$$

where B is a positive constant that depends on the maximum second moments.

Theorem 6.3 *Consider the above 1-hop network with $\omega(t)$ i.i.d. over slots and with arrival rates $(\lambda_1, \ldots, \lambda_L)$. Fix β such that $0 < \beta \leq 1$. Suppose there is an $\epsilon > 0$ such that:*

$$(\lambda_1 + \epsilon, \ldots, \lambda_L + \epsilon) \in \beta\Lambda \tag{6.10}$$

If a (β, C)-approximation is used for all slots t (where $C \geq 0$ is a given constant), and if $\mathbb{E}\{L(\boldsymbol{Q}(0))\} < \infty$, then the network is mean rate stable and strongly stable, with average queue backlog bound:

$$\limsup_{t\to\infty} \frac{1}{t}\sum_{\tau=0}^{t-1}\sum_{l=1}^{L} \mathbb{E}\{Q_l(\tau)\} \leq B/\epsilon$$

where B is the constant from (6.9).

Proof. Fix slot t. Because our decision $I(t)$ yields a (β, C)-approximation for minimizing the final term in the right-hand-side of (6.9), we have:

$$\Delta(\boldsymbol{Q}(t)) \leq B + C + \sum_{l=1}^{L} Q_l(t)\lambda_l - \beta\sum_{l=1}^{L} Q_l(t)\mathbb{E}\left\{\hat{b}_l(I^*(t), S(t))|\boldsymbol{Q}(t)\right\} \tag{6.11}$$

where $I^*(t)$ is any other (possibly randomized) decision in the set $\mathcal{I}_{S(t)}$. Because (6.10) holds, we know that:

$$(\lambda_1/\beta + \epsilon/\beta, \ldots, \lambda_L/\beta + \epsilon/\beta) \in \Lambda$$

Thus, there exists a S-only policy $I^*(t)$ that satisfies:

$$\mathbb{E}\left\{\hat{b}_l(I^*(t), S(t))|\boldsymbol{Q}(t)\right\} = \mathbb{E}\left\{\hat{b}_l(I^*(t), S(t))\right\} \geq \lambda_l/\beta + \epsilon/\beta \;\; \forall l \in \{1, \ldots, L\}$$

where the first equality above holds because $I^*(t)$ is S-only and hence independent of the queue backlogs $\boldsymbol{Q}(t)$. Plugging this policy into the right-hand-side of (6.11) yields:

$$\Delta(\boldsymbol{Q}(t)) \leq B + C + \sum_{l=1}^{L} Q_l(t)\lambda_l - \beta\sum_{l=1}^{L} Q_l(t)(\lambda_l/\beta + \epsilon/\beta) \tag{6.12}$$

$$= B + C - \epsilon\sum_{l=1}^{L} Q_l(t) \tag{6.13}$$

The result then follows by the Lyapunov drift theorem (Theorem 4.1). □

The above theorem can be intuitively interpreted as follows: Any (perhaps approximate) effort to schedule transmissions to maximize the weighted sum of transmission rates translates into good network performance. More concretely, simple greedy algorithms with $\beta = 1/2$ and $C = 0$ (i.e. $(1/2, 0)$-approximation algorithms) exist for networks with *matching constraints* (where links can be simultaneously scheduled if they do not share a common node). Indeed, it can be shown that the *greedy maximal match* algorithm that first selects the largest weight link (breaking ties arbitrarily), then selects the next largest weight link that does not conflict with the previous one, and so on, yields a $(1/2, 0)$-approximation, so that it comes within a factor $\beta = 1/2$ of the max-weight decision (see, for example, (137)). Distributed random access versions of this that produce (β, C) approximations are considered in (154).

Different forms of approximate scheduling, not based on approximating the queue-based max-weight rule, are treated using *maximal matchings* for stable switch scheduling in (183)(102), for stable wireless networks in (184)(104)(103), for utility optimization in (185), and for energy optimization in (186).

CHAPTER 7

Optimization of Renewal Systems

Here we extend the drift-plus-penalty framework to allow optimization over *renewal systems*. In previous chapters, we considered a slotted structure and assumed that every slot t a single random event $\omega(t)$ is observed, a single action $\alpha(t)$ is taken, and the combination of $\alpha(t)$ and $\omega(t)$ generates a vector of attributes (i.e., either penalties or rewards) for that slot. Here, we change the slot structure to a *renewal frame* structure. The frame durations are variable and can depend on the decisions made over the course of the frame. Rather than specifying a single action to take on each frame r, we must specify a *dynamic policy* $\pi[r]$ for the frame. A policy is a contingency plan for making a sequence of decisions, where new random events might take place after each decision in the sequence. This model allows a larger class of problems to be treated, including *Markov Decision Problems*, described in more detail in Section 7.6.2.

An example renewal system is a wireless sensor network that is repeatedly used to perform *sensing tasks*. Assume that each new task starts immediately when the previous task is completed. The duration of each task and the network resources used depend on the policy implemented for that task. Examples of this type are given in Section 7.4 and Exercise 7.1.

7.1 THE RENEWAL SYSTEM MODEL

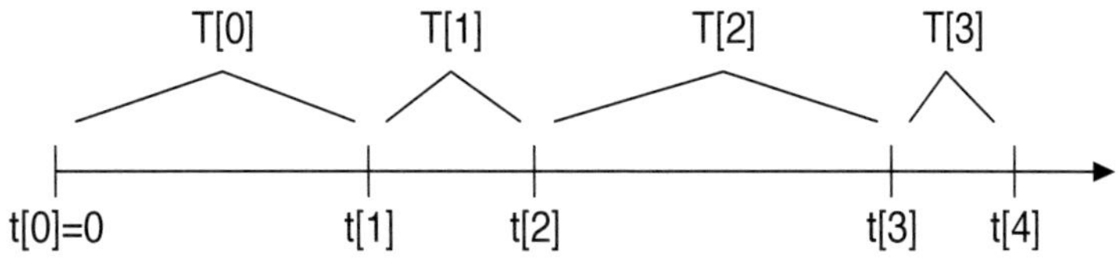

Figure 7.1: An illustration of a sequence of renewal frames.

Consider a dynamic system over the continuous timeline $t \geq 0$ (where t can be a real number). We decompose the timeline into successive *renewal frames*. Renewal frames occur one after the other, and the start of each renewal frame is a time when the system state is "refreshed," which will be made precise below. Define $t[0] = 0$, and let $\{t[0], t[1], t[2], \ldots\}$ be a strictly increasing sequence that represents *renewal events*. For each $r \in \{0, 1, 2, \ldots\}$, the interval of time $[t[r], t[r+1])$ is the

rth renewal frame. Denote $T[r] \triangleq t[r+1] - t[r]$ as the duration of the rth renewal frame (see Fig. 7.1).

At the start of each renewal frame $r \in \{0, 1, 2, \ldots\}$, the controller chooses a *policy* $\pi[r]$ from some abstract policy space $\mathcal{P}$. This policy is implemented over the course of the frame. There may be a sequence of random events during each frame r, and the policy $\pi[r]$ specifies decisions that are made in reaction to these events. The size of the frame $T[r]$ is random and may depend on the policy. Further, the policy on frame r generates a random vector of *penalties* $\boldsymbol{y}[r] = (y_0[r], y_1[r], \ldots, y_L[r])$. We formally write the renewal size $T[r]$ and the penalties $y_l[r]$ as random functions of $\pi[r]$:

$$T[r] = \hat{T}(\pi[r]) \ , \ \ y_l[r] = \hat{y}_l(\pi[r]) \ \forall l \in \{0, 1, \ldots, L\}$$

Thus, given $\pi[r]$, $\hat{T}(\pi[r])$ and $\hat{y}_l(\pi[r])$ are random variables. We make the following renewal assumptions:

- For any policy $\pi \in \mathcal{P}$, the conditional distribution of $(T[r], \boldsymbol{y}[r])$, given $\pi[r] = \pi$, is independent of the events and outcomes from past frames, and is identically distributed for each frame that uses the same policy π.

- The frame sizes $T[r]$ are always strictly positive, and there are finite constants T_{min}, T_{max}, $y_{0,min}$, $y_{0,max}$ such that for all policies $\pi \in \mathcal{P}$, we have:

$$0 < T_{min} \leq \mathbb{E}\left\{\hat{T}(\pi[r])|\pi[r] = \pi\right\} \leq T_{max} \ , \ \ y_{0,min} \leq \mathbb{E}\left\{\hat{y}_0(\pi[r])|\pi[r] = \pi\right\} \leq y_{0,max}$$

- There are finite constants D^2 and $y_{l,max}^2$ for $l \in \{1, \ldots, L\}$ such that for all $\pi \in \mathcal{P}$:

$$\mathbb{E}\left\{\hat{T}(\pi[r])^2|\pi[r] = \pi\right\} \leq D^2 \tag{7.1}$$

$$\mathbb{E}\left\{\hat{y}_l(\pi[r])^2|\pi[r] = \pi\right\} \leq y_{l,max}^2 \ \ \forall l \in \{1, \ldots, L\} \tag{7.2}$$

  That is, second moments are uniformly bounded, regardless of the policy.

In the special case when the system evolves in discrete time with unit time slots, all frame sizes $T[r]$ are positive integers, and $T_{min} = 1$.

7.1.1 THE OPTIMIZATION GOAL

Suppose we have an algorithm that chooses $\pi[r] \in \mathcal{P}$ at the beginning of each frame $r \in \{0, 1, 2, \ldots\}$. Assume temporarily that this algorithm yields well defined frame averages $\overline{T}$ and $\overline{y}_l$ with probability 1, so that:

$$\lim_{R\to\infty} \frac{1}{R}\sum_{r=0}^{R-1} T[r] = \overline{T} \ (w.p.1) \ , \ \ \lim_{R\to\infty} \frac{1}{R}\sum_{r=0}^{R-1} y_l[r] = \overline{y}_l \ (w.p.1) \tag{7.3}$$

We want to design an algorithm that chooses policies $\pi[r]$ over each frame $r \in \{0, 1, 2, \ldots\}$ to solve the following problem:

$$\text{Minimize:} \quad \overline{y}_0/\overline{T} \tag{7.4}$$
$$\text{Subject to:} \quad \overline{y}_l/\overline{T} \leq c_l \;\; \forall l \in \{1, \ldots, L\} \tag{7.5}$$
$$\pi[r] \in \mathcal{P} \;\; \forall r \in \{0, 1, 2, \ldots\} \tag{7.6}$$

where $(c_1, \ldots, c_L)$ are a given collection of real numbers that define time average *cost constraints* for each penalty.

The value $\overline{y}_l/\overline{T}$ represents the *time average penalty* associated with the $y_l[r]$ process. To understand this, note that the time average penalty, sampled at renewal times, is given by:

$$\lim_{R\rightarrow\infty} \frac{\sum_{r=0}^{R-1} y_l[r]}{\sum_{r=0}^{R-1} T[r]} = \frac{\lim_{R\rightarrow\infty} \frac{1}{R}\sum_{r=0}^{R-1} y_l[r]}{\lim_{R\rightarrow\infty} \frac{1}{R}\sum_{r=0}^{R-1} T[r]} = \frac{\overline{y}_l}{\overline{T}}$$

Hence, our goal is to minimize the time average associated with the $y_0[r]$ penalty, subject to the constraint that the time average associated with the $y_l[r]$ process is less than or equal to c_l, for all $l \in \{1, \ldots, L\}$.

As before, we shall find it easier to work with time average *expectations* of the form:

$$\overline{T}[R] \triangleq \frac{1}{R}\sum_{r=0}^{R-1} \mathbb{E}\{T[r]\} \ , \ \ \overline{y}_l[R] \triangleq \frac{1}{R}\sum_{r=0}^{R-1} \mathbb{E}\{y_l[r]\} \;\; \forall l \in \{0, 1, \ldots, L\} \tag{7.7}$$

Under mild boundedness assumptions on $T[r]$ and $y_l[r]$ (for example, when these are deterministically bounded), the Lebesgue dominated convergence theorem ensures that the limiting values of $\overline{T}[R]$ and $\overline{y}_l[R]$ also converge to $\overline{T}$ and $\overline{y}_l$ whenever (7.3) holds (see Exercise 7.9).

7.1.2 OPTIMALITY OVER I.I.D. ALGORITHMS

Define an *i.i.d. algorithm* as one that, at the beginning of each new frame $r \in \{0, 1, 2, \ldots\}$, chooses a policy $\pi[r]$ by independently and probabilistically selecting $\pi \in \mathcal{P}$ according to some distribution that is the same for all frames r. Let $\pi^*[r]$ represent such an i.i.d. algorithm. Then the values $\{\hat{T}(\pi^*[r])\}_{r=0}^{\infty}$ are independent and identically distributed (i.i.d.) over frames, as are $\{\hat{y}_l(\pi^*[r])\}_{r=0}^{\infty}$. Thus, by the law of large numbers, these have well defined averages $\overline{T}^*$ and $\overline{y}_l^*$ with probability 1, where the averages are equal to the expectations over one frame. We say that the problem (7.4)-(7.6) is *feasible* if there is an i.i.d. algorithm $\pi^*[r]$ that satisfies:

$$\frac{\mathbb{E}\left\{\hat{y}_l(\pi^*[r])\right\}}{\mathbb{E}\left\{\hat{T}(\pi^*[r])\right\}} \leq c_l \;\; \forall l \in \{1, \ldots, L\} \tag{7.8}$$

Assuming feasibility, we define $ratio^{opt}$ as the infimum value of the following quantity over all i.i.d. algorithms that meet the constraints (7.8):

$$\frac{\mathbb{E}\left\{\hat{y}_0(\pi^*[r])\right\}}{\mathbb{E}\left\{\hat{T}(\pi^*[r])\right\}}$$

The following lemma is an immediate consequence of these definitions:

Lemma 7.1 *If there is an i.i.d. algorithm that satisfies the feasibility constraints (7.8), then for any $\delta > 0$ there is an i.i.d. algorithm $\pi^*[r]$ that satisfies:*

$$\mathbb{E}\left\{\hat{y}_0(\pi^*[r])\right\} \leq \mathbb{E}\left\{\hat{T}(\pi^*[r])\right\}(ratio^{opt} + \delta) \tag{7.9}$$

$$\mathbb{E}\left\{\hat{y}_l(\pi^*[r])\right\} \leq \mathbb{E}\left\{\hat{T}(\pi^*[r])\right\}c_l \;\; \forall l \in \{1, \ldots, L\} \tag{7.10}$$

The value $ratio^{opt}$ is defined in terms of i.i.d. algorithms. It can be shown that, under mild assumptions, the value $ratio^{opt}$ is *also* the infimum of the objective function in the problem (7.4)-(7.6), which does not restrict to i.i.d. algorithms. This is similar in spirit to Theorems 4.18 and 4.5. However, rather than stating these assumptions and proving this result, we simply use $ratio^{opt}$ as our target, so that we desire to push the time average penalty objective as close as possible to the smallest value that can be achieved over i.i.d. algorithms.

It is often useful to additionally assume that the following "Slater" assumption holds:

Slater Assumption for Renewal Systems: There is a value $\epsilon > 0$ and an i.i.d. algorithm $\pi^*[r]$ such that:

$$\mathbb{E}\left\{\hat{y}_l(\pi^*[r])\right\} \leq \mathbb{E}\left\{\hat{T}(\pi^*[r])\right\}(c_l - \epsilon) \;\; \forall l \in \{1, \ldots, L\} \tag{7.11}$$

7.2 DRIFT-PLUS-PENALTY FOR RENEWAL SYSTEMS

For each $l \in \{1, \ldots, L\}$, define virtual queues $Z_l[r]$ with $Z_l[0] = 0$, and with dynamics as follows:

$$Z_l[r+1] = \max[Z_l[r] + y_l[r] - c_l T[r], 0] \;\; \forall l \in \{1, \ldots, L\} \tag{7.12}$$

Let $\boldsymbol{Z}[r]$ be the vector of queue values, and define the Lyapunov function $L(\boldsymbol{Z}[r])$ by:

$$L(\boldsymbol{Z}[r]) \triangleq \frac{1}{2}\sum_{l=1}^{L} Z_l[r]^2 \tag{7.13}$$

Define the conditional Lyapunov drift $\Delta(\boldsymbol{Z}[r])$ as:

$$\Delta(\boldsymbol{Z}[r]) \triangleq \mathbb{E}\left\{L(\boldsymbol{Z}[r+1]) - L(\boldsymbol{Z}[r]) | \boldsymbol{Z}[r]\right\}$$

Using the same techniques as in previous chapters, it is easy to show that:

$$\Delta(\boldsymbol{Z}[r]) \leq B + \sum_{l=1}^{L} Z_l[r]\mathbb{E}\left\{\hat{y}_l(\pi[r]) - c_l\hat{T}(\pi[r])|\boldsymbol{Z}[r]\right\} \tag{7.14}$$

where B is a finite constant that satisfies the following for all r and all possible $\boldsymbol{Z}[r]$:

$$B \geq \frac{1}{2}\sum_{l=1}^{L}\mathbb{E}\left\{(y_l[r] - c_lT[r])^2|\boldsymbol{Z}[r]\right\} \tag{7.15}$$

Such a finite constant B exists by the boundedness assumptions (7.1)-(7.2). The drift-plus-penalty for frame r thus satisfies:

$$\begin{aligned}\Delta(\boldsymbol{Z}[r]) + V\mathbb{E}\{y_0[r]|\boldsymbol{Z}[r]\} \quad \leq \quad & B + V\mathbb{E}\left\{\hat{y}_0(\pi[r])|\boldsymbol{Z}[r]\right\} + \sum_{l=1}^{L} Z_l[r]\mathbb{E}\left\{\hat{y}_l(\pi[r])|\boldsymbol{Z}[r]\right\} \\ & - \sum_{l=1}^{L} Z_l[r]c_l\mathbb{E}\left\{\hat{T}(\pi[r])|\boldsymbol{Z}[r]\right\} \end{aligned} \tag{7.16}$$

This variable-frame drift methodology was developed in (56)(57) for optimizing delay in networks defined on Markov chains. However, the analysis in (56)(57) used a policy based on minimizing the right-hand-side of the above inequality, which was only shown to be effective for pure feasibility problems (where $\hat{y}_0(\pi[r]) = 0$ for all r) or for problems where the frame durations are independent of the policy (see also Exercise 7.3). Our algorithm below, which can be applied to the general problem, is inspired by the decision rule in (58), which minimizes the ratio of expected drift-plus-penalty over expected frame size.

Renewal-Based Drift-Plus-Penalty Algorithm: At the beginning of each frame $r \in \{0, 1, 2, \ldots\}$, observe $\boldsymbol{Z}[r]$ and do the following:

- Choose a policy $\pi[r] \in \mathcal{P}$ that minimizes the following ratio:

$$\frac{\mathbb{E}\left\{V\hat{y}_0(\pi[r]) + \sum_{l=1}^{L} Z_l[r]\hat{y}_l(\pi[r])|\boldsymbol{Z}[r]\right\}}{\mathbb{E}\left\{\hat{T}(\pi[r])|\boldsymbol{Z}[r]\right\}} \tag{7.17}$$

- Update the virtual queues $Z_l[r]$ by (7.12).

As before, we define a C-additive approximation to the ratio-minimizing decision as follows.

Definition 7.2 A policy $\pi[r]$ is a *C-additive approximation* of the policy that minimizes (7.17) if:

$$\frac{\mathbb{E}\left\{V\hat{y}_0(\pi[r]) + \sum_{l=1}^{L} Z_l[r]\hat{y}_l(\pi[r])|\boldsymbol{Z}[r]\right\}}{\mathbb{E}\left\{\hat{T}(\pi[r])|\boldsymbol{Z}[r]\right\}} \leq C + \inf_{\pi\in\mathcal{P}}\left[\frac{\mathbb{E}\left\{V\hat{y}_0(\pi) + \sum_{l=1}^{L} Z_l[r]\hat{y}_l(\pi)|\boldsymbol{Z}[r]\right\}}{\mathbb{E}\left\{\hat{T}(\pi)|\boldsymbol{Z}[r]\right\}}\right]$$

In particular, if policy $\pi[r]$ is a C-additive approximation, then:

$$\mathbb{E}\left\{V\hat{y}_0(\pi[r]) + \sum_{l=1}^{L} Z_l[r]\hat{y}_l(\pi[r])|\boldsymbol{Z}[r]\right\} \leq CT_{max}$$
$$+\mathbb{E}\left\{\hat{T}(\pi[r])|\boldsymbol{Z}[r]\right\}\frac{\mathbb{E}\left\{V\hat{y}_0(\pi^*[r]) + \sum_{l=1}^{L} Z_l[r]\hat{y}_l(\pi^*[r])\right\}}{\mathbb{E}\left\{\hat{T}(\pi^*[r])\right\}} \tag{7.18}$$

where $\pi^*[r]$ is any i.i.d. algorithm that is chosen in $\mathcal{P}$ and is independent of queues $\boldsymbol{Z}[r]$. In the above inequality, we have used the fact that:

$$\mathbb{E}\left\{\hat{T}(\pi[r])|\boldsymbol{Z}[r]\right\} \leq T_{max}$$

Theorem 7.3 *(Renewal-Based Drift-Plus-Penalty Performance) Assume there is an i.i.d. algorithm $\pi^*[r]$ that satisfies the feasibility constraints (7.8). Suppose we implement the above renewal-based drift-plus-penalty algorithm using a C-additive approximation for all frames r, with initial condition $Z_l[0] = 0$ for all $l \in \{1, \ldots, L\}$. Then:*

a) All queues $Z_l[r]$ are mean rate stable, in that:

$$\lim_{R\to\infty} \frac{\mathbb{E}\{Z_l[R]\}}{R} = 0 \;\; \forall l \in \{1, \ldots, L\}$$

b) For all $l \in \{1, \ldots, L\}$ we have:

$$\limsup_{R\to\infty}(\overline{y}_l[R] - c_l\overline{T}[R]) \leq 0 \quad \textit{and so} \quad \limsup_{R\to\infty} \overline{y}_l[R]/\overline{T}[R] \leq c_l$$

where $\overline{y}_l[R]$ and $\overline{T}[R]$ are defined in (7.7).

c) The penalty process $y_0[r]$ satisfies the following for all $R > 0$:

$$\overline{y}_0[R] - ratio^{opt}\overline{T}[R] \leq \frac{B + CT_{max}}{V}$$

where B is defined in (7.15).

d) If the Slater assumption (7.11) holds for a constant $\epsilon > 0$, then all queues $Z_l[r]$ are strongly stable and satisfy the following for all $R > 0$:

$$\frac{1}{R}\sum_{r=0}^{R-1}\sum_{l=1}^{L}\mathbb{E}\{Z_l[r]\} \leq \frac{VF}{\epsilon T_{min}} \tag{7.19}$$

where the constant F is defined below in (7.22). Further, if for all $l \in \{1, \ldots, L\}$, $y_l[r] - c_l T[r]$ is either deterministically lower bounded or deterministically upper bounded, then queues $Z_l[r]$ are rate stable and:

$$\limsup_{R\to\infty}\left[\frac{\frac{1}{R}\sum_{r=0}^{R-1} y_l[r]}{\frac{1}{R}\sum_{r=0}^{R-1} T[r]}\right] \leq c_l \ \forall l \in \{1, \ldots, L\} \ (w.p.1)$$

Proof. (Theorem 7.3) Because we use a C-additive approximation every frame r, we know that (7.18) holds. Plugging the i.i.d. algorithm $\pi^*[r]$ from (7.18) into the right-hand-side of the drift-plus-penalty inequality (7.16) yields:

$$\begin{aligned}\Delta(\boldsymbol{Z}[r]) + V\mathbb{E}\{y_0[r]|\boldsymbol{Z}[r]\} \ &\leq \ B + CT_{max} + \frac{\mathbb{E}\left\{\hat{T}(\pi[r])|\boldsymbol{Z}[r]\right\}}{\mathbb{E}\left\{\hat{T}(\pi^*[r])\right\}}V\mathbb{E}\left\{\hat{y}_0(\pi^*[r])\right\} \\ +\frac{\mathbb{E}\left\{\hat{T}(\pi[r])|\boldsymbol{Z}[r]\right\}}{\mathbb{E}\left\{\hat{T}(\pi^*[r])\right\}}&\sum_{l=1}^{L} Z_l[r]\mathbb{E}\left\{\hat{y}_l(\pi^*[r])\right\} - \sum_{l=1}^{L} Z_l[r]c_l\mathbb{E}\left\{\hat{T}(\pi[r])|\boldsymbol{Z}[r]\right\} \end{aligned} \quad (7.20)$$

where $\pi^*[r]$ is any policy in $\mathcal{P}$. Now fix $\delta > 0$, and plug into the right-hand-side of (7.20) the policy $\pi^*[r]$ that satisfies (7.9)-(7.10), which makes decisions independent of $\boldsymbol{Z}[r]$, to yield:

$$\Delta(\boldsymbol{Z}[r]) + V\mathbb{E}\{y_0[r]|\boldsymbol{Z}[r]\} \ \leq \ B + CT_{max} + \mathbb{E}\left\{\hat{T}(\pi[r])|\boldsymbol{Z}[r]\right\} V(ratio^{opt} + \delta)$$

The above holds for all $\delta > 0$. Taking a limit as $\delta \to 0$ yields:

$$\Delta(\boldsymbol{Z}[r]) + V\mathbb{E}\{y_0[r]|\boldsymbol{Z}[r]\} \ \leq \ B + CT_{max} + \mathbb{E}\left\{\hat{T}(\pi[r])|\boldsymbol{Z}[r]\right\} Vratio^{opt} \quad (7.21)$$

To prove part (a), we can rearrange (7.21) to yield:

$$\Delta(\boldsymbol{Z}[r]) \leq B + CT_{max} + V\max[ratio^{opt}T_{max}, ratio^{opt}T_{min}] - Vy_{0,min}$$

where we use $\max[ratio^{opt}T_{max}, ratio^{opt}T_{min}]$ because $ratio^{opt}$ may be negative. This proves that all components $Z_l[r]$ are mean rate stable by Theorem 4.1, proving part (a). The first lim sup statement in part (b) follows immediately from mean rate stability of $Z_l[r]$ (via Theorem 2.5(b)). The second lim sup statement in part (b) follows from the first (see Exercise 7.4).

To prove part (c), we take expectations of (7.21) to find:

$$\mathbb{E}\{L(\boldsymbol{Z}[r+1])\} - \mathbb{E}\{L(\boldsymbol{Z}[r])\} + V\mathbb{E}\{y_0[r]\} \ \leq \ B + CT_{max} + \mathbb{E}\left\{\hat{T}(\pi[r])\right\} Vratio^{opt}$$

Summing over $r \in \{0, \ldots, R-1\}$ and dividing by RV yields:

$$\frac{\mathbb{E}\{L(\boldsymbol{Z}[R])\} - \mathbb{E}\{L(\boldsymbol{Z}[0])\}}{RV} + \frac{1}{R}\sum_{r=0}^{R-1}\mathbb{E}\{y_0[r]\} \leq \frac{B + CT_{max}}{V} + ratio^{opt}\frac{1}{R}\sum_{r=0}^{R-1}\mathbb{E}\{T[r]\}$$

Using the definitions of $\overline{y}_0[R]$ and $\overline{T}[R]$ in (7.7) and noting that $\mathbb{E}\{L(\boldsymbol{Z}[R])\} \geq 0$ and $\mathbb{E}\{L(\boldsymbol{Z}[0])\} = 0$ yields:

$$\overline{y}_0[R] \leq \frac{B + CT_{max}}{V} + ratio^{opt}\overline{T}[R]$$

This proves part (c).

Part (d) follows from plugging the policy $\pi^*[r]$ from (7.11) into (7.20) to obtain:

$$\Delta(\boldsymbol{Z}[r]) + V\mathbb{E}\{y_0[r]|\boldsymbol{Z}[r]\} \leq B + CT_{max} + V\frac{\mathbb{E}\left\{\hat{T}(\pi[r])|\boldsymbol{Z}[r]\right\}}{\mathbb{E}\left\{\hat{T}(\pi^*[r])\right\}}y_{0,max} - \epsilon T_{min}\sum_{l=1}^{L} Z_l[r]$$

This can be written in the form:

$$\Delta(\boldsymbol{Z}[r]) \leq VF - \epsilon T_{min}\sum_{l=1}^{L} Z_l[r]$$

where the constant F is defined:

$$F \triangleq \frac{B + CT_{max}}{V} + \max\left[\frac{T_{max}}{T_{min}}y_{0,max}, \frac{T_{min}}{T_{max}}y_{0,max}\right] - y_{0,min} \tag{7.22}$$

Thus, from Theorem 4.1, we have that (7.19) holds, so that all queues $Z_l[r]$ are strongly stable. In the special case when the $y_l[r] - c_lT[r]$ are deterministically bounded, we have by the Strong Stability Theorem (Theorem 2.8) that all queues are rate stable. Thus, by Theorem 2.5(a):

$$\limsup_{R\to\infty}\left[\frac{1}{R}\sum_{r=0}^{R-1} y_l[r] - c_l\frac{1}{R}\sum_{r=0}^{R-1} T[r]\right] \leq 0 \quad (w.p.1)$$

However:

$$\begin{aligned}
\frac{\frac{1}{R}\sum_{r=0}^{R-1} y_l[r]}{\frac{1}{R}\sum_{r=0}^{R-1} T[r]} - c_l &\leq \max\left[\frac{\frac{1}{R}\sum_{r=0}^{R-1} y_l[r]}{\frac{1}{R}\sum_{r=0}^{R-1} T[r]} - c_l, 0\right]\frac{\frac{1}{R}\sum_{r=0}^{R-1} T[r]}{\frac{1}{R}\sum_{r=0}^{R-1} T[r]} \\
&= \max\left[\frac{1}{R}\sum_{r=0}^{R-1} y_l[r] - c_l\frac{1}{R}\sum_{r=0}^{R-1} T[r], 0\right]\frac{1}{\frac{1}{R}\sum_{r=0}^{R-1} T[r]}
\end{aligned} \tag{7.23}$$

Further, because for all $r \in \{1, 2, \ldots\}$ we have $\mathbb{E}\{T[r]|T[0], T[1], \ldots, T[r-1]\} \geq T_{min}$ and $\mathbb{E}\left\{T[r]^2|T[0], T[1], \ldots, T[r-1]\right\} \leq D^2$, from Lemma 4.3 it follows that:

$$\liminf_{R\to\infty}\frac{1}{R}\sum_{r=0}^{R-1} T[r] \geq T_{min} > 0 \quad (w.p.1)$$

and so taking a lim sup of (7.23) yields:

$$\limsup_{R\to\infty}\left[\frac{\frac{1}{R}\sum_{r=0}^{R-1} y_l[r]}{\frac{1}{R}\sum_{r=0}^{R-1} T[r]} - c_l\right] \leq 0 \times \frac{1}{T_{min}} = 0 \ (w.p.1)$$

This proves part (d). □

The above theorem shows that time average penalty can be pushed to within $O(1/V)$ of optimal (for arbitrarily large V). The tradeoff is that the virtual queues are $O(V)$ in size, which affects the time required for the penalties to be close to their required time averages c_l.

7.2.1 ALTERNATE FORMULATIONS

In some cases, we care more about $\overline{y}_l$ itself, rather than $\overline{y}_l/\overline{T}$. Consider the following variation of problem (7.4)-(7.6):

$$\begin{aligned} \text{Minimize:} \quad & \overline{y}_0/\overline{T} \\ \text{Subject to:} \quad & \overline{y}_l \leq 0 \ \forall l \in \{1, \ldots, L\} \\ & \pi[r] \in \mathcal{P} \ \forall r \in \{0, 1, 2, \ldots\} \end{aligned}$$

This changes the constraints from $\overline{y}_l/\overline{T} \leq c_l$ to $\overline{y}_l \leq 0$. However, this is just a special case of the original problem (7.4)-(7.6) with $c_l = 0$.

Now suppose we seek to minimize $\overline{y}_0$, rather than $\overline{y}_0/\overline{T}$. The problem is:

$$\begin{aligned} \text{Minimize:} \quad & \overline{y}_0 \\ \text{Subject to:} \quad & \overline{y}_l/\overline{T} \leq c_l \ \forall l \in \{1, \ldots, L\} \\ & \pi[r] \in \mathcal{P} \ \forall r \in \{0, 1, 2, \ldots\} \end{aligned}$$

This problem has a significantly different structure than (7.4)-(7.6), and it is considerably easier to solve. Indeed, Exercise 7.3 shows that it can be solved by minimizing an expectation every frame, rather than a ratio of expectations.

Finally, we note that Exercise 7.5 explores an alternative algorithm for the original problem (7.4)-(7.6). The alternative uses only a minimum of an expectation every frame, rather than a ratio of expectations.

7.3 MINIMIZING THE DRIFT-PLUS-PENALTY RATIO

We re-write the drift-plus-penalty ratio (7.17) in the following simplified form:

$$\frac{\mathbb{E}\{a(\pi)\}}{\mathbb{E}\{b(\pi)\}}$$

where $a(\pi)$ represents the numerator and $b(\pi)$ the denominator, both expressed as a function of the policy $\pi \in \mathcal{P}$. We note that $T_{max} \geq \mathbb{E}\{b(\pi)\} \geq T_{min} > 0$ for all $\pi \in \mathcal{P}$. Define θ^* as the infimum of the above ratio:

$$\theta^* \triangleq \inf_{\pi \in \mathcal{P}} \left[\frac{\mathbb{E}\{a(\pi)\}}{\mathbb{E}\{b(\pi)\}} \right] \tag{7.24}$$

We want to understand how to find θ^*.

In the special case when $\mathbb{E}\{b(\pi)\}$ does not depend on the policy π (which holds when the expected renewal interval size is the same for all policies), the minimization is achieved by choosing $\pi \in \mathcal{P}$ to minimize $\mathbb{E}\{a(\pi)\}$. This is important because the minimization of an expectation is typically much simpler than a minimization of the ratio of expectations, and it can often be accomplished through *dynamic programming algorithms* (64)(67)(57) and their special cases of *stochastic shortest path algorithms*.

To treat the case when $\mathbb{E}\{b(\pi)\}$ may depend on the policy, we use the following simple but useful lemmas.

Lemma 7.4 *For any policy $\pi \in \mathcal{P}$, we have:*

$$\mathbb{E}\left\{a(\pi) - \theta^* b(\pi)\right\} \geq 0 \tag{7.25}$$

with equality if and only if policy π achieves the infimum ratio $\mathbb{E}\{a(\pi)\} / \mathbb{E}\{b(\pi)\} = \theta^*$.

Proof. By definition of θ^*, we have for any policy $\pi \in \mathcal{P}$:

$$\frac{\mathbb{E}\{a(\pi)\}}{\mathbb{E}\{b(\pi)\}} \geq \inf_{\pi \in \mathcal{P}} \left[\frac{\mathbb{E}\{a(\pi)\}}{\mathbb{E}\{b(\pi)\}} \right] = \theta^*$$

Multiplying both sides by $\mathbb{E}\{b(\pi)\}$ and noting that $\mathbb{E}\{b(\pi)\} > 0$ yields (7.25). That equality holds if and only if $\mathbb{E}\{a(\pi)\} / \mathbb{E}\{b(\pi)\} = \theta^*$ follows immediately. □

Lemma 7.5 *We have:*

$$\inf_{\pi \in \mathcal{P}} \mathbb{E}\left\{a(\pi) - \theta^* b(\pi)\right\} = 0 \tag{7.26}$$

Further, for any value $\theta \in \mathbb{R}$, we have:

$$\inf_{\pi \in \mathcal{P}} \mathbb{E}\{a(\pi) - \theta b(\pi)\} < 0 \quad \textit{if } \theta > \theta^* \tag{7.27}$$

$$\inf_{\pi \in \mathcal{P}} \mathbb{E}\{a(\pi) - \theta b(\pi)\} > 0 \quad \textit{if } \theta < \theta^* \tag{7.28}$$

Proof. To prove (7.26), note from Lemma 7.4 that we have for any policy π:

$$0 \leq \mathbb{E}\left\{a(\pi) - \theta^* b(\pi)\right\} = \mathbb{E}\{b(\pi)\}\left(\frac{\mathbb{E}\{a(\pi)\}}{\mathbb{E}\{b(\pi)\}} - \theta^*\right) \leq T_{max}\left(\frac{\mathbb{E}\{a(\pi)\}}{\mathbb{E}\{b(\pi)\}} - \theta^*\right)$$

Taking infimums over $\pi \in \mathcal{P}$ of the above yields:

$$0 \leq \inf_{\pi \in \mathcal{P}} \mathbb{E}\left\{a(\pi) - \theta^* b(\pi)\right\} \leq T_{max} \inf_{\pi \in \mathcal{P}}\left(\frac{\mathbb{E}\{a(\pi)\}}{\mathbb{E}\{b(\pi)\}} - \theta^*\right) = 0$$

where the final equality uses the definition of θ^* in (7.24). This proves (7.26).

To prove (7.27), suppose that $\theta > \theta^*$. Then:

$$\begin{aligned}
\inf_{\pi \in \mathcal{P}} \mathbb{E}\{a(\pi) - \theta b(\pi)\} &= \inf_{\pi \in \mathcal{P}}\left[\mathbb{E}\left\{a(\pi) - \theta^* b(\pi)\right\} - (\theta - \theta^*)\mathbb{E}\{b(\pi)\}\right] \\
&\leq \inf_{\pi \in \mathcal{P}} \mathbb{E}\left\{a(\pi) - \theta^* b(\pi)\right\} - (\theta - \theta^*)T_{min} \\
&= -(\theta - \theta^*)T_{min} < 0
\end{aligned}$$

where we have used (7.26). This proves (7.27). To prove (7.28), suppose $\theta < \theta^*$. Then:

$$\begin{aligned}
\inf_{\pi \in \mathcal{P}} \mathbb{E}\{a(\pi) - \theta b(\pi)\} &= \inf_{\pi \in \mathcal{P}}\left[\mathbb{E}\left\{a(\pi) - \theta^* b(\pi)\right\} + \mathbb{E}\left\{(\theta^* - \theta)b(\pi)\right\}\right] \\
&\geq \inf_{\pi \in \mathcal{P}} \mathbb{E}\left\{a(\pi) - \theta^* b(\pi)\right\} + (\theta^* - \theta)T_{min} \\
&= (\theta^* - \theta)T_{min} > 0
\end{aligned}$$

□

7.3.1 THE BISECTION ALGORITHM

Lemmas 7.4 and 7.5 show that we can approach the optimal ratio θ^* with a simple *iterative bisection algorithm* that computes infimums of expectations at each step. Specifically, suppose that on stage k of the iteration, we have finite bounds $\theta_{min}^{(k)}$ and $\theta_{max}^{(k)}$ such that we know:

$$\theta_{min}^{(k)} < \theta^* < \theta_{max}^{(k)}$$

Define $\theta_{bisect}^{(k)}$ as:

$$\theta_{bisect}^{(k)} \triangleq (\theta_{max}^{(k)} + \theta_{min}^{(k)})/2$$

We then compute $\inf_{\pi \in \mathcal{P}} \mathbb{E}\left\{a(\pi) - \theta_{bisect}^{(k)} b(\pi)\right\}$. If the result is 0, then $\theta_{bisect}^{(k)} = \theta^*$. If the result is positive then we know $\theta_{bisect}^{(k)} < \theta^*$, and if the result is negative we know $\theta_{bisect}^{(k)} > \theta^*$. We then appropriately adjust our upper and lower bounds for stage $k+1$. The uncertainty interval decreases by a factor of 2 on each stage, and so this algorithm converges exponentially fast to the value θ^*. This is useful because each stage involves minimizing an expectation, rather than a ratio of expectations.

7.3.2 OPTIMIZATION OVER PURE POLICIES

Let $\mathcal{P}^{pure}$ be any finite or countably infinite set of policies that we call *pure policies*:

$$\mathcal{P}^{pure} = \{\pi_1, \pi_2, \pi_3, \ldots\}$$

Let $\mathcal{P}$ be the larger policy space that considers all probabilistic mixtures of pure policies. Specifically, the space $\mathcal{P}$ considers policies that make a randomized decision about which policy $\pi_i \in \mathcal{P}^{pure}$ to use, according to some probabilities $q_i = Pr[\text{Implement policy } \pi_i]$ with $\sum_{i=1}^{\infty} q_i = 1$. It turns out that minimizing the ratio $\mathbb{E}\{a(\pi)\}/\mathbb{E}\{b(\pi)\}$ over $\pi \in \mathcal{P}$ can be achieved by considering only pure policies $\pi \in \mathcal{P}^{pure}$. To see this, define θ^* as the infimum ratio over $\pi \in \mathcal{P}$, and for simplicity, assume that θ^* is achieved by some particular policy $\pi^* \in \mathcal{P}$, which corresponds to a probability distribution $(q_1^*, q_2^*, \ldots)$ for selecting pure policies $(\pi_1, \pi_2, \ldots)$. Then:

$$\begin{aligned}
0 = \mathbb{E}\left\{a(\pi^*) - \theta^* b(\pi^*)\right\} &= \sum_{i=1}^{\infty} q_i^* \mathbb{E}\left\{a(\pi_i) - \theta^* b(\pi_i)\right\} \\
&\geq \sum_{i=1}^{\infty} q_i^* \left[\inf_{\pi \in \mathcal{P}^{pure}} \mathbb{E}\left\{a(\pi) - \theta^* b(\pi)\right\}\right] \\
&= \inf_{\pi \in \mathcal{P}^{pure}} \mathbb{E}\left\{a(\pi) - \theta^* b(\pi)\right\}
\end{aligned}$$

On the other hand, because $\mathcal{P}$ is a larger policy space than $\mathcal{P}^{pure}$, we have:

$$0 = \inf_{\pi \in \mathcal{P}} \mathbb{E}\left\{a(\pi) - \theta^* b(\pi)\right\} \leq \inf_{\pi \in \mathcal{P}^{pure}} \mathbb{E}\left\{a(\pi) - \theta^* b(\pi)\right\}$$

Thus:

$$\inf_{\pi \in \mathcal{P}^{pure}} \mathbb{E}\left\{a(\pi) - \theta^* b(\pi)\right\} = 0$$

which shows that the infimum ratio θ^* can be found over the set of pure policies.

The same result holds more generally: Let $\mathcal{P}^{pure}$ be any (possibly uncountably infinite) set of policies that we call pure policies. Define Ω as the set of all vectors $(\mathbb{E}\{a(\pi)\}, \mathbb{E}\{b(\pi)\})$ that can be achieved by policies $\pi \in \mathcal{P}^{pure}$. Suppose $\mathcal{P}$ is a larger policy space that contains all pure policies and is such that the set of all vectors $(\mathbb{E}\{a(\pi)\}, \mathbb{E}\{b(\pi)\})$ that can be achieved by policies $\pi \in \mathcal{P}$ is equal to the convex hull of Ω, denoted $Conv(\Omega)$.[1] If θ^* is the infimum ratio of $\mathbb{E}\{a(\pi)\}/\mathbb{E}\{b(\pi)\}$ over $\pi \in \mathcal{P}$, then:

$$\begin{aligned}
0 = \inf_{\pi \in \mathcal{P}} \mathbb{E}\left\{a(\pi) - \theta^* b(\pi)\right\} &= \inf_{(a,b) \in Conv(\Omega)} [a - \theta^* b] \\
&= \inf_{(a,b) \in \Omega} [a - \theta^* b] \\
&= \inf_{\pi \in \mathcal{P}^{pure}} \mathbb{E}\left\{a(\pi) - \theta^* b(\pi)\right\}
\end{aligned}$$

[1]The convex hull of a set $\Omega \subseteq \mathbb{R}^k$ (for some integer $k > 0$) is the set of all finite probabilistic mixtures of vectors in Ω. It can be shown that $Conv(\Omega)$ is the set of all expectations $\mathbb{E}\{\boldsymbol{X}\}$ that can be achieved by random vectors $\boldsymbol{X}$ that take values in the set Ω according to any probability distribution that leads to a finite expectation.

where we have used the well known fact that the infimum of a linear function over the convex hull of a set is equal to the infimum over the set itself. Therefore, by Lemma 7.4, it follows that θ^* is also the infimum ratio of $\mathbb{E}\{a(\pi)\}/\mathbb{E}\{b(\pi)\}$ over the smaller set of pure policies $\mathcal{P}^{pure}$.

7.3.3 CAVEAT — FRAMES WITH INITIAL INFORMATION

Suppose at the beginning of each frame r, we observe a vector $\boldsymbol{\eta}[r]$ of *initial information* that influences the penalties and frame size. Assume $\{\boldsymbol{\eta}[r]\}_{r=0}^{\infty}$ is i.i.d. over frames. Each policy $\pi \in \mathcal{P}$ first observes $\boldsymbol{\eta}[r]$ and then chooses a *sub-policy* $\pi' \in \mathcal{P}_{\boldsymbol{\eta}[r]}$ that possibly depends on the observed $\boldsymbol{\eta}[r]$. One might (incorrectly) implement the policy that first observes $\boldsymbol{\eta}[r]$ and then chooses $\pi' \in \mathcal{P}_{\boldsymbol{\eta}[r]}$ that minimizes the ratio of conditional expectations $\mathbb{E}\left\{a(\pi')|\boldsymbol{\eta}[r]\right\}/\mathbb{E}\left\{b(\pi')|\boldsymbol{\eta}[r]\right\}$. This would work if the denominator does not depend on the policy, but it may be incorrect in general. Minimizing the ratio of expectations is not always achieved by the policy that minimizes the ratio of conditional expectations given the observed initial information. For example, suppose there are two possible initial vectors $\boldsymbol{\eta}_1$ and $\boldsymbol{\eta}_2$, both equally likely. Suppose there are two possible policies for each vector:

- Under $\boldsymbol{\eta}_1$: π_{11} gives $[a = 1, b = 1]$, π_{12} gives $[a = 2, b = 1]$.
- Under $\boldsymbol{\eta}_2$: π_{21} gives $[a = 20, b = 10]$, π_{22} gives $[a = .4, b = .1]$.

It can be shown that any achievable $(\mathbb{E}\{a(\pi)\}, \mathbb{E}\{b(\pi)\})$ vector can be achieved by a probabilistic mixture of the following four *pure policies*:

- Pure policy π_1: Choose π_{11} if $\boldsymbol{\eta}[r] = \boldsymbol{\eta}_1$, π_{21} if $\boldsymbol{\eta}[r] = \boldsymbol{\eta}_2$.
- Pure policy π_2: Choose π_{11} if $\boldsymbol{\eta}[r] = \boldsymbol{\eta}_1$, π_{22} if $\boldsymbol{\eta}[r] = \boldsymbol{\eta}_2$.
- Pure policy π_3: Choose π_{12} if $\boldsymbol{\eta}[r] = \boldsymbol{\eta}_1$, π_{21} if $\boldsymbol{\eta}[r] = \boldsymbol{\eta}_2$.
- Pure policy π_4: Choose π_{12} if $\boldsymbol{\eta}[r] = \boldsymbol{\eta}_1$, π_{22} if $\boldsymbol{\eta}[r] = \boldsymbol{\eta}_2$.

Clearly π_{11} minimizes the conditional ratio a/b given $\boldsymbol{\eta}_1$, and π_{21} minimizes the conditional ratio a/b given $\boldsymbol{\eta}_2$. The policy π_1 that chooses π_{11} whenever $\boldsymbol{\eta}_1$ is observed, and π_{21} whenever $\boldsymbol{\eta}_2$ is observed, yields:

$$\frac{\mathbb{E}\{a(\pi_1)\}}{\mathbb{E}\{b(\pi_1)\}} = \frac{(1/2)1 + (1/2)20}{(1/2)1 + (1/2)10} = \frac{10.5}{5.5} \approx 1.909$$

On the other hand, the policy that minimizes the ratio $\mathbb{E}\{a(\pi)\}/\mathbb{E}\{b(\pi)\}$ is the policy π_2, which chooses π_{11} whenever $\boldsymbol{\eta}_1$ is observed, and chooses π_{22} whenever $\boldsymbol{\eta}_2$ is observed:

$$\frac{\mathbb{E}\{a(\pi_2)\}}{\mathbb{E}\{b(\pi_2)\}} = \frac{(1/2)1 + (1/2)0.4}{(1/2)1 + (1/2)0.1} = \frac{.7}{.55} \approx 1.273$$

A correct minimization of the ratio can be obtained as follows: If we happen to know the optimal ratio θ^*, we can use the fact that:

$$0 = \inf_{\pi \in \mathcal{P}} \mathbb{E}\left\{a(\pi) - \theta^* b(\pi)\right\} = \mathbb{E}\left\{\inf_{\pi' \in \mathcal{P}_{\boldsymbol{\eta}[r]}} \mathbb{E}\left\{a(\pi') - \theta^* b(\pi')|\boldsymbol{\eta}[r]\right\}\right\}$$

and so using the policy π^* that first observes $\boldsymbol{\eta}[r]$ and then chooses $\pi' \in \mathcal{P}_{\boldsymbol{\eta}[r]}$ to minimize the conditional expectation $\mathbb{E}\left\{a(\pi') - \theta^* b(\pi')|\boldsymbol{\eta}[r]\right\}$ yields $\mathbb{E}\left\{a(\pi^*) - \theta^* b(\pi^*)\right\} = 0$, which by Lemma 7.4 shows it must also minimize the ratio $\mathbb{E}\left\{a(\pi)\right\}/\mathbb{E}\left\{b(\pi)\right\}$.

If θ^* is unknown, we can compute an approximation of θ^* via the bisection algorithm as follows. At step k, we have $\theta_{bisect}[k]$, and we want to compute:

$$\inf_{\pi\in\mathcal{P}} \mathbb{E}\left\{a(\pi) - \theta_{bisect}[k]b(\pi)\right\} = \mathbb{E}\left\{\inf_{\pi'\in\mathcal{P}_{\boldsymbol{\eta}[r]}} \left[\mathbb{E}\left\{a(\pi') - \theta_{bisect}[k]b(\pi')|\boldsymbol{\eta}[r]\right\}\right]\right\}$$

This can be done by generating a collection of W i.i.d. samples $\{\boldsymbol{\eta}_1, \boldsymbol{\eta}_2, \ldots, \boldsymbol{\eta}_W\}$ (all with the same distribution as $\boldsymbol{\eta}[r]$), computing the infimum conditional expectation for each sample, and then using the law of large numbers to approximate the expectation as follows:

$$\mathbb{E}\left\{\inf_{\pi'\in\mathcal{P}_{\boldsymbol{\eta}[r]}} \left[\mathbb{E}\left\{a(\pi') - \theta_{bisect}[k]b(\pi')|\boldsymbol{\eta}[r]\right\}\right]\right\} \approx$$
$$\frac{1}{W}\sum_{w=1}^{W} \inf_{\pi'\in\mathcal{P}_{\boldsymbol{\eta}_w}} \mathbb{E}\left\{a(\pi') - \theta_{bisect}[k]b(\pi')|\boldsymbol{\eta}[r] = \boldsymbol{\eta}_w\right\} \triangleq val(\theta_{bisect}[k]) \tag{7.29}$$

For a given frame r, the same samples $\{\boldsymbol{\eta}_1, \ldots, \boldsymbol{\eta}_W\}$ should be used for each step of the bisection routine. This ensures the stage-r approximation function $val(\theta)$ uses the same samples and is thus non-increasing in θ, important for the bisection to work properly (see Exercise 7.2). However, new samples should be used on each frame. If it is difficult to generate new i.i.d. samples $\{\boldsymbol{\eta}_1, \ldots, \boldsymbol{\eta}_W\}$ on each frame (possibly because the distribution of $\boldsymbol{\eta}[r]$ is unknown), we can use W *past values* of $\boldsymbol{\eta}[r]$. There is a subtle issue here because these past values are not independent of the queue backlogs $Z_l[r]$ that are part of the $a(\pi)$ function. However, using these past values can still be shown to work via a *delayed-queue* argument given in the *max-weight learning* theory of (166).

7.4 TASK PROCESSING EXAMPLE

Consider a network of L wireless nodes that collaboratively process tasks and report the results to a receiver. There are an infinite sequence of tasks $\{Task[0], Task[1], Task[2], \ldots\}$ that are performed back-to-back, and the starting time of task $r \in \{0, 1, 2, \ldots\}$ is considered to be the start of renewal frame r. At the beginning of each task $r \in \{0, 1, 2, \ldots\}$, the network observes a vector $\boldsymbol{\eta}[r]$ of *task information*. We assume $\{\boldsymbol{\eta}[r]\}_{r=0}^{\infty}$ is i.i.d. over tasks with an unknown distribution. Every task must be processed using one of K *pure policies* $\mathcal{P}^{pure} = \{\pi_1, \pi_2, \ldots, \pi_K\}$. The frame size $T[r]$, task processing utility $g[r]$, and energy expenditures $y_l[r]$ for each node $l \in \{1, \ldots, L\}$ are *deterministic functions* of $\boldsymbol{\eta}[r]$ and $\pi[r]$:

$$T[r] = \hat{T}(\boldsymbol{\eta}[r], \pi[r]) \;,\; g[r] = \hat{g}(\boldsymbol{\eta}[r], \pi[r]) \;,\; y_l[r] = \hat{y}_l(\boldsymbol{\eta}[r], \pi[r])$$

Let p_{av} be a positive constant. The goal is design an algorithm to solve:

$$\begin{array}{ll} \text{Maximize:} & \overline{g}/\overline{T} \\ \text{Subject to:} & \overline{y}_l/\overline{T} \leq p_{av} \ \forall l \in \{1, \ldots, L\} \\ & \pi[r] \in \mathcal{P}^{pure} \ \forall r \in \{0, 1, 2, \ldots\} \end{array}$$

Example Problem:

a) State the renewal-based drift-plus-penalty algorithm for this problem.

b) Assume that the frame size is independent of the policy, so that $\hat{T}(\boldsymbol{\eta}[r], \pi[r]) = \hat{T}(\boldsymbol{\eta}[r])$. Show that minimization of the ratio of expectations can be done without bisection, by solving a single deterministic problem every slot.

c) Assume the general case when the frame size depends on the policy. Suppose the optimal ratio value $\theta^*[r]$ is known for frame r. State the deterministic problem to solve every slot, with the structure of minimizing $a(\pi) - \theta^*[r]b(\pi)$ as in Section 7.3.3.

d) Describe the bisection algorithm that obtains an estimate of $\theta^*[r]$ for part (c). Assume we have W past values of initial information $\{\boldsymbol{\eta}[r], \boldsymbol{\eta}[r-1], \ldots, \boldsymbol{\eta}[r-W+1]\}$, and that we know $\theta_{min} \leq \theta^*[r] \leq \theta_{max}$ for some constants θ_{min} and θ_{max}.

Solution:

a) Create virtual queues $Z_l[r]$ for each $l \in \{1, \ldots, L\}$ as follows:

$$Z_l[r+1] = \max[Z_l[r] + \hat{y}_l(\boldsymbol{\eta}[r], \pi[r]) - \hat{T}(\boldsymbol{\eta}[r], \pi[r])p_{av}, 0] \tag{7.30}$$

Every frame $r \in \{0, 1, 2, \ldots\}$, observe $\boldsymbol{\eta}[r]$ and $\boldsymbol{Z}[r]$ and do the following:

- Choose $\pi[r] \in \mathcal{P}^{pure}$ to minimize:

$$\frac{\mathbb{E}\left\{-V\hat{g}(\boldsymbol{\eta}[r], \pi[r]) + \sum_{l=1}^{L} Z_l[r]\hat{y}_l(\boldsymbol{\eta}[r], \pi[r])|\boldsymbol{Z}[r]\right\}}{\mathbb{E}\left\{\hat{T}(\boldsymbol{\eta}[r], \pi[r])|\boldsymbol{Z}[r]\right\}} \tag{7.31}$$

- Update queues $Z_l[r]$ according to (7.30).

b) If $\mathbb{E}\left\{\hat{T}(\boldsymbol{\eta}[r], \pi[r])|\boldsymbol{Z}[r]\right\}$ does not depend on the policy, it suffices to minimize the numerator in (7.31). This is done by observing $\boldsymbol{\eta}[r]$ and $\boldsymbol{Z}[r]$ and choosing the policy $\pi[r] \in \mathcal{P}^{pure}$ as the one that minimizes:

$$-V\hat{g}(\boldsymbol{\eta}[r], \pi[r]) + \sum_{l=1}^{L} Z_l[r]\hat{y}_l(\boldsymbol{\eta}[r], \pi[r])$$

c) If $\theta^*[r]$ is known, then we observe $\boldsymbol{\eta}[r]$ and $\boldsymbol{Z}[r]$ and choose the policy $\pi[r] \in \mathcal{P}^{pure}$ as the one that minimizes:

$$-V\hat{g}(\boldsymbol{\eta}[r], \pi[r]) + \sum_{l=1}^{L} Z_l[r]\hat{y}_l(\boldsymbol{\eta}[r], \pi[r]) - \theta^*[r]\hat{T}(\boldsymbol{\eta}[r], \pi[r])$$

d) Fix a particular frame r. Let $\theta_{min}^{(k)}$ and $\theta_{max}^{(k)}$ be the bounds on $\theta^*[r]$ for step k of the bisection, where $\theta_{min}^{(0)} = \theta_{min}$ and $\theta_{max}^{(k)} = \theta_{max}$. Define $\theta_{bisect}^{(k)} = (\theta_{min}^{(k)} + \theta_{max}^{(k)})/2$. Define $\{\boldsymbol{\eta}_1, \ldots \boldsymbol{\eta}_W\}$ as the W samples to be used. Define the function $val(\theta)$ as follows:

$$val(\theta) = \frac{1}{W}\sum_{i=1}^{W}\left[\min_{\pi \in \mathcal{P}^{pure}}\left[-V\hat{g}(\boldsymbol{\eta}_i, \pi) + \sum_{l=1}^{L} Z_l[r]\hat{y}_l(\boldsymbol{\eta}_i, \pi) - \theta\hat{T}(\boldsymbol{\eta}_i, \pi)\right]\right] \tag{7.32}$$

Note that computing $val(\theta)$ involves W separate minimizations. Note also that $val(\theta)$ is non-increasing in θ (see Exercise 7.2). Now compute $val(\theta_{bisect}^{(k)})$:

- If $val(\theta_{bisect}^{(k)}) = 0$, we are done and we declare $\theta_{bisect}^{(k)}$ as our estimate of $\theta^*[r]$.
- If $val(\theta_{bisect}^{(k)}) > 0$, then define $\theta_{min}^{(k+1)} = \theta_{bisect}^{(k)}, \theta_{max}^{(k+1)} = \theta_{max}^{(k)}$.
- If $val(\theta_{bisect}^{(k)}) < 0$, then define $\theta_{min}^{(k+1)} = \theta_{min}^{(k)}, \theta_{max}^{(k+1)} = \theta_{bisect}^{(k)}$.

Then proceed with the iterations until our error bounds are sufficiently low. Note that this algorithm requires $val(\theta_{min}^{(0)}) \geq 0 \geq val(\theta_{max}^{(0)})$, which should be checked before the iterations begin. If this is violated, we simply increase $\theta_{max}^{(0)}$ and/or decrease $\theta_{min}^{(0)}$.

7.5 UTILITY OPTIMIZATION FOR RENEWAL SYSTEMS

Now consider a renewal system that generates both a penalty vector $\boldsymbol{y}[r] = (y_1[r], \ldots, y_L[r])$ and an attribute vector $\boldsymbol{x}[r] = (x_1[r], \ldots, x_M[r])$. These are random functions of the policy $\pi[r]$ implemented on frame r:

$$x_m[r] = \hat{x}_m(\pi[r]) \;\;,\;\; y_l[r] = \hat{y}_l(\pi[r]) \;\forall m \in \{1, \ldots, M\}, l \in \{1, \ldots, L\}$$

The frame size $T[r]$ is also a random function of the policy as before: $T[r] = \hat{T}(\pi[r])$. We make the same assumptions as before, including that second moments of $\hat{x}_m(\pi[r])$ are uniformly bounded regardless of the policy, and that the conditional distribution of $(T[r], \boldsymbol{y}[r], \boldsymbol{x}[r])$, given $\pi[r] = \pi$, is independent of events on previous frames, and is identically distributed on each frame that uses the same policy π. Let $T_{min}, T_{max}, x_{m,min}, x_{m,max}$ be finite constants such that for all policies $\pi \in \mathcal{P}$ and all $m \in \{1, \ldots, M\}$, we have:

$$0 < T_{min} \leq \mathbb{E}\left\{\hat{T}(\pi[r])|\pi[r] = \pi\right\} \leq T_{max} \;\;,\;\; x_{m,min} \leq \mathbb{E}\left\{\hat{x}(\pi[r])|\pi[r] = \pi\right\} \leq x_{m,max}$$

Under a particular algorithm for choosing policies $\pi[r]$ over frames $r \in \{0, 1, 2, \ldots\}$, define $\overline{T}[R]$, $\overline{y}_l[R], \overline{x}_m[R]$ for $R > 0$ by:

$$\overline{T}[R] \triangleq \frac{1}{R}\sum_{r=0}^{R-1}\mathbb{E}\{T[r]\} \;\;,\;\; \overline{y}_l[R] \triangleq \frac{1}{R}\sum_{r=0}^{R-1}\mathbb{E}\{y_l[r]\} \;\;,\;\; \overline{x}_m[R] \triangleq \frac{1}{R}\sum_{r=0}^{R-1}\mathbb{E}\{x_m[r]\}$$

Define $\overline{T}$, $\overline{y}_l$, $\overline{x}_m$ as the limiting values of $\overline{T}[R]$, $\overline{y}_l[R]$, $\overline{x}_m[R]$, assuming temporarily that the limit exists. For each $m \in \{1, \ldots, M\}$, define $\gamma_{m,min}$ and $\gamma_{m,max}$ by:

$$\gamma_{m,min} \triangleq \min\left[\frac{x_{m,min}}{T_{min}}, \frac{x_{m,min}}{T_{max}}\right] \quad , \quad \gamma_{m,max} \triangleq \max\left[\frac{x_{m,max}}{T_{min}}, \frac{x_{m,max}}{T_{max}}\right]$$

It is clear that for all $m \in \{1, \ldots, M\}$ and all $R > 0$, we have:

$$\gamma_{m,min} \leq \frac{\overline{x}_m[R]}{\overline{T}[R]} \leq \gamma_{m,max} \quad , \quad \gamma_{m,min} \leq \frac{\overline{x}_m}{\overline{T}} \leq \gamma_{m,max} \tag{7.33}$$

Let $\phi(\boldsymbol{\gamma})$ be a continuous, concave, and entrywise non-decreasing function of vector $\boldsymbol{\gamma} = (\gamma_1, \ldots, \gamma_M)$ over the rectangle $\boldsymbol{\gamma} \in \mathcal{R}$, where:

$$\mathcal{R} = \{(\gamma_1, \ldots, \gamma_M) | \gamma_{m,min} \leq \gamma_m \leq \gamma_{m,max} \ \forall m \in \{1, \ldots, M\}\} \tag{7.34}$$

Consider the following problem:

$$\text{Maximize:} \quad \phi(\overline{\boldsymbol{x}}/\overline{T}) \tag{7.35}$$

$$\text{Subject to:} \quad \overline{y}_l/\overline{T} \leq c_l \ \forall l \in \{1, \ldots, L\} \tag{7.36}$$

$$\pi[r] \in \mathcal{P} \ \forall r \in \{0, 1, 2, \ldots\} \tag{7.37}$$

To transform this problem to one that has the structure given in Section 7.1.1, we define auxiliary variables $\boldsymbol{\gamma}[r] = (\gamma_1[r], \ldots, \gamma_M[r])$ that are chosen in the rectangle $\mathcal{R}$ every frame r. We then define a new penalty $y_0[r]$ as follows:

$$y_0[r] \triangleq -T[r]\phi(\boldsymbol{\gamma}[r])$$

Now consider the following transformed (and equivalent) problem:

$$\text{Maximize:} \quad \overline{T\phi(\boldsymbol{\gamma})}/\overline{T} \tag{7.38}$$

$$\text{Subject to:} \quad \overline{x}_m \geq \overline{T\gamma_m} \ \forall m \in \{1, \ldots, M\} \tag{7.39}$$

$$\overline{y}_l/\overline{T} \leq c_l \ \forall l \in \{1, \ldots, L\} \tag{7.40}$$

$$\boldsymbol{\gamma}[r] \in \mathcal{R} \ \forall r \in \{0, 1, 2, \ldots\} \tag{7.41}$$

$$\pi[r] \in \mathcal{P} \ \forall r \in \{0, 1, 2, \ldots\} \tag{7.42}$$

where:

$$\overline{T\phi(\boldsymbol{\gamma})} \triangleq \lim_{R\to\infty} \frac{1}{R} \sum_{r=0}^{R-1} \mathbb{E}\{T[r]\phi(\boldsymbol{\gamma}[r])\} = -\overline{y}_0$$

$$\overline{T\gamma_m} \triangleq \lim_{R\to\infty} \frac{1}{R} \sum_{r=0}^{R-1} \mathbb{E}\{T[r]\gamma_m[r]\} \ \forall m \in \{1, \ldots, M\}$$

That the problems (7.35)-(7.37) and (7.38)-(7.42) are equivalent is proven in Exercise 7.7 using the fact:

$$\overline{T\phi(\boldsymbol{\gamma})}/\overline{T} \leq \phi(\overline{T\boldsymbol{\gamma}}/\overline{T})$$

This fact is a variation on Jensen's inequality and is proven in the following lemma.

Lemma 7.6 *Let $\phi(\boldsymbol{\gamma})$ be any continuous and concave (not necessarily non-decreasing) function defined over $\boldsymbol{\gamma} \in \mathcal{R}$, where $\mathcal{R}$ is defined in (7.34).*

(a) Let $(T, \boldsymbol{\gamma})$ be a random vector that takes values in the set $\{(T, \boldsymbol{\gamma})|T > 0, \boldsymbol{\gamma} \in \mathcal{R}\}$ according to any joint distribution that satisfies $0 < \mathbb{E}\{T\} < \infty$. Then:

$$\frac{\mathbb{E}\{T\phi(\boldsymbol{\gamma})\}}{\mathbb{E}\{T\}} \leq \phi\left(\frac{\mathbb{E}\{T\boldsymbol{\gamma}\}}{\mathbb{E}\{T\}}\right)$$

(b) Let $(T[r], \boldsymbol{\gamma}[r])$ be a sequence of random vectors of the type specified in part (a), for $r \in \{0, 1, 2, \ldots\}$. Then for any integer $R > 0$:

$$\frac{\frac{1}{R}\sum_{r=0}^{R-1} T[r]\phi(\boldsymbol{\gamma}[r])}{\frac{1}{R}\sum_{r=0}^{R-1} T[r]} \leq \phi\left(\frac{\frac{1}{R}\sum_{r=0}^{R-1} T[r]\boldsymbol{\gamma}[r]}{\frac{1}{R}\sum_{r=0}^{R-1} T[r]}\right) \tag{7.43}$$

$$\frac{\frac{1}{R}\sum_{r=0}^{R-1} \mathbb{E}\{T[r]\phi(\boldsymbol{\gamma}[r])\}}{\frac{1}{R}\sum_{r=0}^{R-1} \mathbb{E}\{T[r]\}} \leq \phi\left(\frac{\frac{1}{R}\sum_{r=0}^{R-1} \mathbb{E}\{T[r]\boldsymbol{\gamma}[r]\}}{\frac{1}{R}\sum_{r=0}^{R-1} \mathbb{E}\{T[r]\}}\right) \tag{7.44}$$

and thus $\overline{T\phi(\boldsymbol{\gamma})}/\overline{T} \leq \phi(\overline{T\boldsymbol{\gamma}}/\overline{T})$.

Proof. Part (b) follows easily from part (a) (see Exercise 7.6). Here we prove part (a). Let $\{(T[r], \boldsymbol{\gamma}[r])\}_{r=0}^{\infty}$ be an i.i.d. sequence of random vectors, each with the same distribution as $(T, \boldsymbol{\gamma})$. Define $t_0 = 0$, and for integers $R > 0$ define $t_R = \sum_{r=0}^{R-1} T[r]$. Let interval $[t_r, t_{r+1})$ represent the rth frame. Define $\hat{\gamma}(t)$ to take the value $\boldsymbol{\gamma}[r]$ if t is in the rth frame, so that:

$$\hat{\boldsymbol{\gamma}}(t) = \boldsymbol{\gamma}[r] \quad \text{if } t \in [t_r, t_{r+1})$$

We thus have for any integer $R > 0$:

$$\frac{1}{t_R}\int_0^{t_R} \phi(\hat{\boldsymbol{\gamma}}(t))dt = \frac{\sum_{r=0}^{R-1} T[r]\phi(\boldsymbol{\gamma}[r])}{\sum_{r=0}^{R-1} T[r]} = \frac{\frac{1}{R}\sum_{r=0}^{R-1} T[r]\phi(\boldsymbol{\gamma}[r])}{\frac{1}{R}\sum_{r=0}^{R-1} T[r]} \tag{7.45}$$

On the other hand, by Jensen's inequality for the concave function $\phi(\boldsymbol{\gamma})$:

$$\frac{1}{t_R}\int_0^{t_R} \phi(\hat{\boldsymbol{\gamma}}(t))dt \leq \phi\left(\frac{1}{t_R}\int_0^{t_R} \hat{\boldsymbol{\gamma}}(t)dt\right) = \phi\left(\frac{\frac{1}{R}\sum_{r=0}^{R-1} T[r]\boldsymbol{\gamma}[r]}{\frac{1}{R}\sum_{r=0}^{R-1} T[r]}\right) \tag{7.46}$$

Taking limits of (7.45) as $R \to \infty$ and using the law of large numbers yields:

$$\lim_{R\to\infty} \frac{1}{t_R} \int_0^{t_R} \phi(\hat{\boldsymbol{\gamma}}(t))dt = \frac{\mathbb{E}\{T\phi(\boldsymbol{\gamma})\}}{\mathbb{E}\{T\}} \quad (w.p.1)$$

Taking limits of (7.46) as $R \to \infty$ and using the law of large numbers and continuity of $\phi(\boldsymbol{\gamma})$ yields:

$$\lim_{R\to\infty} \frac{1}{t_R} \int_0^{t_R} \phi(\hat{\boldsymbol{\gamma}}(t))dt \leq \phi\left(\frac{\mathbb{E}\{T\boldsymbol{\gamma}\}}{\mathbb{E}\{T\}}\right) \quad (w.p.1)$$

□

7.5.1 THE UTILITY OPTIMAL ALGORITHM FOR RENEWAL SYSTEMS

To solve (7.38)-(7.42), we enforce the constraints $\overline{x}_m \geq \overline{T\gamma_m}$ and $\overline{y}_l/\overline{T} \leq c_l$ with virtual queues $Z_l[r]$ and $G_m[r]$ for $l \in \{1, \ldots, L\}$ and $m \in \{1, \ldots, M\}$:

$$Z_l[r+1] = \max[Z_l[r] + y_l[r] - T[r]c_l, 0] \quad (7.47)$$
$$G_m[r+1] = \max[G_m[r] + T[r]\gamma_m[r] - x_m[r], 0] \quad (7.48)$$

Note that the constraint $\overline{x}_m \geq \overline{T\gamma_m}$ is equivalent to $\overline{T\gamma_m - x_m} \leq 0$, which is the same as $\overline{p}_m/\overline{T} \leq 0$ for $p_m[r] \triangleq T[r]\gamma_m[r] - x_m[r]$. Hence, the transformed problem fits the general renewal framework (7.4)-(7.6). Using $y_0[r] = -T[r]\phi(\boldsymbol{\gamma}[r])$, the algorithm then observes $\boldsymbol{Z}[r], \boldsymbol{G}[r]$ at the beginning of each frame $r \in \{0, 1, 2, \ldots, \}$ and chooses a policy $\pi[r] \in \mathcal{P}$ and auxiliary variables $\boldsymbol{\gamma}[r] \in \mathcal{R}$ to minimize:

$$\frac{-V\mathbb{E}\left\{\hat{T}(\pi[r])\phi(\boldsymbol{\gamma}[r])|\boldsymbol{Z}[r], \boldsymbol{G}[r]\right\}}{\mathbb{E}\left\{\hat{T}(\pi[r])|\boldsymbol{Z}[r], \boldsymbol{G}[r]\right\}} + \frac{\mathbb{E}\left\{\sum_{l=1}^{L} Z_l[r]\hat{y}_l(\pi[r]) + \sum_{m=1}^{M} G_m[r][\hat{T}(\pi[r])\gamma_m[r] - \hat{x}(\pi[r])]|\boldsymbol{Z}[r], \boldsymbol{G}[r]\right\}}{\mathbb{E}\left\{\hat{T}(\pi[r])|\boldsymbol{Z}[r], \boldsymbol{G}[r]\right\}}$$

This minimization can be simplified by separating out the terms that use auxiliary variables. The expression to minimize is thus:

$$\frac{\mathbb{E}\left\{\hat{T}(\pi[r])[-V\phi(\boldsymbol{\gamma}[r]) + \sum_{m=1}^{M} G_m[r]\gamma_m[r]]|\boldsymbol{Z}[r], \boldsymbol{G}[r]\right\}}{\mathbb{E}\left\{\hat{T}(\pi[r])|\boldsymbol{Z}[r], \boldsymbol{G}[r]\right\}} + \frac{\mathbb{E}\left\{\sum_{l=1}^{L} Z_l[r]\hat{y}_l(\pi[r]) - \sum_{m=1}^{M} G_m[r]\hat{x}(\pi[r])|\boldsymbol{Z}[r], \boldsymbol{G}[r]\right\}}{\mathbb{E}\left\{\hat{T}(\pi[r])|\boldsymbol{Z}[r], \boldsymbol{G}[r]\right\}}$$

Clearly, the $\boldsymbol{\gamma}[r]$ variables can be optimized separately to minimize the first term, making the frame size in the numerator and denominator of the first term cancel. The resulting algorithm is thus: Observe $\boldsymbol{Z}[r]$ and $\boldsymbol{G}[r]$ at the beginning of each frame $r \in \{0, 1, 2, \ldots\}$, and perform the following:

- (Auxiliary Variables) Choose $\boldsymbol{\gamma}[r]$ to solve:

$$\begin{array}{ll}\text{Maximize:} & V\phi(\boldsymbol{\gamma}[r]) - \sum_{m=1}^{M} G_m[r]\gamma_m[r] \\ \text{Subject to:} & \gamma_{m,min} \leq \gamma_m[r] \leq \gamma_{m,max} \ \ \forall m \in \{1, \ldots, M\}\end{array}$$

- (Policy Selection) Choose $\pi[r] \in \mathcal{P}$ to minimize the following:

$$\frac{\mathbb{E}\left\{\sum_{l=1}^{L} Z_l[r]\hat{y}_l(\pi[r]) - \sum_{m=1}^{M} G_m[r]\hat{x}(\pi[r])|\boldsymbol{Z}[r], \boldsymbol{G}[r]\right\}}{\mathbb{E}\left\{\hat{T}(\pi[r])|\boldsymbol{Z}[r], \boldsymbol{G}[r]\right\}}$$

- (Virtual Queue Updates) At the end of frame r, update $\boldsymbol{Z}[r]$ and $\boldsymbol{G}[r]$ by (7.47) and (7.48).

The auxiliary variable update has the same structure as that given in Chapter 5, and it is a deterministic optimization that reduces to M optimizations of single variable functions if $\phi(\boldsymbol{\gamma})$ has the form $\phi(\boldsymbol{\gamma}) = \sum_{m=1}^{M} \phi_m(\gamma_m)$. The policy selection stage is a minimization of a ratio of expectations, and it can be solved with the techniques given in Section 7.3.

7.6 DYNAMIC PROGRAMMING EXAMPLES

This section presents more complex renewal system examples that involve the theory of dynamic programming. Readers unfamiliar with dynamic programming can skip this section, and are referred to (64) for a coverage of that theory. Readers familiar with dynamic programming can peruse these examples.

7.6.1 DELAY-LIMITED TRANSMISSION EXAMPLE

Here we present an example similar to the delay-limited transmission system developed for cooperative communication in (71), although we remove the cooperative component for simplicity. Consider a system with L wireless transmitters that deliver data to a common receiver. Time is slotted with unit size, and all frames are fixed to T slots, where T is a positive integer. At the beginning of each frame $r \in \{0, 1, 2, \ldots\}$, new packets arrive for transmission. These packets must be delivered within the T slot frame $\tau \in \{rT, \ldots, (r+1)T - 1\}$, or they are dropped at the end of the frame. Let $\boldsymbol{A}[r] = (A_1[r], \ldots, A_L[r])$ be the vector of new packet arrivals, treated as *initial information* about frame r. Assume that $\boldsymbol{A}[r]$ is i.i.d. over frames. On each slot τ of the T-slot frame, at most one transmitter $l \in \{1, \ldots, L\}$ is allowed to transmit, and it can transmit at most a single packet. Let $Q_l(\tau)$ represent the (integer) queue size for transmitter l on slot τ. Then for frame $r \in \{0, 1, \ldots\}$, we have:

$$\begin{aligned} Q_l(rT) &= A_l[r] \\ Q_l(rT+v) &= A_l[r] - \sum_{\tau=rT}^{rT+v-1} 1_l(\tau) \quad, \forall v \in \{1,\ldots,T-1\} \end{aligned}$$

where $1_l(\tau)$ is an indicator function that is 1 if transmitter l successfully delivers a packet on slot τ, and is 0 otherwise.

The success of each packet transmission depends on the power that was used. Let $\boldsymbol{p}(\tau) = (p_1(\tau),\ldots,p_L(\tau))$ represent the power allocation vector on each slot τ in the T-slot frame. This vector is chosen every slot τ subject to the constraints:

$$\begin{aligned} 0 \leq p_l(\tau) \leq p_{max} & \quad \forall l \in \{1,\ldots,L\}, \forall \tau \\ p_l(\tau) = 0 \ \text{ if } Q_l(\tau) = 0 & \quad \forall l \in \{1,\ldots,L\}, \forall \tau \\ p_l(\tau)p_m(\tau) = 0 & \quad \forall l \neq m, \forall \tau \end{aligned}$$

The third constraint above ensures at most one transmitter can send on any given slot. Transmission successes are conditionally independent of past history given the transmission power used, with success probability for each $l \in \{1,\ldots,L\}$ given by:

$$q_l(p) \triangleq Pr[\text{transmitter } l \text{ is successful on slot } \tau | p_l(\tau) = p, Q_l(\tau) > 0]$$

We assume that $q_l(0) = 0$ for all $l \in \{1,\ldots,L\}$. Define $D_l[r]$ and $y_l[r]$ as the total packets delivered and total energy expended by transmitter l on frame r:

$$\begin{aligned} D_l[r] &\triangleq \sum_{\tau=rT}^{rT+T-1} 1_l(\tau) \ \forall l \in \{1,\ldots,L\} \\ y_l[r] &\triangleq \sum_{\tau=rT}^{rT+T-1} p_l(\tau) \ \forall l \in \{1,\ldots,L\} \end{aligned}$$

The goal is to maximize a weighted sum of throughput subject to average power constraints:

$$\begin{aligned} \text{Maximize:} & \quad \sum_{l=1}^{L} w_l \overline{D}_l/T \\ \text{Subject to:} & \quad \overline{y}_l/T \leq p_{av} \ \forall l \in \{1,\ldots,L\} \\ & \quad \pi[r] \in \mathcal{P} \ \forall r \in \{0,1,2,\ldots\} \end{aligned}$$

where $\{w_l\}_{l=1}^{L}$ are a given collection of positive weights, p_{av} is a given constant power constraint, $\overline{D}_l$ and $\overline{y}_l$ are the average delivered data and energy expenditure by transmitter l on one frame, and $\mathcal{P}$ is the policy space that conforms to the above transmission constraints over the frame. This problem fits the standard renewal form given in Section 7.1 with $c_l = p_{av}$ for all $l \in \{1,\ldots,L\}$, and:

$$y_0[r] \triangleq -\sum_{l=1}^{L} w_l \sum_{\tau=rT}^{rT+T-1} 1_l(\tau)$$

We thus form virtual queues $Z_l[r]$ for each $l \in \{1, \ldots, L\}$, with updates:

$$Z_l[r+1] = \max\left[Z_l[r] + \sum_{\tau=rT}^{rT+T-1} p_l(\tau) - p_{av}T, 0\right] \tag{7.49}$$

Then perform the following:

- For every frame r, observe $\boldsymbol{A}[r]$ and make actions over the course of the frame to solve:

$$\begin{array}{lll}\text{Maximize:} & & \sum_{l=1}^{L}\sum_{\tau=rT}^{rT+T-1} \mathbb{E}\{Vw_l 1_l(\tau) - Z_l[r]p_l(\tau)|\boldsymbol{Z}[r], \boldsymbol{A}[r]\} \\ \text{Subject to:} & (1) & 0 \leq p_l(\tau) \leq p_{max} \quad \forall l, \forall \tau \in \{rT, \ldots, rT+T-1\} \\ & (2) & p_l(\tau) = 0 \text{ if } Q_l(\tau) = 0 \quad \forall l, \forall \tau \in \{rT, \ldots, rT+T-1\} \\ & (3) & p_l(\tau)p_m(\tau) = 0 \quad \forall l \neq m, \forall \tau \in \{rT, \ldots, rT+T-1\}\end{array}$$

- Update $Z_l[r]$ according to (7.49).

The above uses the fact that the desired ratio (7.17) in this case has a constant denominator T, and hence it suffices to minimize the numerator, which can be achieved by minimizing the conditional expectation given both the $\boldsymbol{Z}[r]$ and $\boldsymbol{A}[r]$ values. Using iterated expectations, the expression to be maximized can be re-written:

$$\sum_{l=1}^{L}\sum_{\tau=rT}^{rT+T-1} \mathbb{E}\{Vw_l q_l(p_l(\tau)) - Z_l[r]p_l(\tau)|\boldsymbol{Z}[r], \boldsymbol{A}[r]\}$$

The problem can be solved as a *dynamic program* (64). Specifically, we can start backwards and define $J_T(\boldsymbol{Q})$ as the optimal reward in the final stage T (corresponding to slot $\tau = rT + T - 1$) given that $\boldsymbol{Q}(rT + T - 1) = \boldsymbol{Q}$:

$$J_T(\boldsymbol{Q}) \triangleq \max_{l|Q_l>0}\left[\max_{\{p|0\leq p\leq p_{max}\}}[Vw_l q_l(p) - Z_l[r]p]\right]$$

This function $J_T(\boldsymbol{Q})$ is computed for all integer vectors $\boldsymbol{Q}$ that satisfy $\boldsymbol{0} \leq \boldsymbol{Q} \leq \boldsymbol{A}[r]$. Then define $J_{T-1}(\boldsymbol{Q})$ as the optimal expected sum reward in the last two stages $\{T-1, T\}$, given that $\boldsymbol{Q}(rT + T - 2) = \boldsymbol{Q}$:

$$J_{T-1}(\boldsymbol{Q}) \triangleq \max_{l|Q_l>0}\left[\max_{\{p|0\leq p\leq p_{max}\}}[Vw_l q_l(p) - Z_l[r]p + q_l(p)J_T(\boldsymbol{Q} - \boldsymbol{e}_l) + (1 - q_l(p))J_T(\boldsymbol{Q})]\right]$$

where $\boldsymbol{e}_l$ is a vector that is zero in all entries $j \neq l$, and is 1 in entry l. The function $J_{T-1}(\boldsymbol{Q})$ is also computed for all $\boldsymbol{Q}$ that satisfy $\boldsymbol{0} \leq \boldsymbol{Q} \leq \boldsymbol{A}[r]$. In general, we have for stages $k \in \{1, \ldots, T-1\}$ the following recursive equation:

$$J_k(\boldsymbol{Q}) \triangleq \max_{l|Q_l>0}\left[\max_{\{p|0\leq p\leq p_{max}\}}[Vw_l q_l(p) - Z_l[r]p + q_l(p)J_{k+1}(\boldsymbol{Q} - \boldsymbol{e}_l) + (1 - q_l(p))J_{k+1}(\boldsymbol{Q})]\right]$$

The value $J_1(\boldsymbol{Q})$ represents the expected total reward over frame r under the optimal policy, given that $\boldsymbol{Q}(rT) = \boldsymbol{Q}$. The optimal action to take at each stage k corresponds to the transmitter l and the power level p that achieves the maximum in the computation of $J_k(\boldsymbol{Q})$.

For a modified problem where power allocations are restricted to $p_l(\tau) \in \{0, p_{max}\}$, it can be shown the problem has a simple greedy solution: On each slot τ of frame r, consider the set of links l such that $Q_l(\tau) > 0$, and transmit over the link l in this set that has the largest positive $Vw_l q_l(p_{max}) - Z_l[r]p_{max}$ value, breaking ties arbitrarily and choosing not to transmit over any link if none of these values are positive.

7.6.2 MARKOV DECISION PROBLEM FOR MINIMUM DELAY SCHEDULING

Here we consider a Markov decision problem involving queueing delay, from (56)(57). Consider a 2-queue wireless downlink in slotted time $t \in \{0, 1, 2, \ldots\}$. Packets arrive randomly every slot, and the controller can transmit a packet from at most one queue per slot. Let $Q_i(t)$ be the (integer) number of packets in queue i on slot t, for $i \in \{1, 2\}$. We assume the queues have a finite buffer of 10 packets, so that packets arriving when the $Q_i(t) = 10$ are dropped. To enforce a renewal structure, let $\chi(t)$ be an independent process of i.i.d. Bernoulli variables with $Pr[\chi(t) = 1] = \delta$, for some *renewal probability* $\delta > 0$. The contents of both queues are emptied whenever $\chi(t) = 1$, so that queueing dynamics are given by:

$$Q_i(t+1) = \begin{cases} \min[Q_i(t) + A_i(t), 10] - 1_i(t) & \text{if } \chi(t) = 0 \\ 0 & \text{if } \chi(t) = 1 \end{cases}$$

where $1_i(t)$ is an indicator function that is 1 if a packet is successfully transmitted from queue i on slot t (and is 0 otherwise), and $A_i(t)$ is the (integer) number of new packet arrivals to queue i. The maximum packet loss rate due to forced renewals is thus 20δ, which can be made arbitrarily small with a small choice of $\delta > 0$. We assume the controller knows the value of $\chi(t)$ at the beginning of each slot. We have two choices of a renewal definition: (i) Define a renewal event on slot t whenever $(Q_1(t), Q_2(t)) = (0, 0)$, (ii) Define a renewal event on slot t whenever $\chi(t-1) = 1$. The first definition has shorter renewal frames, but the frames sizes depend on the control actions. This would require minimizing a ratio of expectations every slot. The second definition has frame sizes that are independent of the control actions, and have mean $1/\delta$. For simplicity, we use the second definition.

Let $g_i(t)$ be the number of packets dropped from queue i on slot t:

$$g_i(t) = \begin{cases} A_i(t)1\{Q_i(t) = 10\} & \text{if } \chi(t) = 0 \\ Q_i(t) + A_i(t) - 1_i(t) & \text{if } \chi(t) = 1 \end{cases}$$

where $1\{Q_i(t) = 10\}$ is an indicator function that is 1 if $Q_i(t) = 10$, and 0 otherwise.

Assume the processes $A_1(t)$ and $A_2(t)$ are independent of each other. $A_1(t)$ is i.i.d. Bernoulli with $Pr[A_1(t) = 1] = \lambda_1$, and $A_2(t)$ is i.i.d. Bernoulli with $Pr[A_2(t) = 1] = \lambda_2$. Every slot, the controller chooses a queue for transmission by selecting a power allocation vector $(p_1(t), p_2(t))$ subject to the constraints:

$$\begin{aligned} 0 \le p_i(t) \le p_{max} \ , \quad p_1(t)p_2(t) = 0 \ \forall i \in \{1, 2\}, \forall t \\ p_i(t) = 0 \ , \text{if } Q_i(t) = 0 \ \forall i \in \{1, 2\}, \forall t \end{aligned}$$

where p_{max} is a given maximum power level. Let $\mathcal{P}(\boldsymbol{Q})$ denote the set of all power vectors that satisfy these constraints. Transmission successes are independent of past history given the power level used, with probabilities:

$$q_i(p) \triangleq Pr[1_i(t) = 1 | Q_i(t) > 0, p_i(t) = p]$$

Assume that $q_1(0) = q_2(0) = 0$.

The goal is to minimize the time average rate of packet drops $\overline{g}_1 + \overline{g}_2$ subject to an average power constraint p_{av} and an average delay constraint of 3 slots for all non-dropped packets in each queue: $\overline{W}_1 \le 3$, $\overline{W}_2 \le 3$. Specifically, define $\tilde{\lambda}_i = \lambda_i - \overline{g}_i$ as the throughput of queue i. By Little's Theorem (129), we have $\overline{Q}_i = \tilde{\lambda}_i \overline{W}_i$, and so the delay constraints can be transformed to $\overline{Q}_i \le 3\tilde{\lambda}_i$, which is equivalent to $\overline{Q}_i - 3(\lambda_i - \overline{g}_i) \le 0$.

Let $t[r]$ be the slot that starts renewal frame $r \in \{0, 1, 2, \ldots\}$ (where $t[0] = 0$), and let $T[r]$ represent the number of slots in the rth renewal frame. Thus, we have constraints:

$$\begin{aligned} \lim_{R\to\infty} \frac{1}{R} \sum_{r=0}^{R-1} \sum_{\tau=t[r]}^{t[r]+T[r]-1} [Q_1(\tau) - 3(A_1(\tau) - g_1(\tau))] &\le 0 \\ \lim_{R\to\infty} \frac{\frac{1}{R}\sum_{r=0}^{R-1} \sum_{\tau=t[r]}^{t[r]+T[r]-1} [p_1(\tau) + p_2(\tau)]}{\frac{1}{R}\sum_{r=0}^{R-1} T[r]} &\le p_{av} \end{aligned}$$

Following the renewal system framework, we define virtual queues $Z_1[r]$, $Z_2[r]$, $Z_p[r]$:

$$Z_1[r+1] = \max\left[Z_1[r] + \sum_{\tau=t[r]}^{t[r]+T[r]-1} [Q_1(\tau) - 3(A_1(\tau) - g_1(\tau))], 0 \right] \quad (7.50)$$

$$Z_2[r+1] = \max\left[Z_2[r] + \sum_{\tau=t[r]}^{t[r]+T[r]-1} [Q_2(\tau) - 3(A_2(\tau) - g_2(\tau))], 0 \right] \quad (7.51)$$

$$Z_p[r+1] = \max\left[Z_p[r] + \sum_{\tau=t[r]}^{t[r]+T[r]-1} [p_1(\tau) + p_2(\tau) - p_{av}], 0 \right] \quad (7.52)$$

Making the queues $Z_1[r]$ and $Z_2[r]$ rate stable ensures the desired delay constraints are satisfied, and making queue $Z_p[r]$ rate stable ensures the power constraint is satisfied. We thus have the following algorithm, which only minimizes the numerator in the ratio of expectations because the denominator is independent of the policy:

- At the beginning of each frame r, observe $\boldsymbol{Z}[r] = [Z_1[r], Z_2[r], Z_p[r]]$ and make power allocation decisions to minimize the following expression over the frame:

$$\mathbb{E}\left\{\sum_{\tau=t[r]}^{t[r]+T[r]-1} f(\boldsymbol{p}(\tau), \boldsymbol{A}(\tau), \boldsymbol{Q}(\tau), \boldsymbol{Z}[r]) \middle| \boldsymbol{Z}[r]\right\}$$

where $f(\boldsymbol{p}(\tau), \boldsymbol{A}(\tau), \boldsymbol{Q}(\tau), \boldsymbol{Z}[r])$ is defined:

$$\begin{aligned} f(\boldsymbol{p}(\tau), \boldsymbol{A}(\tau), \boldsymbol{Q}(\tau), \boldsymbol{Z}[r]) &\triangleq V(g_1(\tau) + g_2(\tau)) + \sum_{i=1}^{2} Z_i[r][Q_i(\tau) + 3g_i(\tau)] \\ &\quad + Z_p[r][p_1(\tau) + p_2(\tau)] \end{aligned}$$

- Update the virtual queues $\boldsymbol{Z}[r]$ by (7.50)-(7.52).

The minimization in the above algorithm can be solved by dynamic programming. Specifically, given queues $\boldsymbol{Z}[r] = \boldsymbol{Z}$ that start the frame, define $J_{\boldsymbol{Z}}(\boldsymbol{Q})$ as the optimal cost until the end of a renewal frame, given the initial queue backlog is $\boldsymbol{Q}$. Then $J_{\boldsymbol{Z}}(\boldsymbol{0})$ is the value of the expression to be minimized. We have (56)(57):

$$\begin{aligned} J_{\boldsymbol{Z}}(\boldsymbol{Q}) &= \delta\mathbb{E}_{\boldsymbol{A}}\left\{\inf_{\boldsymbol{p}\in\mathcal{P}(\boldsymbol{Q})} f(\boldsymbol{p}, \boldsymbol{A}, \boldsymbol{Q}, \boldsymbol{Z})|\chi = 1, \boldsymbol{Q}, \boldsymbol{Z}\right\} \\ &\quad + (1-\delta)\mathbb{E}_{\boldsymbol{A}}\left\{\inf_{\boldsymbol{p}\in\mathcal{P}(\boldsymbol{Q})}\left[f(\boldsymbol{p}, \boldsymbol{A}, \boldsymbol{Q}, \boldsymbol{Z}) + h(\boldsymbol{p}, \boldsymbol{A}, \boldsymbol{Q}, \boldsymbol{Z})\right]|\chi = 0, \boldsymbol{Q}, \boldsymbol{Z}\right\} \end{aligned} \tag{7.53}$$

where $h(\boldsymbol{p}, \boldsymbol{A}, \boldsymbol{Q}, \boldsymbol{Z})$ is defined:

$$h(\boldsymbol{p}, \boldsymbol{A}, \boldsymbol{Q}, \boldsymbol{Z}) \triangleq J_{\boldsymbol{Z}}(\min[\boldsymbol{Q}+\boldsymbol{A}, 10])(1-q(\boldsymbol{p})) + J_{\boldsymbol{Z}}(\min[\boldsymbol{Q}+\boldsymbol{A}, 10] - \boldsymbol{e}(\boldsymbol{p}))q(\boldsymbol{p})$$

where:

$$q(\boldsymbol{p}) = \begin{cases} q_1(p_1) & \text{if } p_1 > 0 \\ q_2(p_2) & \text{if } p_1 = 0 \end{cases} \quad , \quad \boldsymbol{e}(p) = \begin{cases} (1,0) & \text{if } p_1 > 0 \\ (0,1) & \text{if } p_1 = 0 \end{cases}$$

The equation (7.53) must be solved to find $J_{\boldsymbol{Z}}(\boldsymbol{Q})$ for all $\boldsymbol{Q} \in \{0, 1, \ldots, 10\} \times \{0, 1, \ldots, 10\}$.

Define $\Psi(J)$ as an operator that takes a function $J(\boldsymbol{Q})$ (for $\boldsymbol{Q} \in \{0, 1, \ldots, 10\} \times \{0, 1, \ldots, 10\}$) and maps it to another such function via the right-hand-side of (7.53). Then (7.53) reduces to:

$$J_{\boldsymbol{Z}}(\boldsymbol{Q}) = \Psi(J_{\boldsymbol{Z}}(\boldsymbol{Q}))$$

and hence the desired $J_{\boldsymbol{Z}}(\boldsymbol{Q})$ is a *fixed point* of the $\Psi(\cdot)$ operator. It can be shown that $\Psi(\cdot)$ is a *contraction* with an appropriate definition of distance (67)(57), and so the fixed point is unique and can be obtained by iteration of the $\Psi(\cdot)$ operator starting with any initial function $J^{(0)}(\boldsymbol{Q})$ (such as $J^{(0)}(\boldsymbol{Q}) = 0$):

$$J^{(0)}(\boldsymbol{Q}) = 0 \ , \ J^{(i+1)}(\boldsymbol{Q}) = \Psi(J^{(i)}(\boldsymbol{Q})) \ \forall i \in \{0, 1, 2, \ldots\}$$

Then $\lim_{i\to\infty} J^{(i)}(\boldsymbol{Q})$ solves the fixed point equation and hence is equal to the desired $J_Z(\boldsymbol{Q})$ function. While this then needs to be recomputed for the next frame (because the queue $\boldsymbol{Z}[r]$ change), the change in these queues over one frame is bounded and the resulting $J_Z(\boldsymbol{Q})$ function for frame r is already a good approximation for this function on frame $r+1$. Thus, the initial value of the iteration can be the final value found in the previous frame.

Iteration of the $\Psi(J)$ operator requires knowledge of the $\boldsymbol{A}(t)$ distribution to compute the desired expectations. In this case of independent Bernoulli inputs, this involves knowing only two scalars λ_1 and λ_2. However, for larger problems when the random events every slot can be a large vector, the expectations can be accurately approximated by averaging over past samples, as in (7.29). See (57) for an analysis of the error bounds in this technique.

See also (61)(60)(59) for alternative approximations to the Markov Decision Problem for wireless queueing delay. A detailed treatment of stochastic shortest path problems and approximations is found in (67). Approximate dynamic programming methods that approximate value functions with simpler functions can be found in (68)(187)(67)(69). Recent work in (62)(63) combines Markov Decision theory and approximate value functions for treatment of energy and delay optimization in wireless systems.

7.7 EXERCISES

Exercise 7.1. (Deterministic Task Processing) Suppose N network nodes cooperate to process a sequence of tasks. A new task is started when the previous task ends, and we label the tasks $r \in \{0, 1, 2, \ldots\}$. For each new task r, the network controller makes a decision about which single node $n[r]$ will process the task, and what modality $m[r]$ will be used in the processing. Assume there are M possible modalities, each with different durations and energy expenditures. The task r decision is $\pi[r] = (n[r], m[r])$, where $n[r] \in \{1, \ldots, N\}$ and $m[r] \in \{1, \ldots, M\}$. Define $T(n, m)$ and $\beta(n, m)$ as the duration of time and the energy expenditure, respectively, required for node n to process a task using modality m. Assume that $T(n, m) \geq 0$ and $\beta(n, m) \geq 0$ for all n, m. Let $e_n[r]$ represent the energy expended by node $n \in \{1, \ldots, N\}$ during task r:

$$e_n[r] = \begin{cases} \beta(n[r], m[r]) & \text{if } n[r] = n \\ 0 & \text{if } n[r] \neq n \end{cases}$$

We want to maximize the task processing rate subject to average power constraints at each node:

$$\begin{array}{lll} \text{Maximize:} & & 1/\overline{T} \\ \text{Subject to:} & 1) & \overline{e}_n/\overline{T} \leq p_{n,av} \quad , \forall n \in \{1, \ldots, N\} \\ & 2) & n[r] \in \{1, \ldots, N\}, m[r] \in \{1, \ldots, M\} \quad , \forall r \in \{0, 1, 2, \ldots\} \end{array}$$

where $p_{n,av}$ is the average power constraint for node $n \in \{1, \ldots, N\}$. State the renewal-based drift-plus-penalty algorithm of Section 7.2 for this problem. Note that there is no randomness here, and so the ratio of expectations to be minimized on each frame becomes a ratio of deterministic functions.

Exercise 7.2. (Non-Increasing Property of $val(\theta)$). Consider the $val(\theta)$ function in (7.32). Suppose that $\theta_1 \leq \theta_2$.

a) Argue that for all $\boldsymbol{\eta}_i, \pi, Z_l[r]$, we have:

$$-V\hat{g}(\boldsymbol{\eta}_i, \pi) + \sum_{l=1}^{L} Z_l[r]\hat{y}_l(\boldsymbol{\eta}_i, \pi) - \theta_1\hat{T}(\boldsymbol{\eta}_i, \pi) \geq -V\hat{g}(\boldsymbol{\eta}_i, \pi) + \sum_{l=1}^{L} Z_l[r]\hat{y}_l(\boldsymbol{\eta}_i, \pi) - \theta_2\hat{T}(\boldsymbol{\eta}_i, \pi)$$

b) Prove that $val(\theta_1) \geq val(\theta_2)$.

Exercise 7.3. (An Alternative Algorithm with Modified Objective) Consider the system of Section 7.1. However, suppose we desire a solution to the following modified problem:

$$\begin{array}{ll} \text{Minimize:} & \overline{y}_0 \\ \text{Subject to:} & \overline{y}_l/\overline{T} \leq c_l \;\; \forall l \in \{1, \ldots, L\} \\ & \pi[r] \in \mathcal{P} \;\; \forall r \in \{0, 1, 2, \ldots\} \end{array}$$

This differs from (7.4)-(7.6) because we seek to minimize $\overline{y}_0$ rather than $\overline{y}_0/\overline{T}$. Define the same virtual queues $\boldsymbol{Z}[r]$ in (7.12). Note that (7.16) still applies. Consider the algorithm that, every frame r, observes $\boldsymbol{Z}[r]$ and chooses a policy $\pi[r] \in \mathcal{P}$ to minimize the right-hand-side of (7.16). It then updates $\boldsymbol{Z}[r]$ by (7.12) at the end of the frame. Assume there is an i.i.d. algorithm $\pi^*[r]$ that yields:

$$\mathbb{E}\left\{\hat{y}_0(\pi^*[r])\right\} = y_0^{opt} \tag{7.54}$$

$$\mathbb{E}\left\{\hat{y}_l(\pi^*[r])\right\} \leq \mathbb{E}\left\{\hat{T}(\pi^*[r])\right\} c_l \;\; \forall l \in \{1, \ldots, L\} \tag{7.55}$$

a) Plug the i.i.d. algorithm $\pi^*[r]$ into the right-hand-side of (7.16) to show that $\Delta(\boldsymbol{Z}[r]) \leq F$ for some finite constant F, and hence all queues are mean rate stable so that:

$$\limsup_{R\to\infty}[\overline{y}_l[R] - c_l\overline{T}[R]] \leq 0$$

b) Again plug the i.i.d. algorithm $\pi^*[r]$ into the right-hand-side of (7.16), and use iterated expectations and telescoping sums to prove:

$$\limsup_{R\to\infty} \overline{y}_0[R] \leq y_0^{opt} + B/V$$

Exercise 7.4. (Manipulating limits) Suppose that $\limsup_{R\to\infty}[\overline{y}_l[R] - c_l\overline{T}[R]] \leq 0$, where $0 < T_{min} \leq \overline{T}[R] \leq T_{max}$ for all $R > 0$.

a) Argue that for all integers $R > 0$:

$$\frac{\overline{y}_l[R]}{\overline{T}[R]} - c_l \leq \max\left[0, \frac{\overline{y}_l[R]}{\overline{T}[R]} - c_l\right]\frac{\overline{T}[R]}{T_{min}} = \max\left[0, \overline{y}_l[R] - c_l\overline{T}[R]\right]\frac{1}{T_{min}}$$

b) Take limits of the inequality in (a) to conclude that:

$$\limsup_{R\to\infty} \frac{\overline{y}_l[R]}{\overline{T}[R]} \leq c_l$$

Exercise 7.5. (An Alternative Algorithm with Time Averaging) Consider the optimization problem (7.4)-(7.6) for a renewal system with frame sizes $T[r]$ that depend on the policy $\pi[r]$. Define $\theta[0] = 0$. For each stage $r \in \{1, 2, \ldots, \}$ define $\theta[r]$ by:

$$\theta[r] \triangleq \frac{\frac{1}{r}\sum_{k=0}^{r-1} y_0[k]}{\frac{1}{r}\sum_{k=0}^{r-1} T[k]} \tag{7.56}$$

so that $\theta[r]$ is the empirical time average of the penalty to be minimized over the first r frames. Consider the following modified algorithm, which does not require the multi-step bisection phase, but makes assumptions about convergence:

- Every frame r, observe $\theta[r]$, $\boldsymbol{Z}[r]$, and choose a policy $\pi[r] \in \mathcal{P}$ to minimize:

$$\mathbb{E}\left\{V[\hat{y}_0(\pi[r]) - \theta[r]\hat{T}(\pi[r])] + \sum_{l=1}^{L} Z_l[r][\hat{y}_l(\pi[r]) - c_l\hat{T}(\pi[r])]|\boldsymbol{Z}[r], \theta[r]\right\}$$

- Update $\theta[r]$ by (7.56) and update $\boldsymbol{Z}[r]$ by (7.12).

To analyze this algorithm, we assume that there are constants $\theta, \overline{T}, \overline{y}_0$ such that, with probability 1:

$$\lim_{R\to\infty}\theta[R] = \theta \ , \ \lim_{R\to\infty}\frac{1}{R}\sum_{r=0}^{R-1} T[r] = \overline{T} \ , \ \lim_{R\to\infty}\frac{1}{R}\sum_{r=0}^{R-1} y_0[r] = \overline{y}_0 \tag{7.57}$$

We further assume there is an i.i.d. algorithm $\pi^*[r]$ that satisfies (7.9)-(7.10) with $\delta = 0$.

a) Use (7.14) to complete the right-hand-side of the following inequality:

$$\Delta(\boldsymbol{Z}[r]) + V\mathbb{E}\{y_0[r] - \theta[r]T[r]|\boldsymbol{Z}[r]\} \leq B + \cdots$$

b) Assume $\mathbb{E}\{L(\boldsymbol{Z}[0])\} = 0$. Plug the i.i.d. algorithm $\pi^*[r]$ from (7.9)-(7.10) into the right-hand-side of part (a) to prove that $\Delta(\boldsymbol{Z}[r]) \leq F$ for some constant F, and so all queues are mean rate stable. Use iterated expectations and the law of telescoping sums to conclude that for any $R > 0$:

$$\mathbb{E}\left\{\frac{1}{R}\sum_{r=0}^{R-1}[y_0[r] - \theta[r]T[r]]\right\} \leq \mathbb{E}\left\{\hat{T}(\pi^*[r])\right\}\left[ratio^{opt} - \frac{1}{R}\sum_{r=0}^{R-1}\mathbb{E}\{\theta[r]\}\right] + B/V$$

c) Argue from (7.56) and (7.57) that, with probability 1:

$$\lim_{R\to\infty}\frac{1}{R}\sum_{r=0}^{R-1}[y_0[r] - \theta[r]T[r]] = 0 \ , \ \lim_{R\to\infty}\frac{1}{R}\sum_{r=0}^{R-1}\theta[r] = \theta$$

d) Assume that:

$$\lim_{R\to\infty} \mathbb{E}\left\{\frac{1}{R}\sum_{r=0}^{R-1}[y_0[r] - \theta[r]T[r]]\right\} = 0 \ , \ \lim_{R\to\infty} \frac{1}{R}\sum_{r=0}^{R-1}\mathbb{E}\{\theta[r]\} = \theta$$

This can be justified via part (c) together with the Lebesgue Dominated convergence theorem, provided that mild additional boundedness assumptions on the processes are introduced. Use this with part (b) to prove:

$$\theta = \lim_{R\to\infty} \frac{\frac{1}{R}\sum_{r=0}^{R-1} y_0[r]}{\frac{1}{R}\sum_{r=0}^{R-1} T[r]} \leq ratio^{opt} + \frac{B}{\mathbb{E}\left\{\hat{T}(\pi^*[r])\right\} V} \quad (w.p.1)$$

Exercise 7.6. (Variation on Jensen's Inequality) Assume the result of Lemma 7.6(a).

a) Let $\{T[0], T[1], \ldots, T[R-1]\}$, $\{\boldsymbol{\gamma}[0], \boldsymbol{\gamma}[1], \ldots, \boldsymbol{\gamma}[R-1]\}$ be deterministic sequences. Prove (7.43) by defining X as a random integer that is uniform over $\{0, \ldots, R-1\}$ and defining the random vector $(T[X], \boldsymbol{\gamma}[X])$.

b) Prove (7.44) by considering $\{T[0], T[1], \ldots, T[R-1]\}$, $\{\boldsymbol{\gamma}[0], \boldsymbol{\gamma}[1], \ldots, \boldsymbol{\gamma}[R-1]\}$ as *random* sequences that are independent of X.

Exercise 7.7. (Equivalence of the Transformed Problem)

a) Suppose that $\pi^\star[r]$, $\boldsymbol{\gamma}^\star[r]$ solve (7.38)-(7.42), yielding $\overline{\gamma}_m^\star$, $\overline{T}^\star$, $\overline{y}_l^\star$, $\overline{T^\star\phi(\boldsymbol{\gamma}^\star)}$, $\overline{T^\star\boldsymbol{\gamma}^\star}$. Use the fact that $\phi(\overline{T^\star\boldsymbol{\gamma}^\star}/\overline{T}^\star) \geq \overline{T^\star\phi(\gamma^\star)}/\overline{T}^\star$ to show that the same policy $\pi^\star[r]$ satisfies the feasibility constraints (7.36)-(7.37) and yields $\phi(\overline{\boldsymbol{x}}^\star/\overline{T}^\star) \geq \overline{T^\star\phi(\boldsymbol{\gamma}^\star)}/\overline{T}^\star$.

b) Suppose that $\pi^*[r]$ is an algorithm that solves (7.35)-(7.37), yielding $\overline{\boldsymbol{x}}^*$, $\overline{T}^*$, and $\overline{y}_l^*$. Show that the optimal value of (7.38) is greater than or equal to $\phi(\overline{\boldsymbol{x}}^*/\overline{T}^*)$. Hint: Use the same policy $\pi^*[r]$, and use the constant $\boldsymbol{\gamma}[r] = \overline{\boldsymbol{x}}^*/\overline{T}^*$ for all $r \in \{0, 1, 2, \ldots\}$, noting from (7.33) that this is in $\mathcal{R}$.

Exercise 7.8. (Utility Optimization with Delay-Limited Scheduling) Modify the example in Section 7.6.1 to treat the problem of maximizing the utility function $\sum_{l=1}^{L}\log(1 + \overline{D}_l/T)$, rather than maximizing $\sum_{l=1}^{L} w_l\overline{D}_l/T$.

Exercise 7.9. (A simple form of Lebesgue Dominated Convergence) Let $\{f[r]\}_{r=0}^{\infty}$ be an infinite sequence of random variables. Suppose there are finite constants f_{min} and f_{max} such that the random variables deterministically satisfy $f_{min} \leq f[r] \leq f_{max}$ for all $r \in \{0, 1, 2, \ldots\}$. Suppose there is a finite constant $\overline{f}$ such that:

$$\lim_{R\to\infty} \frac{1}{R}\sum_{r=0}^{R-1} f[r] = \overline{f} \quad (w.p.1)$$

We will show that $\lim_{R\to\infty} \frac{1}{R}\sum_{r=0}^{R-1} \mathbb{E}\{f[r]\} = \overline{f}$.

a) Fix $\epsilon > 0$. Argue that for any integer $R > 0$:

$$\mathbb{E}\left\{\frac{1}{R}\sum_{r=0}^{R-1} f[r]\right\} \leq (\overline{f}+\epsilon)Pr\left[\frac{1}{R}\sum_{r=0}^{R-1} f[r] \leq \overline{f}+\epsilon\right] + f_{max}\,Pr\left[\frac{1}{R}\sum_{r=0}^{R-1} f[r] > \overline{f}+\epsilon\right]$$

b) Argue that for any $\epsilon > 0$:

$$\lim_{R\to\infty} Pr\left[\frac{1}{R}\sum_{r=0}^{R-1} f[r] > \overline{f}+\epsilon\right] = 0$$

Use this with part (a) to conclude that for all $\epsilon > 0$:

$$\lim_{R\to\infty} \frac{1}{R}\sum_{r=0}^{R-1} \mathbb{E}\{f[r]\} \leq \overline{f}+\epsilon$$

Conclude that the left-hand-side in the above inequality is less than or equal to $\overline{f}$.

c) Make a similar argument to show $\lim_{t\to\infty} \frac{1}{R}\sum_{r=0}^{R-1} \mathbb{E}\{f[r]\} \geq \overline{f}$.

CHAPTER 8

Conclusions

This text has presented a theory for optimizing time averages in stochastic networks. The tools of Lyapunov drift and Lyapunov optimization were developed to solve these problems. Our focus was on communication and queueing networks, including networks with wireless links and mobile devices. The theory can be used for networks with a variety of goals and functionalities, such as networks with:

- Network coding capabilities (see Exercise 4.12 and (188)(189)(190)).
- Dynamic data compression (see Exercise 4.14 and (191)(165)(143)).
- Multi-input, multi-output (MIMO) antenna capabilities (162)(192)(193).
- Multi-receiver diversity (154).
- Cooperative combining (194)(71).
- Hop count minimization (155).
- Economic considerations (195)(153).

Lyapunov optimization theory also has applications to a wide array of other problems, including (but not limited to):

- Stock market trading (40)(41).
- Product assembly plants (196)(175)(197)(198).
- Energy allocation for smart grids (159).

This text has included several representative simulation results for 1-hop networks (see Chapter 3). Further simulation and experimentation results for Lyapunov based algorithms in single-hop and multi-hop networks can be found in (54)(55)(199)(200)(201)(202)(203)(154)(142)(42).

We have highlighted the simplicity of Lyapunov drift and Lyapunov optimization, emphasizing that it only uses techniques of (see Chapters 1 and 3): (i) Telescoping sums, (ii) Iterated expectations, (iii) Opportunistically minimizing an expectation, (iv) Jensen's inequality. Further, the drift-plus-penalty algorithm of Lyapunov optimization theory is analyzed with the following simple framework:

1. Define a Lyapunov function as the sum of squares of queue backlog.

2. Compute a bound on the drift-plus-penalty by squaring the queueing equation. The bound typically has the form:

$$\Delta(\boldsymbol{\Theta}(t)) + V\mathbb{E}\{penalty(t)|\boldsymbol{\Theta}(t)\} \leq$$
$$B + V\mathbb{E}\{penalty(t)|\boldsymbol{\Theta}(t)\} + \sum_{n=1}^{N} \Theta_n(t)\mathbb{E}\{h_n(t)|\boldsymbol{\Theta}(t)\}$$

   where B is a constant that bounds second moments of the processes, $\boldsymbol{\Theta}(t) = (\Theta_1(t), \ldots, \Theta_n(t))$ is a general vector of (possibly virtual) queues, and $h_n(t)$ is the arrival-minus-departure value for queue $\Theta_n(t)$ on slot t.

3. Design the policy to minimize the right-hand-side of the above drift-plus-penalty bound.

4. Conclude that, under this algorithm, the drift-plus-penalty is bounded by plugging any other policy into the right-hand-side:

$$\Delta(\boldsymbol{\Theta}(t)) + V\mathbb{E}\{penalty(t)|\boldsymbol{\Theta}(t)\} \leq$$
$$B + V\mathbb{E}\left\{penalty^*(t)|\boldsymbol{\Theta}(t)\right\} + \sum_{n=1}^{N} \Theta_n(t)\mathbb{E}\left\{h_n^*(t)|\boldsymbol{\Theta}(t)\right\}$$

5. Plug an ω-only policy $\alpha^*(t)$ into the right-hand-side, one that is known to exist (although it would be hard to compute) that satisfies all constraints and yields a greatly simplified drift-plus-penalty expression on the right-hand-side.

Also important in this theory is the use of *virtual queues* to transform time average inequality constraints into queue stability problems, and *auxiliary variables* for the case of optimizing convex functions of time averages. The drift-plus-penalty framework was also shown to hold for optimization of non-convex functions of time averages, and for optimization over renewal systems.

The resulting min-drift (or "max-weight") algorithms can be very complex for general problems, particularly for wireless networks with interference. However, we have seen that low complexity approximations can be used to provide good performance. Further, for interference networks without time-variation, methods that take a longer time to find the max-weight solution (either by a deterministic or randomized search) were seen to provide full throughput and throughput-utility optimality with arbitrarily low per-timeslot computation complexity, provided that we let convergence time and/or delay increase (possibly non-polynomially) to infinity. Simple distributed Carrier Sense Multiple Access (CSMA) implementations are often possible (and provably throughput optimal) for these networks via the Jiang-Walrand theorem, which hints at deeper connections with Lyapunov optimization, max-weight theory, C-additive approximations, maximum entropy solutions, randomized algorithms, and Markov chain steady state theory.

Bibliography

[1] F. Kelly. Charging and rate control for elastic traffic. *European Transactions on Telecommunications*, vol. 8, no. 1 pp. 33-37, Jan.-Feb. 1997. DOI: 10.1002/ett.4460080106 3, 98

[2] F.P. Kelly, A.Maulloo, and D. Tan. Rate control for communication networks: Shadow prices, proportional fairness, and stability. *Journ. of the Operational Res. Society*, vol. 49, no. 3, pp. 237-252, March 1998. DOI: 10.2307/3010473 3, 7, 98, 104

[3] J. Mo and J. Walrand. Fair end-to-end window-based congestion control. *IEEE/ACM Transactions on Networking*, vol. 8, no. 5, Oct. 2000. DOI: 10.1109/90.879343 3, 128

[4] L. Massoulié and J. Roberts. Bandwidth sharing: Objectives and algorithms. *IEEE/ACM Transactions on Networking*, vol. 10, no. 3, pp. 320-328, June 2002. DOI: 10.1109/TNET.2002.1012364 3

[5] A. Tang, J. Wang, and S. Low. Is fair allocation always inefficient. *Proc. IEEE INFOCOM*, March 2004. DOI: 10.1109/INFCOM.2004.1354479 3, 98, 128

[6] B. Radunovic and J.Y. Le Boudec. Rate performance objectives of multihop wireless networks. *IEEE Transactions on Mobile Computing*, vol. 3, no. 4, pp. 334-349, Oct.-Dec. 2004. DOI: 10.1109/TMC.2004.45 3, 128

[7] L. Tassiulas and A. Ephremides. Stability properties of constrained queueing systems and scheduling policies for maximum throughput in multihop radio networks. *IEEE Transactions on Automatic Control*, vol. 37, no. 12, pp. 1936-1948, Dec. 1992. DOI: 10.1109/9.182479 6, 49, 113, 138

[8] L. Tassiulas and A. Ephremides. Dynamic server allocation to parallel queues with randomly varying connectivity. *IEEE Transactions on Information Theory*, vol. 39, no. 2, pp. 466-478, March 1993. DOI: 10.1109/18.212277 6, 10, 24, 49, 66

[9] P. R. Kumar and S. P. Meyn. Stability of queueing networks and scheduling policies. *IEEE Trans. on Automatic Control*, vol.40,.n.2, pp.251-260, Feb. 1995. DOI: 10.1109/9.341782 6

[10] N. McKeown, A. Mekkittikul, V. Anantharam, and J. Walrand. Achieving 100% throughput in an input-queued switch. *IEEE Transactions on Communications*, vol. 47, no. 8, August 1999. 6, 88

[11] E. Leonardi, M. Mellia, F. Neri, and M. Ajmone Marsan. Bounds on average delays and queue size averages and variances in input-queued cell-based switches. *Proc. IEEE INFOCOM*, 2001. DOI: 10.1109/INFCOM.2001.916303 6

[12] L. Tassiulas. Scheduling and performance limits of networks with constantly changing topology. *IEEE Transactions on Information Theory*, vol. 43, no. 3, pp. 1067-1073, May 1997. DOI: 10.1109/18.568722 6

[13] N. Kahale and P. E. Wright. Dynamic global packet routing in wireless networks. *Proc. IEEE INFOCOM*, 1997. DOI: 10.1109/INFCOM.1997.631182 6

[14] M. Andrews, K. Kumaran, K. Ramanan, A. Stolyar, P. Whiting, and R. Vijaykumar. Providing quality of service over a shared wireless link. *IEEE Communications Magazine*, vol. 39, no.2, pp.150-154, Feb. 2001. DOI: 10.1109/35.900644 6

[15] M. J. Neely, E. Modiano, and C. E Rohrs. Dynamic power allocation and routing for time varying wireless networks. *IEEE Journal on Selected Areas in Communications*, vol. 23, no. 1, pp. 89-103, January 2005. DOI: 10.1109/JSAC.2004.837349 6, 24, 56, 113

[16] B. Awerbuch and T. Leighton. A simple local-control approximation algorithm for multicommodity flow. *Proc. 34th IEEE Conf. on Foundations of Computer Science*, Oct. 1993. DOI: 10.1109/SFCS.1993.366841 6

[17] M. J. Neely. *Dynamic Power Allocation and Routing for Satellite and Wireless Networks with Time Varying Channels*. PhD thesis, Massachusetts Institute of Technology, LIDS, 2003. 6, 8, 11, 49, 105, 119, 120, 128, 134, 145

[18] M. J. Neely, E. Modiano, and C. Li. Fairness and optimal stochastic control for heterogeneous networks. *Proc. IEEE INFOCOM*, March 2005. DOI: 10.1109/INFCOM.2005.1498453 6, 49, 105, 134

[19] M. J. Neely, E. Modiano, and C. Li. Fairness and optimal stochastic control for heterogeneous networks. *IEEE/ACM Transactions on Networking*, vol. 16, no. 2, pp. 396-409, April 2008. DOI: 10.1109/TNET.2007.900405 6, 105, 145

[20] M. J. Neely. Energy optimal control for time varying wireless networks. *Proc. IEEE INFOCOM*, March 2005. DOI: 10.1109/INFCOM.2005.1497924 6, 49

[21] M. J. Neely. Energy optimal control for time varying wireless networks. *IEEE Transactions on Information Theory*, vol. 52, no. 7, pp. 2915-2934, July 2006. DOI: 10.1109/TIT.2006.876219 6, 28, 38, 49, 56, 83, 84

[22] L. Georgiadis, M. J. Neely, and L. Tassiulas. Resource allocation and cross-layer control in wireless networks. *Foundations and Trends in Networking*, vol. 1, no. 1, pp. 1-149, 2006. DOI: 10.1561/1300000001 xi, 7, 8, 24, 49, 105, 110, 111, 113, 145

[23] S. H. Low and D. E. Lapsley. Optimization flow control, i: Basic algorithm and convergence. *IEEE/ACM Transactions on Networking*, vol. 7 no. 6, pp. 861-875, Dec. 1999. DOI: 10.1109/90.811451 7, 104, 109

[24] S. H. Low. A duality model of TCP and queue management algorithms. *IEEE Trans. on Networking*, vol. 11, no. 4, pp. 525-536, August 2003. DOI: 10.1109/TNET.2003.815297 7

[25] L. Xiao, M. Johansson, and S. Boyd. Simultaneous routing and resource allocation for wireless networks. *Proc. of the 39th Annual Allerton Conf. on Comm., Control, Comput.*, Oct. 2001. 7

[26] L. Xiao, M. Johansson, and S. P. Boyd. Simultaneous routing and resource allocation via dual decomposition. *IEEE Transactions on Communications*, vol. 52, no. 7, pp. 1136-1144, July 2004. DOI: 10.1109/TCOMM.2004.831346 7

[27] J. W. Lee, R. R. Mazumdar, and N. B. Shroff. Downlink power allocation for multi-class cdma wireless networks. *Proc. IEEE INFOCOM*, 2002. DOI: 10.1109/INFCOM.2002.1019399 7

[28] M. Chiang. Balancing transport and physical layer in wireless multihop networks: Jointly optimal congestion control and power control. *IEEE Journal on Selected Areas in Communications*, vol. 23, no. 1, pp. 104-116, Jan. 2005. DOI: 10.1109/JSAC.2004.837347 7

[29] M. Chiang, S. H. Low, A. R. Calderbank, and J. C. Doyle. Layering as optimization decomposition: A mathematical theory of network architectures. *Proceedings of the IEEE*, vol. 95, no. 1, Jan. 2007. DOI: 10.1109/JPROC.2006.887322 7, 104, 109

[30] R. Cruz and A. Santhanam. Optimal routing, link scheduling, and power control in multi-hop wireless networks. *Proc. IEEE INFOCOM*, April 2003. DOI: 10.1109/INFCOM.2003.1208720 7

[31] X. Lin and N. B. Shroff. Joint rate control and scheduling in multihop wireless networks. *Proc. of 43rd IEEE Conf. on Decision and Control, Paradise Island, Bahamas*, Dec. 2004. 7, 8, 109

[32] R. Agrawal and V. Subramanian. Optimality of certain channel aware scheduling policies. *Proc. 40th Annual Allerton Conference on Communication , Control, and Computing, Monticello, IL*, Oct. 2002. 7, 119

[33] H. Kushner and P. Whiting. Asymptotic properties of proportional-fair sharing algorithms. *Proc. of 40th Annual Allerton Conf. on Communication, Control, and Computing*, 2002. 7, 119

[34] A. Stolyar. Maximizing queueing network utility subject to stability: Greedy primal-dual algorithm. *Queueing Systems*, vol. 50, no. 4, pp. 401-457, 2005. DOI: 10.1007/s11134-005-1450-0 7, 119

[35] A. Stolyar. Greedy primal-dual algorithm for dynamic resource allocation in complex networks. *Queueing Systems*, vol. 54, no. 3, pp. 203-220, 2006. DOI: 10.1007/s11134-006-0067-2 7

[36] Q. Li and R. Negi. Scheduling in wireless networks under uncertainties: A greedy primal-dual approach. *Arxiv Technical Report: arXiv:1001:2050v2*, June 2010. 8, 119

[37] L. Huang and M. J. Neely. Delay reduction via lagrange multipliers in stochastic network optimization. *Proc. of 7th Intl. Symposium on Modeling and Optimization in Mobile, Ad Hoc, and Wireless Networks (WiOpt)*, June 2009. DOI: 10.1109/WIOPT.2009.5291609 8, 10, 69, 71, 113

[38] M. J. Neely. Universal scheduling for networks with arbitrary traffic, channels, and mobility. *Proc. IEEE Conf. on Decision and Control (CDC)*, Atlanta, GA, Dec. 2010. 8, 77, 81, 102, 112, 119

[39] M. J. Neely. Universal scheduling for networks with arbitrary traffic, channels, and mobility. *ArXiv technical report*, arXiv:1001.0960v1, Jan. 2010. 8, 77, 81, 102, 107

[40] M. J. Neely. Stock market trading via stochastic network optimization. *Proc. IEEE Conference on Decision and Control (CDC)*, Atlanta, GA, Dec. 2010. 8, 77, 179

[41] M. J. Neely. Stock market trading via stochastic network optimization. *ArXiv Technical Report*, arXiv:0909.3891v1, Sept. 2009. 8, 77, 179

[42] M. J. Neely and R. Urgaonkar. Cross layer adaptive control for wireless mesh networks. *Ad Hoc Networks (Elsevier)*, vol. 5, no. 6, pp. 719-743, August 2007. DOI: 10.1016/j.adhoc.2007.01.004 8, 102, 112, 119, 179

[43] M. J. Neely. Stochastic network optimization with non-convex utilities and costs. *Proc. Information Theory and Applications Workshop (ITA)*, Feb. 2010. DOI: 10.1109/ITA.2010.5454100 8, 116, 117, 118

[44] A. Eryilmaz and R. Srikant. Fair resource allocation in wireless networks using queue-length-based scheduling and congestion control. *Proc. IEEE INFOCOM*, March 2005. DOI: 10.1109/INFCOM.2005.1498459 8

[45] A. Eryilmaz and R. Srikant. Fair resource allocation in wireless networks using queue-length-based scheduling and congestion control. *IEEE/ACM Transactions on Networking*, vol. 15, no. 6, pp. 1333-1344, Dec. 2007. DOI: 10.1109/TNET.2007.897944 8, 69, 71

[46] J. W. Lee, R. R. Mazumdar, and N. B. Shroff. Opportunistic power scheduling for dynamic multiserver wireless systems. *IEEE Transactions on Wireless Communications*, vol. 5, no.6, pp. 1506-1515, June 2006. DOI: 10.1109/TWC.2006.1638671 8

[47] V. Tsibonis, L. Georgiadis, and L. Tassiulas. Exploiting wireless channel state information for throughput maximization. *IEEE Transactions on Information Theory*, vol. 50, no. 11, pp. 2566-2582, Nov. 2004. DOI: 10.1109/TIT.2004.836687 8

[48] V. Tsibonis, L. Georgiadis, and L. Tassiulas. Exploiting wireless channel state information for throughput maximization. *Proc. IEEE INFOCOM*, April 2003. DOI: 10.1109/TIT.2004.836687 8

[49] X. Liu, E. K. P. Chong, and N. B. Shroff. A framework for opportunistic scheduling in wireless networks. *Computer Networks*, vol. 41, no. 4, pp. 451-474, March 2003. DOI: 10.1016/S1389-1286(02)00401-2 8

[50] R. Berry and R. Gallager. Communication over fading channels with delay constraints. *IEEE Transactions on Information Theory*, vol. 48, no. 5, pp. 1135-1149, May 2002. DOI: 10.1109/18.995554 8, 9, 67

[51] M. J. Neely. Optimal energy and delay tradeoffs for multi-user wireless downlinks. *IEEE Transactions on Information Theory*, vol. 53, no. 9, pp. 3095-3113, Sept. 2007. DOI: 10.1109/TIT.2007.903141 8, 10, 67, 71

[52] M. J. Neely. Super-fast delay tradeoffs for utility optimal fair scheduling in wireless networks. *IEEE Journal on Selected Areas in Communications, Special Issue on Nonlinear Optimization of Communication Systems*, vol. 24, no. 8, pp. 1489-1501, Aug. 2006. DOI: 10.1109/JSAC.2006.879357 8, 10, 67, 71

[53] M. J. Neely. Intelligent packet dropping for optimal energy-delay tradeoffs in wireless downlinks. *IEEE Transactions on Automatic Control*, vol. 54, no. 3, pp. 565-579, March 2009. DOI: 10.1109/TAC.2009.2013652 8, 10, 67, 71

[54] S. Moeller, A. Sridharan, B. Krishnamachari, and O. Gnawali. Routing without routes: The backpressure collection protocol. *Proc. 9th ACM/IEEE Intl. Conf. on Information Processing in Sensor Networks (IPSN)*, April 2010. DOI: 10.1145/1791212.1791246 8, 10, 71, 72, 113, 179

[55] L. Huang, S. Moeller, M. J. Neely, and B. Krishnamachari. LIFO-backpressure achieves near optimal utility-delay tradeoff. *Arxiv Technical Report, arXiv:1008.4895v1*, August 2010. 8, 10, 72, 113, 179

[56] M. J. Neely. Stochastic optimization for Markov modulated networks with application to delay constrained wireless scheduling. *Proc. IEEE Conf. on Decision and Control (CDC)*, Shanghai, China, Dec. 2009. DOI: 10.1109/CDC.2009.5400270 8, 9, 153, 171, 173

[57] M. J. Neely. Stochastic optimization for Markov modulated networks with application to delay constrained wireless scheduling. *ArXiv Technical Report, arXiv:0905.4757v1*, May 2009. 8, 9, 153, 158, 171, 173, 174

[58] C.-P. Li and M. J. Neely. Network utility maximization over partially observable markovian channels. *Arxiv Technical Report: arXiv:1008.3421v1*, Aug. 2010. 8, 153

[59] F. J. Vázquez Abad and V. Krishnamurthy. Policy gradient stochastic approximation algorithms for adaptive control of constrained time varying Markov decision processes. *Proc. IEEE Conf. on Decision and Control*, Dec. 2003. DOI: 10.1109/CDC.2003.1273053 8, 174

[60] D. V. Djonin and V. Krishnamurthy. q-learning algorithms for constrained Markov decision processes with randomized monotone policies: Application to mimo transmission control. *IEEE Transactions on Signal Processing*, vol. 55, no. 5, pp. 2170-2181, May 2007. DOI: 10.1109/TSP.2007.893228 8, 9, 174

[61] N. Salodkar, A. Bhorkar, A. Karandikar, and V. S. Borkar. An on-line learning algorithm for energy efficient delay constrained scheduling over a fading channel. *IEEE Journal on Selected Areas in Communications*, vol. 26, no. 4, pp. 732-742, May 2008. DOI: 10.1109/JSAC.2008.080514 8, 9, 174

[62] F. Fu and M. van der Schaar. A systematic framework for dynamically optimizing multi-user video transmission. *IEEE Journal on Selected Areas in Communications*, vol. 28, no. 3, pp. 308-320, April 2010. DOI: 10.1109/JSAC.2010.100403 8, 9, 174

[63] F. Fu and M. van der Schaar. Decomposition principles and online learning in cross-layer optimization for delay-sensitive applications. *IEEE Trans. Signal Processing*, vol. 58, no. 3, pp. 1401-1415, March 2010. DOI: 10.1109/TSP.2009.2034938 8, 9, 174

[64] D. P. Bertsekas. *Dynamic Programming and Optimal Control, vols. 1 and 2*. Athena Scientific, Belmont, Mass, 1995. 8, 158, 168, 170

[65] E. Altman. *Constrained Markov Decision Processes*. Boca Raton, FL, Chapman and Hall/CRC Press, 1999. 8

[66] S. Ross. *Introduction to Probability Models*. Academic Press, 8th edition, Dec. 2002. 8, 12, 27, 76

[67] D. P. Bertsekas and J. N. Tsitsiklis. *Neuro-Dynamic Programming*. Athena Scientific, Belmont, Mass, 1996. 8, 158, 173, 174

[68] W. B. Powell. *Approximate Dynamic Programming: Solving the Curses of Dimensionality*. John Wiley & Sons, 2007. DOI: 10.1002/9780470182963 8, 174

[69] S. Meyn. *Control Techniques for Complex Networks*. Cambridge University Press, 2008. 8, 174

[70] D. Tse and S. Hanly. Multi-access fading channels: Part ii: Delay-limited capacities. *IEEE Transactions on Information Theory*, vol. 44, no. 7, pp. 2816-2831, Nov. 1998. DOI: 10.1109/18.737514 9, 135

[71] R. Urgaonkar and M. J. Neely. Delay-limited cooperative communication with reliability constraints in wireless networks. *Proc. IEEE INFOCOM*, April 2009. DOI: 10.1109/INFCOM.2009.5062187 9, 135, 168, 179

[72] A. Mekkittikul and N. McKeown. A starvation free algorithm for achieving 100% throughput in an input-queued switch. *Proc. ICCN*, pp. 226-231, 1996. 9

[73] A. L. Stolyar and K. Ramanan. Largest weighted delay first scheduling: Large deviations and optimality. *Annals of Applied Probability*, vol. 11, no. 1, pp. 1-48, 2001. DOI: 10.1214/aoap/998926986 9, 11

[74] M. Andrews, K. Kumaran, K. Ramanan, A. Stolyar, R. Vijaykumar, and P. Whiting. Scheduling in a queueing system with asynchronously varying service rates. *Probability in the Engineering and Informational Sciences*, vol. 18, no. 2, pp. 191-217, April 2004. DOI: 10.1017/S0269964804182041 9

[75] S. Shakkottai and A. Stolyar. Scheduling for multiple flows sharing a time-varying channel: The exponential rule. *American Mathematical Society Translations, series 2*, vol. 207, 2002. 9

[76] M. J. Neely. Delay-based network utility maximization. *Proc. IEEE INFOCOM*, March 2010. DOI: 10.1109/INFCOM.2010.5462097 9, 120, 122

[77] A. Fu, E. Modiano, and J. Tsitsiklis. Optimal energy allocation for delay-constrained data transmission over a time-varying channel. *Proc. IEEE INFOCOM*, 2003. 9

[78] M. Zafer and E. Modiano. Optimal rate control for delay-constrained data transmission over a wireless channel. *IEEE Transactions on Information Theory*, vol. 54, no. 9, pp. 4020-4039, Sept. 2008. DOI: 10.1109/TIT.2008.928249 9

[79] M. Zafer and E. Modiano. Minimum energy transmission over a wireless channel with deadline and power constraints. *IEEE Transactions on Automatic Control*, vol. 54, no. 12, pp. 2841-2852, December 2009. DOI: 10.1109/TAC.2009.2034202 9

[80] M. Goyal, A. Kumar, and V. Sharma. Power constrained and delay optimal policies for scheduling transmission over a fading channel. *Proc. IEEE INFOCOM*, April 2003. DOI: 10.1109/INFCOM.2003.1208683 9

[81] A. Wierman, L. L. H. Andrew, and A. Tang. Power-aware speed scaling in processor sharing systems. *Proc. IEEE INFOCOM*, Rio de Janeiro, Brazil, April 2009. DOI: 10.1109/INFCOM.2009.5062123 9

[82] E. Uysal-Biyikoglu, B. Prabhakar, and A. El Gamal. Energy-efficient packet transmission over a wireless link. *IEEE/ACM Trans. Networking*, vol. 10, no. 4, pp. 487-499, Aug. 2002. DOI: 10.1109/TNET.2002.801419 9

[83] M. Zafer and E. Modiano. A calculus approach to minimum energy transmission policies with quality of service guarantees. *Proc. IEEE INFOCOM*, March 2005. DOI: 10.1109/INFCOM.2005.1497922 9

[84] M. Zafer and E. Modiano. A calculus approach to energy-efficient data transmission with quality-of-service constraints. *IEEE/ACM Transactions on Networking*, vol. 17, no. 13, pp. 898-911, June 2009. DOI: 10.1109/TNET.2009.2020831 9

[85] W. Chen, M. J. Neely, and U. Mitra. Energy-efficient transmissions with individual packet delay constraints. *IEEE Transactions on Information Theory*, vol. 54, no. 5, pp. 2090-2109, May 2008. DOI: 10.1109/TIT.2008.920344 9

[86] W. Chen, U. Mitra, and M. J. Neely. Energy-efficient scheduling with individual packet delay constraints over a fading channel. *Wireless Networks*, vol. 15, no. 5, pp. 601-618, July 2009. DOI: 10.1007/s11276-007-0093-y 9

[87] M. A. Khojastepour and A. Sabharwal. Delay-constrained scheduling: Power efficiency, filter design, and bounds. *Proc. IEEE INFOCOM*, March 2004. DOI: 10.1109/INFCOM.2004.1354603 9

[88] B. Hajek. Optimal control of two interacting service stations. *IEEE Transactions on Automatic Control*, vol. 29, no. 6, pp. 491-499, June 1984. DOI: 10.1109/TAC.1984.1103577 9

[89] S. Sarkar. Optimum scheduling and memory management in input queued switches with finite buffer space. *Proc. IEEE INFOCOM*, April 2003. DOI: 10.1109/INFCOM.2003.1208973 9

[90] A. Tarello, J. Sun, M. Zafer, and E. Modiano. Minimum energy transmission scheduling subject to deadline constraints. *ACM Wireless Networks*, vol. 14, no. 5, pp. 633-645, 2008. DOI: 10.1007/s11276-006-0005-6 9

[91] B. Sadiq, S. Baek, and Gustavo de Veciana. Delay-optimal opportunistic scheduling and approximations: the log rule. *Proc. IEEE INFOCOM*, April 2009. DOI: 10.1109/INFCOM.2009.5062088 9

[92] B. Sadiq and G. de Veciana. Optimality and large deviations of queues under the pseudo-log rule opportunistic scheduling. *46th Annual Allerton Conference on Communication, Control, and Computing, Monticello, IL*, Sept. 2008. DOI: 10.1109/ALLERTON.2008.4797636 9, 11

[93] A. L. Stolyar. Large deviations of queues sharing a randomly time-varying server. *Queueing Systems Theory and Applications*, vol. 59, no. 1, pp. 1-35, 2008. DOI: 10.1007/s11134-008-9072-y 9, 11

[94] A. Ganti, E. Modiano, and J. N. Tsitsiklis. Optimal transmission scheduling in symmetric communication models with intermittent connectivity. *IEEE Transactions on Information Theory*, vol. 53, no. 3, pp. 998-1008, March 2007. DOI: 10.1109/TIT.2006.890695 10

[95] E. M. Yeh and A. S. Cohen. Delay optimal rate allocation in multiaccess fading communications. *Proc. Allerton Conference on Communication, Control, and Computing, Monticello, IL*, 2004. 10

[96] E. M. Yeh. *Multiaccess and Fading in Communication Networks*. PhD thesis, Massachusetts Institute of Technology, Laboratory for Information and Decision Systems (LIDS), 2001. 10

[97] S. Kittipiyakul and T. Javidi. Delay-optimal server allocation in multi-queue multi-server systems with time-varying connectivities. *IEEE Transactions on Information Theory*, vol. 55, no. 5, pp. 2319-2333, May 2009. DOI: 10.1109/TIT.2009.2016051 10

[98] A. Ephremides, P. Varaiya, and J. Walrand. A simple dynamic routing problem. *IEEE Transactions on Automatic Control*, vol. AC-25, no.4, pp. 690-693, Aug. 1980. 10

[99] M. J. Neely, E. Modiano, and Y.-S. Cheng. Logarithmic delay for $n \times n$ packet switches under the crossbar constraint. *IEEE Transactions on Networking*, vol. 15, no. 3, pp. 657-668, June 2007. DOI: 10.1109/TNET.2007.893876 10, 11, 37

[100] M. J. Neely. Order optimal delay for opportunistic scheduling in multi-user wireless uplinks and downlinks. *IEEE/ACM Transactions on Networking*, vol. 16, no. 5, pp. 1188-1199, October 2008. DOI: 10.1109/TNET.2007.909682 10, 24, 37

[101] M. J. Neely. Delay analysis for max weight opportunistic scheduling in wireless systems. *IEEE Transactions on Automatic Control*, vol. 54, no. 9, pp. 2137-2150, Sept. 2009. DOI: 10.1109/TAC.2009.2026943 10, 11, 24, 37

[102] S. Deb, D. Shah, and S. Shakkottai. Fast matching algorithms for repetitive optimization: An application to switch scheduling. *Proc. of 40th Annual Conference on Information Sciences and Systems (CISS), Princeton, NJ*, March 2006. DOI: 10.1109/CISS.2006.286659 10, 37, 147

[103] M. J. Neely. Delay analysis for maximal scheduling with flow control in wireless networks with bursty traffic. *IEEE Transactions on Networking*, vol. 17, no. 4, pp. 1146-1159, August 2009. DOI: 10.1109/TNET.2008.2008232 10, 11, 37, 147

[104] X. Wu, R. Srikant, and J. R. Perkins. Scheduling efficiency of distributed greedy scheduling algorithms in wireless networks. *IEEE Transactions on Mobile Computing*, vol. 6, no. 6, pp. 595-605, June 2007. DOI: 10.1109/TMC.2007.1061 11, 37, 147

[105] J. G. Dai and B. Prabhakar. The throughput of data switches with and without speedup. *Proc. IEEE INFOCOM*, 2000. DOI: 10.1109/INFCOM.2000.832229 11, 37

[106] J. M. Harrison and J. A. Van Mieghem. Dynamic control of brownian networks: State space collapse and equivalent workload formulations. *The Annals of Applied Probability*, vol. 7(3), pp. 747-771, Aug. 1997. DOI: 10.1214/aoap/1034801252 11

[107] S. Shakkottai, R. Srikant, and A. Stolyar. Pathwise optimality of the exponential scheduling rule for wireless channels. *Advances in Applied Probability*, vol. 36, no. 4, pp. 1021-1045, Dec. 2004. DOI: 10.1239/aap/1103662957 11

[108] A. L. Stolyar. Maxweight scheduling in a generalized switch: State space collapse and workload minimization in heavy traffic. *Annals of Applied Probability, pp. 1-53*, 2004. DOI: 10.1214/aoap/1075828046 11

[109] D. Shah and D. Wischik. Optimal scheduling algorithms for input-queued switches. *Proc. IEEE INFOCOM*, 2006. DOI: 10.1109/INFOCOM.2006.238 11

[110] I. Keslassy and N. McKeown. Analysis of scheduling algorithms that provide 100% throughput in input-queued switches. *Proc. 39th Annual Allerton Conf. on Communication, Control, and Computing*, Oct. 2001. 11

[111] T. Ji, E. Athanasopoulou, and R. Srikant. Optimal scheduling policies in small generalized switches. *Proc. IEEE INFOCOM, Rio De Janeiro, Brazil*, 2009. DOI: 10.1109/INFCOM.2009.5062259 11

[112] V. J. Venkataramanan and X. Lin. Structural properties of ldp for queue-length based wireless scheduling algorithms. *Proc. of 45th Annual Allerton Conference on Communication, Control, and Computing, Monticello, Illinois*, September 2007. 11

[113] D. Bertsimas, I. C. Paschalidis, and J. N. Tsitsiklis. Large deviations analysis of the generalized proccssor sharing policy. *Queueing Systems*, vol. 32, pp. 319-349, 1999. DOI: 10.1023/A:1019151423773 11

[114] D. Bertsimas, I. C. Paschalidis, and J. N. Tsitsiklis. Asymptotic buffer overflow probabilities in multiclass multiplexers: An optimal control approach. *IEEE Transactions on Automatic Control*, vol. 43, no. 3, pp. 315-335, March 1998. DOI: 10.1109/9.661587 11

[115] S. Bodas, S. Shakkottai, L. Ying, and R. Srikant. Scheduling in multi-channel wireless networks: Rate function optimality in the small-buffer regime. *Proc. ACM SIGMETRICS/Performance Conference*, June 2009. DOI: 10.1145/1555349.1555364 11

[116] P. Gupta and P. R. Kumar. The capacity of wireless networks. *IEEE Transactions on Information Theory*, vol. 46, no. 2, pp. 388-404, March 2000. DOI: 10.1109/18.825799 11

[117] M. Grossglauser and D. Tse. Mobility increases the capacity of ad-hoc wireless networks. *IEEE/ACM Trans. on Networking*, vol. 10, no. 4, pp. 477-486, August 2002. DOI: 10.1109/TNET.2002.801403 11

[118] M. J. Neely and E. Modiano. Capacity and delay tradeoffs for ad-hoc mobile networks. *IEEE Transactions on Information Theory*, vol. 51, no. 6, pp. 1917-1937, June 2005. DOI: 10.1109/TIT.2005.847717 11

[119] S. Toumpis and A. J. Goldsmith. Large wireless networks under fading, mobility, and delay constraints. *Proc. IEEE INFOCOM*, 2004. DOI: 10.1109/INFCOM.2004.1354532 12

[120] X. Lin and N. B. Shroff. Towards achieving the maximum capacity in large mobile wireless networks. *Journal of Communications and Networks, Special Issue on Mobile Ad Hoc Wireless Networks*, vol. 6, no. 4, December 2004. 12

[121] X. Lin and N. B. Shroff. The fundamental capacity-delay tradeoff in large mobile ad hoc networks. *Purdue University Tech. Report*, 2004. 12

[122] A. El Gamal, J. Mammen, B. Prabhakar, and D. Shah. Optimal throughput-delay scaling in wireless networks – part 1: The fluid model. *IEEE Transactions on Information Theory*, vol. 52, no. 6, pp. 2568-2592, June 2006. DOI: 10.1109/TIT.2006.874379 12

[123] G. Sharma, R. Mazumdar, and N. Shroff. Delay and capacity trade-offs in mobile ad-hoc networks: A global perspective. *Proc. IEEE INFOCOM*, April 2006. DOI: 10.1109/INFOCOM.2006.144 12

[124] X. Lin, G. Sharma, R. R. Mazumdar, and N. B. Shroff. Degenerate delay-capacity trade-offs in ad hoc networks with brownian mobility. *IEEE Transactions on Information Theory*, vol. 52, no. 6, pp. 2777-2784, June 2006. DOI: 10.1109/TIT.2006.874544 12

[125] N. Bansal and Z. Liu. Capacity, delay and mobility in wireless ad-hoc networks. *Proc. IEEE INFOCOM*, April 2003. DOI: 10.1109/INFCOM.2003.1208990 12

[126] L. Ying, S. Yang, and R. Srikant. Optimal delay-throughput tradeoffs in mobile ad hoc networks. *IEEE Transactions on Information Theory*, vol. 54, no. 9, pp. 4119-4143, Sept. 2008. DOI: 10.1109/TIT.2008.928247 12

[127] Z. Kong, E. M. Yeh, and E. Soljanin. Coding improves the throughput-delay trade-off in mobile wireless networks. *Proceedings of the International Symposium on Information Theory, Seoul, Korea*, June 2009. 12

[128] Z. Kong, E. M. Yeh, and E. Soljanin. Coding improves the throughput-delay trade-off in mobile wireless networks. *IEEE Transactions on Information Theory*, to appear. DOI: 10.1109/ISIT.2009.5205277 12

[129] D. P. Bertsekas and R. Gallager. *Data Networks*. New Jersey: Prentice-Hall, Inc., 1992. 12, 19, 25, 27, 37, 48, 109, 128, 144, 172

[130] R. Gallager. *Discrete Stochastic Processes*. Kluwer Academic Publishers, Boston, 1996. 12, 27, 50, 74, 76

[131] F. P. Kelly. *Reversibility and Stochastic Networks*. Wiley, Chichester, 1979. 12, 27, 144

[132] S. Ross. *Stochastic Processes*. John Wiley & Sons, Inc., New York, 1996. 12, 74

[133] D. P. Bertsekas, A. Nedic, and A. E. Ozdaglar. *Convex Analysis and Optimization*. Boston: Athena Scientific, 2003. 12, 67

[134] S. Boyd and L. Vandenberghe. *Convex Optimization*. Cambridge University Press, 2004. 12, 67

[135] R. T. Rockafellar. *Convex Analysis*. Princeton University Press, 1996. 12

[136] M. J. Neely. Stability and capacity regions for discrete time queueing networks. *ArXiv Technical Report: arXiv:1003.3396v1*, March 2010. 18, 19, 56, 102

[137] R. Urgaonkar and M. J. Neely. Opportunistic scheduling with reliability guarantees in cognitive radio networks. *IEEE Transactions on Mobile Computing*, vol. 8, no. 6, pp. 766-777, June 2009. DOI: 10.1109/TMC.2009.38 28, 145, 147

[138] M. J. Neely. Queue stability and probability 1 convergence via lyapunov optimization. *Arxiv Technical Report, arXiv:1008.3519*, August 2010. 50, 51

[139] O. Kallenberg. *Foundations of Modern Probability, 2nd ed., Probability and its Applications*. Springer-Verlag, 2002. 50

[140] D. Williams. *Probability with Martingales*. Cambridge Mathematical Textbooks, Cambridge University Press, 1991. 50

[141] Y. V. Borovskikh and V. S. Korolyuk. *Martingale Approximation*. VSP BV, The Netherlands, 1997. 50

[142] M. J. Neely and R. Urgaonkar. Opportunism, backpressure, and stochastic optimization with the wireless broadcast advantage. *Asilomar Conference on Signals, Systems, and Computers, Pacific Grove, CA*, Oct. 2008. DOI: 10.1109/ACSSC.2008.5074815 70, 71, 179

[143] M. J. Neely and A. Sharma. Dynamic data compression with distortion constraints for wireless transmission over a fading channel. *arXiv:0807.3768v1*, July 24, 2008. 70, 71, 84, 89, 179

[144] L. Huang and M. J. Neely. Max-weight achieves the exact $[O(1/V), O(V)]$ utility-delay tradeoff under Markov dynamics. *Arxiv Technical Report, arXiv:1008.0200*, August 2010. 74, 77

[145] P. Billingsley. *Probability Theory and Measure, 2nd edition.* New York: John Wiley & Sons, 1986. 76, 92

[146] M. J. Neely. Distributed and secure computation of convex programs over a network of connected processors. *DCDIS Conf., Guelph, Ontario,* July 2005. 81

[147] L. Tassiulas and A. Ephremides. Throughput properties of a queueing network with distributed dynamic routing and flow control. *Advances in Applied Probability*, vol. 28, pp. 285-307, 1996. DOI: 10.2307/1427922 86

[148] Y. Wu, P. A. Chou, and S-Y Kung. Information exchange in wireless networks with network coding and physical-layer broadcast. *Conference on Information Sciences and Systems, Johns Hopkins University*, March 2005. 87

[149] E. Leonardi, M. Mellia, M. A. Marsan, and F. Neri. Optimal scheduling and routing for maximizing network throughput. *IEEE/ACM Transactions on Networking*, vol. 15, no. 6, Dec. 2007. DOI: 10.1109/TNET.2007.896486 104, 107

[150] Y. Li, A. Papachristodoulou, and M. Chiang. Stability of congestion control schemes with delay sensitive traffic. *Proc. IEEE ACC, Seattle, WA,* June 2008. DOI: 10.1109/ACC.2008.4586779 104, 108, 109

[151] J. K. MacKie-Mason and H. R. Varian. Pricing congestible network resources. *IEEE Journal on Selected Areas in Communications*, vol. 13, no. 7, September 1995. DOI: 10.1109/49.414634 109

[152] M. J. Neely and E. Modiano. Convexity in queues with general inputs. *IEEE Transactions on Information Theory*, vol. 51, no. 2, pp. 706-714, Feb. 2005. DOI: 10.1109/TIT.2004.840859 109

[153] M. J. Neely. Optimal pricing in a free market wireless network. *Wireless Networks*, vol. 15, no. 7, pp. 901-915, October 2009. DOI: 10.1007/s11276-007-0083-0 112, 179

[154] M. J. Neely and R. Urgaonkar. Optimal backpressure routing in wireless networks with multi-receiver diversity. *Ad Hoc Networks (Elsevier)*, vol. 7, no. 5, pp. 862-881, July 2009. DOI: 10.1016/j.adhoc.2008.07.009 113, 132, 145, 147, 179

[155] L. Ying, S. Shakkottai, and A. Reddy. On combining shortest-path and backpressure routing over multihop wireless networks. *Proc. IEEE INFOCOM*, 2009. DOI: 10.1109/INFCOM.2009.5062086 113, 179

[156] J.W. Lee, R. R. Mazumdar, and N. B. Shroff. Non-convex optimization and rate control for multi-class services in the internet. *IEEE/ACM Trans. on Networking*, vol. 13, no. 4, pp. 827-840, Aug. 2005. DOI: 10.1109/TNET.2005.852876 116

[157] M. Chiang. Nonconvex optimization of communication systems. *Advances in Mechanics and Mathematics, Special volume on Strang's 70th Birthday, Springer*, vol. 3, 2008. 116

[158] W.-H. Wang, M. Palaniswami, and S. H. Low. Application-oriented flow control: Fundamentals, algorithms, and fairness. *IEEE/ACM Transactions on Networking*, vol. 14, no. 6, Dec. 2006. DOI: 10.1109/TNET.2006.886318 116

[159] M. J. Neely, A. S. Tehrani, and A. G. Dimakis. Efficient algorithms for renewable energy allocation to delay tolerant consumers. *1st IEEE International Conference on Smart Grid Communications*, 2010. 120, 122, 179

[160] L. Tassiulas and S. Sarkar. Maxmin fair scheduling in wireless ad hoc networks. *IEEE Journal on Selected Areas in Communications, Special Issue on Ad Hoc Networks*, vol. 23, no. 1, pp. 163-173, Jan. 2005. 128

[161] H. Shirani-Mehr, G. Caire, and M. J. Neely. Mimo downlink scheduling with non-perfect channel state knowledge. *IEEE Transactions on Communications*, vol. 58, no. 7, pp. 2055-2066, July 2010. DOI: 10.1109/TCOMM.2010.07.090377 129, 132

[162] M. Kobayashi, G. Caire, and D. Gesbert. Impact of multiple transmit antennas in a queued SDMA/TDMA downlink. In *Proc. of 6th IEEE Workshop on Signal Processing Advances in Wireless Communications (SPAWC)*, June 2005. DOI: 10.1109/SPAWC.2005.1506198 132, 179

[163] C. Li and M. J. Neely. Energy-optimal scheduling with dynamic channel acquisition in wireless downlinks. *IEEE Transactions on Mobile Computing*, vol. 9, no. 4, pp. 527-539, April 2010. DOI: 10.1109/TMC.2009.140 132

[164] A. Gopalan, C. Caramanis, and S. Shakkottai. On wireless scheduling with partial channel-state information. *Allerton Conf. on Comm., Control, and Computing*, Sept. 2007. 132

[165] M. J. Neely. Dynamic data compression for wireless transmission over a fading channel. *Proc. Conference on Information Sciences and Systems (CISS), invited paper, Princeton*, March 2008. DOI: 10.1109/CISS.2008.4558703 132, 179

[166] M. J. Neely. Max weight learning algorithms with application to scheduling in unknown environments. *arXiv:0902.0630v1*, Feb. 2009. 132, 162

[167] D. Shah and M. Kopikare. Delay bounds for approximate maximum weight matching algorithms for input queued switches. *Proc. IEEE INFOCOM*, June 2002. DOI: 10.1109/INFCOM.2002.1019350 140

[168] M. J. Neely, E. Modiano, and C. E. Rohrs. Tradeoffs in delay guarantees and computation complexity for $n \times n$ packet switches. *Proc. of Conf. on Information Sciences and Systems (CISS), Princeton*, March 2002. 140, 141

[169] L. Tassiulas. Linear complexity algorithms for maximum throughput in radio networks and input queued switches. *Proc. IEEE INFOCOM*, 1998. DOI: 10.1109/INFCOM.1998.665071 140, 141

[170] E. Modiano, D. Shah, and G. Zussman. Maximizing throughput in wireless networks via gossiping. *Proc. ACM SIGMETRICS / IFIP Performance'06*, June 2006. DOI: 10.1145/1140103.1140283 141

[171] D. Shah, D. N. C. Tse, and J. N. Tsitsiklis. Hardness of low delay network scheduling. *under submission.* 141

[172] L. Jiang and J. Walrand. A distributed csma algorithm for throughput and utility maximization in wireless networks. *Proc. Allerton Conf. on Communication, Control, and Computing*, Sept. 2008. DOI: 10.1109/ALLERTON.2008.4797741 141, 142, 144

[173] S. Rajagopalan and D. Shah. Reversible networks, distributed optimization, and network scheduling: What do they have in common? *Proc. Conf. on Information Sciences and Systems (CISS)*, 2008. 141, 144

[174] T. M. Cover and J. A. Thomas. *Elements of Information Theory.* New York: John Wiley & Sons, Inc., 1991. DOI: 10.1002/0471200611 143

[175] L. Jiang and J. Walrand. Scheduling and congestion control for wireless and processing networks. *Synthesis Lectures on Communication Networks*, vol. 3, no. 1, pp. 1-156, 2010. DOI: 10.2200/S00270ED1V01Y201008CNT006 144, 179

[176] L. Jiang and J. Walrand. Convergence and stability of a distributed csma algorithm for maximal network throughput. *Proc. IEEE Conference on Decision and Control (CDC), Shanghai, China*, December 2009. DOI: 10.1109/CDC.2009.5400349 144

[177] J. Ni, B. Tan, and R. Srikant. Q-csma: Queue length based csma/ca algorithms for achieving maximum throughput and low delay in wireless networks. *ArXive Technical Report: arXiv:0901.2333v4*, Dec. 2009. 144

[178] G. Louth, M. Mitzenmacher, and F. Kelly. Computational complexity of loss networks. *Theoretical Computer Science*, vol. 125, pp. 45-59, 1994. DOI: 10.1016/0304-3975(94)90216-X 144

[179] J. Ni and S. Tatikonda. A factor graph modelling of product-form loss and queueing networks. *43rd Allerton Conference on Communication, Control, and Computing (Monticello, IL)*, September 2005. 144

[180] M. Luby and E. Vigoda. Fast convergence of the glauber dynamics for sampling independent sets: Part i. *International Computer Science Institute, Berkeley, CA, Technical Report TR-99-002*,

Jan. 1999.
DOI: 10.1002/(SICI)1098-2418(199910/12)15:3/4%3C229::AID-RSA3%3E3.0.CO;2-X 144

[181] D. Randall and P. Tetali. Analyzing glauber dynamics by comparison of Markov chains. *Lecture Notes in Computer Science, Proc. of the 3rd Latin American Symposium on Theoretical Informatics*, vol. 1380:pp. 292–304, 1998. DOI: 10.1063/1.533199 144

[182] L. Bui, E. Eryilmaz, R. Srikant, and X. Wu. Joint asynchronous congestion control and distributed scheduling for multi-hop wireless networks. *Proc. IEEE INFOCOM*, 2006. DOI: 10.1109/INFOCOM.2006.210 145

[183] D. Shah. Maximal matching scheduling is good enough. *Proc. IEEE Globecom*, Dec. 2003. DOI: 10.1109/GLOCOM.2003.1258788 147

[184] P. Chaporkar, K. Kar, X. Luo, and S. Sarkar. Throughput and fairness guarantees through maximal scheduling in wireless networks. *IEEE Trans. on Information Theory*, vol. 54, no. 2, pp. 572-594, Feb. 2008. DOI: 10.1109/TIT.2007.913537 147

[185] X. Lin and N. B. Shroff. The impact of imperfect scheduling on cross-layer rate control in wireless networks. *Proc. IEEE INFOCOM*, 2005. DOI: 10.1109/INFCOM.2005.1498460 147

[186] L. Lin, X. Lin, and N. B. Shroff. Low-complexity and distributed energy minimization in multi-hop wireless networks. *Proc. IEEE INFOCOM*, 2007.
DOI: 10.1109/TNET.2009.2032419 147

[187] C. C. Moallemi, S. Kumar, and B. Van Roy. Approximate and data-driven dynamic programming for queuing networks. Submitted for publication, 2008. 174

[188] T. Ho, M. Médard, J. Shi, M. Effros, and D. R. Karger. On randomized network coding. *Proc. of* 41st *Annual Allerton Conf. on Communication, Control, and Computing*, Oct. 2003. 179

[189] A. Eryilmaz and D. S. Lun. Control for inter-session network coding. *Proc. Information Theory and Applications Workshop (ITA)*, Jan./Feb. 2007. 179

[190] X. Yan, M. J. Neely, and Z. Zhang. Multicasting in time varying wireless networks: Cross-layer dynamic resource allocation. *Proc. IEEE International Symposium on Information Theory (ISIT)*, June 2007. DOI: 10.1109/ISIT.2007.4557630 179

[191] A. Sharma, L. Golubchik, R. Govindan, and M. J. Neely. Dynamic data compression in multi-hop wireless networks. *Proc. SIGMETRICS*, 2009. DOI: 10.1145/1555349.1555367 179

[192] C. Swannack, E. Uysal-Biyikoglu, and G. Wornell. Low complexity multiuser scheduling for maximizing throughput in the mimo broadcast channel. *Proc. of 42nd Allerton Conf. on Communication, Control, and Computing*, September 2004. 179

[193] H. Shirani-Mehr, G. Caire, and M. J. Neely. Mimo downlink scheduling with non-perfect channel state knowledge. *IEEE Transactions on Communications*, to appear. DOI: 10.1109/TCOMM.2010.07.090377 179

[194] E. M. Yeh and R. A. Berry. Throughput optimal control of cooperative relay networks. *IEEE Transactions on Information Theory: Special Issue on Models, Theory, and Codes for Relaying and Cooperation in Communication Networks*, vol. 53, no. 10, pp. 3827-3833, October 2007. DOI: 10.1109/TIT.2007.904978 179

[195] L. Huang and M. J. Neely. The optimality of two prices: Maximizing revenue in a stochastic communication system. *IEEE/ACM Transactions on Networking*, vol. 18, no. 2, pp. 406-419, April 2010. DOI: 10.1109/TNET.2009.2028423 179

[196] L. Jiang and J. Walrand. Stable and utility-maximizing scheduling for stochastic processing networks. *Allerton Conference on Communication, Control, and Computing*, 2009. DOI: 10.1109/ALLERTON.2009.5394870 179

[197] M. J. Neely and L. Huang. Dynamic product assembly and inventory control for maximum profit. *Proc. IEEE Conf. on Decision and Control (CDC)*, Atlanta, GA, Dec. 2010. 179

[198] M. J. Neely and L. Huang. Dynamic product assembly and inventory control for maximum profit. *ArXiv Technical Report, arXiv:1004.0479v1*, April 2010. 179

[199] A. Warrier, S. Ha, P. Wason, I. Rhee, and J. H. Kim. Diffq: Differential backlog congestion control for wireless multi-hop networks. *Conference on Sensor, Mesh and Ad Hoc Communications and Networks (SECON), San Francisco, US*, 2008. DOI: 10.1109/SAHCN.2008.78 179

[200] A. Warrier, S. Janakiraman, S. Ha, and I. Rhee. Diffq: Practical differential backlog congestion control for wireless networks. *Proc. IEEE INFOCOM, Rio de Janeiro, Brazil*, 2009. DOI: 10.1109/INFCOM.2009.5061929 179

[201] A. Sridharan, S. Moeller, and B. Krishnamachari. Making distributed rate control using lyapunov drifts a reality in wireless sensor networks. *6th Intl. Symposium on Modeling and Optimization in Mobile, Ad Hoc, and Wireless Networks (WiOpt)*, April 2008. DOI: 10.4108/ICST.WIOPT2008.3205 179

[202] U. Akyol, M. Andrews, P. Gupta, J. Hobby, I. Saniee, and A. Stolyar. Joint scheduling and congestion control in mobile ad-hoc networks. *Proc. IEEE INFOCOM*, 2008. DOI: 10.1109/INFOCOM.2008.111 179

[203] B. Radunović, C. Gkantsidis, D. Gunawardena, and P. Key. Horizon: Balancing tcp over multiple paths in wireless mesh network. *Proc. ACM Mobicom*, 2008. DOI: 10.1145/1409944.1409973 179

Author's Biography

MICHAEL J. NEELY

Michael J. Neely received B.S. degrees in both Electrical Engineering and Mathematics from the University of Maryland, College Park, in 1997. He then received a 3 year Department of Defense NDSEG Fellowship for graduate study at the Massachusetts Institute of Technology, where he completed the M.S. degree in 1999 and the Ph.D. in 2003, both in Electrical Engineering. He joined the faculty of Electrical Engineering at the University of Southern California in 2004, where he is currently an Associate Professor. His research interests are in the areas of stochastic network optimization and queueing theory, with applications to wireless networks, mobile ad-hoc networks, and switching systems. Michael received the NSF Career award in 2008 and the Viterbi School of Engineering Junior Research Award in 2009. He is a member of Tau Beta Pi and Phi Beta Kappa.

Printed in the United States
by Baker & Taylor Publisher Services